Lecture Notes in Computer Science

## Lecture Notes in Artificial Intelligence  14909
Founding Editor

Jörg Siekmann

Series Editors

Randy Goebel, *University of Alberta, Edmonton, Canada*
Wolfgang Wahlster, *DFKI, Berlin, Germany*
Zhi-Hua Zhou, *Nanjing University, Nanjing, China*

The series Lecture Notes in Artificial Intelligence (LNAI) was established in 1988 as a topical subseries of LNCS devoted to artificial intelligence.

The series publishes state-of-the-art research results at a high level. As with the LNCS mother series, the mission of the series is to serve the international R & D community by providing an invaluable service, mainly focused on the publication of conference and workshop proceedings and postproceedings.

Yaxin Bi · Anne-Laure Jousselme ·
Thierry Denoeux
Editors

# Belief Functions: Theory and Applications

8th International Conference, BELIEF 2024
Belfast, UK, September 2–4, 2024
Proceedings

*Editors*
Yaxin Bi
Ulster University
Belfast, UK

Anne-Laure Jousselme
CS Group - France
La Garde, France

Thierry Denoeux
University of Technology of Compiègne
Compiègne, France

ISSN 0302-9743　　　　　　　ISSN 1611-3349　(electronic)
Lecture Notes in Artificial Intelligence
ISBN 978-3-031-67976-6　　　ISBN 978-3-031-67977-3　(eBook)
https://doi.org/10.1007/978-3-031-67977-3

LNCS Sublibrary: SL7 – Artificial Intelligence

© The Editor(s) (if applicable) and The Author(s), under exclusive license
to Springer Nature Switzerland AG 2024

This work is subject to copyright. All rights are solely and exclusively licensed by the Publisher, whether the whole or part of the material is concerned, specifically the rights of translation, reprinting, reuse of illustrations, recitation, broadcasting, reproduction on microfilms or in any other physical way, and transmission or information storage and retrieval, electronic adaptation, computer software, or by similar or dissimilar methodology now known or hereafter developed.
The use of general descriptive names, registered names, trademarks, service marks, etc. in this publication does not imply, even in the absence of a specific statement, that such names are exempt from the relevant protective laws and regulations and therefore free for general use.
The publisher, the authors and the editors are safe to assume that the advice and information in this book are believed to be true and accurate at the date of publication. Neither the publisher nor the authors or the editors give a warranty, expressed or implied, with respect to the material contained herein or for any errors or omissions that may have been made. The publisher remains neutral with regard to jurisdictional claims in published maps and institutional affiliations.

This Springer imprint is published by the registered company Springer Nature Switzerland AG
The registered company address is: Gewerbestrasse 11, 6330 Cham, Switzerland

If disposing of this product, please recycle the paper.

*Professor Arnaud Martin*

*To the memory of Arnaud Martin, founder of the Belief Functions and Applications Society, president from 2010 to 2012, and treasurer from 2012 to 2017.*

# Preface

The theory of belief functions, introduced by Arthur P. Dempster and Glenn Shafer in the 1960s and 1970s, is now well established as a general framework for reasoning under uncertainty, with well-understood connections to other frameworks such as random sets, possibility and imprecise probability theories. In the last 40 years, it has been applied to a wide range of problems in statistics, computer science and engineering. In particular, in recent years, it has inspired many important contributions to statistical inference and machine learning.

The series of biennial International Conferences on Belief Functions (BELIEF), sponsored by the Belief Functions and Applications Society (BFAS), is dedicated to the confrontation of ideas, the reporting of recent achievements, and the presentation of the wide range of applications of this theory. The first edition of this conference series was held in Brest, France, in 2010. Later editions were held in Compiègne, France in 2012, Oxford, UK in 2014, Prague, Czech Republic in 2016, again in Compiègne, France in 2018, Shanghai, China in 2021 and in Paris, France, in 2022.

The 8th International Conference on Belief Functions (BELIEF 2024) was held in Belfast, UK, September 2–4, 2024. This volume contains the proceedings of BELIEF 2024 composed of 30 accepted papers from the 36 submissions, each reviewed by either two or three peers in a single-blind review process. Original contributions address theoretical issues including continuous belief functions, random fuzzy sets, measures of uncertainty and conflict, as well as methods for solving various problems in machine learning, information fusion, statistical inference and optimization.

We would like to thank all the people who made this volume and this conference possible: all contributing authors, the organizers and the Program Committee members. We are especially grateful to our three invited speakers, Zhi-Hua Zhou (Nanjing University, China), for his talk *A Preliminary Exploration to Learnware*, Prakash P. Shenoy (University of Kansas, USA), for his talk *Knowing What You Don't Know: Making Inferences in Incomplete Bayesian Networks* and Frédéric Pichon (Artois University, France), for his talk *Reliability and dependence in information fusion*. We would also like to thank the Belief Functions and Applications Society and the School of Computing at Ulster University, UK for sponsoring the event, Ulster University for hosting the event, the International Journal of Approximate Reasoning and Elsevier. Furthermore, we would like to thank the editors of the Springer-Verlag series Lecture Notes in Artificial Intelligence (LNCS/LNAI) and Springer-Verlag for their dedication to the production of this volume.

June 2024

Yaxin Bi
Anne-Laure Jousselme
Thierry Denoeux

# Organization

## General Chairs

Yaxin Bi — Ulster University, UK
Anne-Laure Jousselme — CS Group, France

## Steering Committee

Thierry Denoeux — Université de Technologie de Compiègne, France
David Mercier — Université d'Artois, France
Frédéric Pichon — Université d'Artois, France

## Publicity Co-chairs

David Mercier — Université d'Artois, France
Anna Jurek-Loughrey — Queen's University Belfast, UK
Fuyuan Xiao — Chonqing University, China

## Publication Chair

Adrian Moore — Ulster University, UK

## Local Organization Committee

Naveed Khan — Ulster University, UK
Jorge Martinez Carracedo — Ulster University, UK
Colin Shewell — Ulster University, UK
Matias Garcia-Constantino — Ulster University, UK
Fiachra Merwick — Ulster University, UK
Maja Pavlovic — Ulster University, UK
Maurice Mulvenna — Ulster University, UK
Jun Liu — Ulster University, UK

## Program Committee

| | |
|---|---|
| Emanuel Aldea | Paris-Saclay University, France |
| Violaine Antoine | Limos, France |
| Alessandro Antonucci | IDSIA, Switzerland |
| Salem Benferhat | CRIL CNRS University of Artois, France |
| Yaxin Bi | Ulster University, UK |
| Andrea Campagner | IRCCS Ospedale Galeazzi Sant'Ambrogio, Italy |
| Leonardo Cella | Wake Forest University, USA |
| Davide Ciucci | University of Milano-Bicocca, Italy |
| Fabio Cuzzolin | Oxford Brookes University, UK |
| Thierry Denoeux | Université de technologie de Compiègne, France |
| Sébastien Destercke | Université de technologie de Compiègne, France |
| Zied Elouedi | ISG Tunis, Tunisia |
| Chao Fu | Hefei University of Technology, China |
| Glenn Ivan Hawe | University of Ulster, UK |
| Nam Huynh | Japan Advanced Institute of Science and Technology, Japan |
| Lianmeng Jiao | Northwestern Polytechnical University, China |
| Anne-Laure Jousselme | CS Group, France |
| Václav Kratochvíl | Academy of Sciences of the Czech Republic, Czech Republic |
| Sylvie Le Hégarat-Mascle | Université Paris-Saclay, France |
| Eric Lefèvre | University of Artois, France |
| Marie-Jeanne Lesot | LIP6, France |
| Xinde Li | Southeast University, China |
| Liping Liu | University of Akron, USA |
| Zhunga Liu | Northwestern Polytechnical University, China |
| Liyao Ma | University of Jinan, China |
| Ryan Martin | North Carolina State University, USA |
| David Mercier | University of Artois, France |
| Enrique Miranda | University of Oviedo, Spain |
| Serafin Moral | University of Granada Spain |
| Frédéric Pichon | University of Artois, France |
| Benjamin Quost | Université de technologie de Compiègne, France |
| Emmanuel Ramasso | FEMTO-ST, France |
| Sébastien Ramel | Université d'Artois, France |
| Roger Reynaud | Université Paris-Saclay, France |
| Prakash Shenoy | University of Kansas School of Business, USA |
| Zhi-gang Su | Southeast University, China |
| Kuang Zhou | Northwestern Polytechnical University, China |

# Contents

**Machine Learning**

Deep Evidential Clustering of Images .................................... 3
  Loïc Guiziou, Emmanuel Ramasso, Sébastien Thibaud,
  and Sébastien Denneulin

Incremental Belief-Peaks Evidential Clustering ........................... 13
  Chaoyu Gong, Sihan Wang, and Zhi-gang Su

Imprecise Deep Networks for Uncertain Image Classification ................ 22
  Chuanqi Liu, Zuowei Zhang, Zechao Liu, Liangbo Ning, and Zhunga Liu

Dempster-Shafer Credal Probabilistic Circuits ............................ 31
  David Ricardo Montalván Hernández, Thomas Krak,
  and Cassio de Campos

Uncertainty Quantification in Regression Neural Networks Using
Likelihood-Based Belief Functions ........................................ 40
  Thierry Denœux

An Evidential Time-to-Event Prediction Model Based on Gaussian
Random Fuzzy Numbers .................................................... 49
  Ling Huang, Yucheng Xing, Thierry Denœux, and Mengling Feng

Object Hallucination Detection in Large Vision Language Models
via Evidential Conflict ................................................. 58
  Zhekun Liu, Tao Huang, Rui Wang, and Liping Jing

Multi-oversampling with Evidence Fusion for Imbalanced Data
Classification .......................................................... 68
  Hongpeng Tian, Zuowei Zhang, Zhunga Liu, and Jingwei Zuo

An Evidence-Based Framework For Heterogeneous Electronic Health
Records: A Case Study In Mortality Prediction ........................... 78
  Yucheng Ruan, Ling Huang, Qianyi Xu, and Mengling Feng

Conflict Management in a Distance to Prototype-Based Evidential Deep
Learning ............................................................. 87
    Mihreteab Negash Geletu, Dǎnuţ-Vasile Giurgi,
    Thomas Josso-Laurain, Maxime Devanne,
    Jean-Philippe Lauffenburger, and Jean Dezert

A Novel Privacy Preserving Framework for Training Dempster-Shafer
Theory-Based Evidential Deep Neural Network ........................... 98
    Anh-Tu Tran, Van-Nam Huynh, and Viet-Hung Dang

**Statistical Inference**

Large-Sample Theory for Inferential Models: A Possibilistic
Bernstein–von Mises Theorem .......................................... 111
    Ryan Martin and Jonathan P. Williams

Variational Approximations of Possibilistic Inferential Models ............... 121
    Leonardo Cella and Ryan Martin

Decision Theory via Model-Free Generalized Fiducial Inference .............. 131
    Jonathan P. Williams and Yang Liu

Which Statistical Hypotheses are Afflicted with False Confidence? ........... 140
    Ryan Martin

Algebraic Expression for the Relative Likelihood-Based Evidential
Prediction of an Ordinal Variable ....................................... 150
    Frédéric Pichon and Sébastien Ramel

**Information Fusion and Optimization**

Why Combining Belief Functions on Quantum Circuits? .................... 161
    Qianli Zhou, Hao Luo, Éloi Bossé, and Yong Deng

SHADED: Shapley Value-Based Deceptive Evidence Detection in Belief
Functions ........................................................... 171
    Haifei Zhang

A Novel Optimization-Based Combination Rule for Dempster-Shafer
Theory .............................................................. 180
    Hasan Ihsan Turhan and Tugba Tanaydin

Fusing Independent Inferential Models in a Black-Box Manner .............. 189
    Leonardo Cella

Optimization Under Severe Uncertainty: a Generalized Minimax Regret
Approach for Problems with Linear Objectives .............................. 197
   *Tuan-Anh Vu, Sohaib Afifi, Éric Lefèvre, and Frédéric Pichon*

**Measures of Uncertainty, Conflict and Distances**

A Mean Distance Between Elements of Same Class for Rich Labels .......... 207
   *Arthur Hoarau, Constance Thierry, Jean-Christophe Dubois,
and Yolande Le Gall*

Threshold Functions and Operations in the Theory of Evidence .............. 216
   *Alexander Lepskiy*

Mutual Information and Kullback-Leibler Divergence
in the Dempster-Shafer Theory ........................................... 225
   *Prakash P. Shenoy*

An OWA-Based Distance Measure for Ordered Frames of Discernment ....... 234
   *Xiong Zhao, Liyao Ma, Yiyang Wang, and Shuhui Bi*

Automated Hierarchical Conflict Reduction for Crowdsourced Annotation
Tasks Using Belief Functions ............................................. 244
   *Constance Thierry, David Gross-Amblard, Yolande Le Gall,
and Jean-Christophe Dubois*

**Continuous Belief Functions, Logics, Computation**

Gamma Belief Functions .................................................. 255
   *Liping Liu*

Combination of Dependent Gaussian Random Fuzzy Numbers ................ 264
   *Thierry Denœux*

A 3-Valued Logical Foundation for Evidential Reasoning .................... 273
   *Chunlai Zhou*

Accelerated Dempster Shafer Using Tensor Train Representation ............. 283
   *Duc P. Truong, Erik Skau, Cassandra L. Armstrong, and Kari Sentz*

**Author Index** ......................................................... 293

# Machine Learning

# Deep Evidential Clustering of Images

Loïc Guiziou[1]([✉]), Emmanuel Ramasso[1], Sébastien Thibaud[1],
and Sébastien Denneulin[2]

[1] Department of Applied Mechanics, FEMTO-ST Institute – UMR CNRS 6174 - ENSMM/UFC/UTBM, 25000 Besançon, France
{loic.guiziou,emmanuel.ramasso}@femto-st.fr,
sebastien.denneulin@safrangroup.com
[2] SAFRAN CERAMICS, 33185 Le Haillan, France

**Abstract.** This paper presents a new image clustering method (DEEM) based on convolutional neural networks and the theory of belief functions used to encode uncertainty between clusters. The algorithm learns to generate mass functions for a given image through a training process that minimises a loss between the conflict computed from pairs of images and their dissimilarities. DEEM extends NN-EVCLUS and provides a gateway to the entire realm of deep learning, capitalising on all its advancements. It enables the full exploitation of the benefits offered by customisable layers, sophisticated optimisation algorithms, and other state-of-the-art techniques. DEEM can learn from the data itself, without requiring external labels but we can incorporate prior on labels if available as proposed in NN-EVCLUS. The first results are shown on the MNIST dataset (digit recognition).

**Keywords:** Belief functions · Unsupervised learning · Semi-supervised learning · Image clustering

## 1 Introduction

Clustering is a data analysis technique wherein data points are grouped based on their inherent similarities. The groups or clusters, are formed such that data points within the same cluster are more closely related to each other than to those in other clusters. Clusters can then be used to get insights in the data for further classification or interpretation.

Managing uncertainty in clustering is an important topic because real-world data often contains noise, outliers, and ambiguities, leading to uncertainty in the assignment of data points to clusters. By addressing uncertainty, clustering algorithms can produce more reliable and interpretable results, ensuring that the identified clusters accurately represent the underlying structure of the data.

In order to represent uncertainty, various formalisms can be used such as fuzzy sets [3,18], probability theory [7,15] or belief functions [4,13]. Using belief functions, doubt between clusters is explicitly represented. In practice, doubt allows the end-user to visualise the contours of clusters which helps in making informed decisions and drawing meaningful insights from the clustering process.

When it comes to image clustering, there are two primary approaches. On one hand, features can be extracted from the input, such as color histograms or merged pixel blobs [10,22]. On the other hand, the image itself can be treated as an input vector for conventional clustering methods. Despite the latter not fully considering the image structure, they can still yield satisfactory results. Both input methods may involve the use of deep clustering methods [17]. However, convolutional neural networks (CNN) are a class of method that leverages data structure for clustering. While often employed in supervised settings, recent applications of CNN have explored unsupervised and self-supervised learning. One common approach consists in a CNN serving as an encoder for the data, producing an output suitable for clustering methods like k-means. In the self-supervised setting, the clusters are used as a feedback to train the encoder like in [6]. The head of the CNN is then coupled with a MLP trained with the pseudo-labels using a discriminative loss.

Our approach, DEEM, diverges from traditional encoders as it directly provides cluster membership from input images, where the uncertainty is encoded by belief functions. Moreover, the option of partially or weakly supervised training is inherently integrated into the loss function, as detailed in the following section. DEEM is based on a recent method called NN-EVCLUS proposed by T. Denoeux [8]. Initially developed for clustering feature vectors with shallow networks, NN-EVCLUS is feature dependent, which requires to pay particular attention to feature relevance. DEEM handles image inputs which outstrips feature limitations. In addition, images can be processed by particular networks such as CNN, which have already shown great performance in supervised learning [12]. To our knowledge, this is the first neural network-based clustering method able to generate belief functions from images. By fully exploiting the benefits offered by customisable layers, sophisticated optimisation algorithms, and other state-of-the-art techniques, DEEM can represent a valuable method for image clustering under uncertainty and be used as a mass function generator.

The presentation of the method is described in Sect. 2 and Sect. 3 presents the results.

## 2 Method

### 2.1 Background on Belief Functions

A mass function on a finite set $\Omega$ is defined as a mapping of each element of the power set $2^\Omega$ onto $[0,1]$:

$$m : 2^\Omega \mapsto [0,1] \\ A \to m^\Omega(A) \text{ s.c. } \sum_A m^\Omega(A) = 1, m^\Omega(A) \geq 0 \qquad (1)$$

When $m^\Omega(S) > 0$, $A$ is called a *focal set*, and if $m^\Omega(\emptyset) = 0$ then the basic belief assignment (BBA) is said *normal*. A mass function can be transformed into several other functions which allows the end-user to get insights about the content of a mass and to make easier some computations [20,21]. One of these functions is the plausibility defined as

$$Pl(A) = \sum_{B \cap A \neq \emptyset} m(B), \forall B \subseteq \Omega \qquad (2)$$

As a particular case, the *contour function* related to $m$ is defined as $pl : \Omega \mapsto [0,1]$. It maps each singleton $\omega \in \Omega$ to its plausibility: $pl(\omega) = Pl(\{\omega\})$.

Given two mass functions, $m_1$ and $m_2$, defined on the same frame of discernment $\Omega$, their combination through the conjunctive rule is

$$(m_1 \cap m_2)(C) = \sum_{A \cap B = C} m_1(A) m_2(B), \forall C \subseteq \Omega \qquad (3)$$

From this combination, a conflict can arise when the intersection is empty between $A$ and $B$:

$$\kappa = (m_1 \cap m_2)(\emptyset) = \sum_{A \cap B = \emptyset} m_1(A) m_2(B) \qquad (4)$$

This conflict has been used in several algorithms based on belief functions such as target association [1] or the evidential hidden Markov model [16] since it is related to the likelihood. It has a very important role in NN-EVCLUS as shown subsequently.

## 2.2 NN-EVCLUS for Feature Vector Clustering

NN-EVCLUS relies on a shallow neural network parameterised by $\boldsymbol{\theta}$ taking as inputs a $d$-dimensional feature vector $\boldsymbol{x}_i \in \Re^d$ and generating a vector of masses $\boldsymbol{m}_i$. Together, the BBA form a collection called credal partition $m = \{m_1, ..., m_N\}$ for $N$ objects [9]. The number of outputs in the last layer of the network depends on the complexity of the mass that the end-user considered as necessary to represent the uncertainty on clusters. Classically, the singletons, the pairs and the whole frame are often sufficient. The remaining description does not depend on this limitation but due to the calculation complexity, adding more subsets will considerably increase the computation time.

The core idea of NN-EVCLUS is to quantify the conflict between each pair of mass functions $(\boldsymbol{m}_i, \boldsymbol{m}_j)$ and to use this conflict to update the network parameters. Therefore, NN-EVCLUS has some analogy with the siamese network (SN) [5] since the network parameters are fixed for a given pair of inputs $(\boldsymbol{x}_i, \boldsymbol{x}_j)$. In a SN, the true labels of individual inputs are not known, but we have to know whether each pair represents a similar (e.g. label "1") or a dissimilar (label "0") object. Based on this prior knowledge, the parameters of a SN are updated according to the cross-entropy loss or a contractive loss.

In NN-EVCLUS, the loss function is slightly different and the method unsupervised. It is defined as

$$\mathcal{L} = \sum_{i=1}^{N} \sum_{j>i} (\delta_{ij} - \kappa_{ij}(\boldsymbol{\theta}))^2 \qquad (5)$$

where $\kappa_{ij}(\boldsymbol{\theta})$ is the conflict between the mass functions $\boldsymbol{m}_i$ and $\boldsymbol{m}_j$ generated after the forward pass using $\boldsymbol{x}_i$ and $\boldsymbol{x}_j$ as inputs and with network parameters $\boldsymbol{\theta}$. The value $\delta_{ij}$ is the dissimilarity between the two inputs. Dissimilarity can be computed in many ways. In the simplest case, we will use the Euclidean distance. The values are then scaled between 0 and 1 to be comparable to conflict. This approach presupposes that the greater the distance between images, the more likely the masses are to be in conflict. The main idea behind this loss is that if the network generates two conflicting masses while the inputs are similar, then the parameters should be updated in consequence. Conversely, if two inputs are similar and generate two masses with limited conflict, then the network behaves properly.

### 2.3 DEEM for Image Clustering

An implementation of the shallow version of NN-EVCLUS was done in R by T. Denoeux [8]. It requires inputs as feature vectors as defined in the original version. The implementation includes the possibility to tune up to three layers with an arbitrary number of neurons and using the ReLU activation function in all layers. Our first implementation was done in Matlab, with the possibility to manage any number of layers, to use all possible activation functions and various optimisers.

We then explored the possibility to improve this implementation in order to manage images as inputs and to change the loss. Therefore, the shallow network was replaced by a CNN. These networks, trained in an unsupervised manner, can extract useful visual features and representations that can be used for other tasks [6]. Implementation was done in both Matlab and Pytorch with GPU compatibility. The main contribution is to give the possibility to perform image clustering in an end-to-end manner without the need of manual feature extraction. Here, the objective of the CNN is to output the evidential partition of the images and does not aim to transform the images into features as auto-encoders do. As with the previous method, the output masses enable clustering and the calculation of the loss function, which depends on the conflict.

Thus, the input data consists of vectorised and normalised raw images. Given the large amount of data typically involved in general image databases, the implementation relies on minibatches. These minibatches are then processed by the CNN, which outputs the masses of the power set, including singletons, pairs, the empty set, and the universal set. Given the complexity of the computation depending on the number of elements and the number of focal sets, it is difficult to add higher-order intersections to the calculation. For 10 classes, this results in an output of 57 masses. The conflict is then extracted from these masses and integrated into the loss function. In the DEEM code, it is possible to change the loss function easily, but the configuration used for the tests remains the same Mean Squared Error (MSE) loss as in Eq. 5. Conversely to the general approach for image clustering which is based on prediction of the cluster assignments used as pseudo-labels in a discriminative loss, DEEM relies on the MSE loss defined in NN-EVCLUS. The way the dissimilarity $\delta_{ij}$ is computed is critical, in particular for image since they have a structure conversely to standard features. If natural images are considered, several dissimilarity measures proposed in the past can be used [11].

In the original publication, T. Denoeux provided the gradient of the loss with respect to all parameters. These gradients can be useful according to the way the optimisation is implemented. In our case, when considering images, the convolutional layers particularly make these computations and implementation more difficult. Therefore, the gradients were found by automatic differentiation (AD) [2] which allows the end-user to easily configure the network as desired for a given application. The associated optimiser is Adam, which, once again, can be easily swapped with others such as Stochastic Gradient Descent with Momentum (SGDM) or Root Mean Squared Propagation (RMSProp). The DEEM code also allows for real-time calculation of ARI and other evaluators. This calculation is performed by evaluating the predicted clusters against the expected ones at each iteration. In our case, clustering is done by assigning the class of the subgroup where the plausibility function is the strongest. Thus, the plausibility calculation is performed on the 10 sets, with the winning cluster being the one with the highest value. Finally, it is difficult to define the stopping criterion for training the network because the loss varies greatly, and a threshold condition would not be adequate. For now, the number of iterations is the stopping criterion for training.

## 3 Results

### 3.1 Clustering Performances on Digits

The network's efficiency was initially evaluated using the MNIST standard dataset, which is made of images of the ten digits. One of the primary challenges involved configuring the network architecture and find a set of hyperparameters. The convergence outcomes are significantly influenced by the selection of layers and optimisation rules applied to each layer. One effective network configuration is illustrated in Fig. 1.

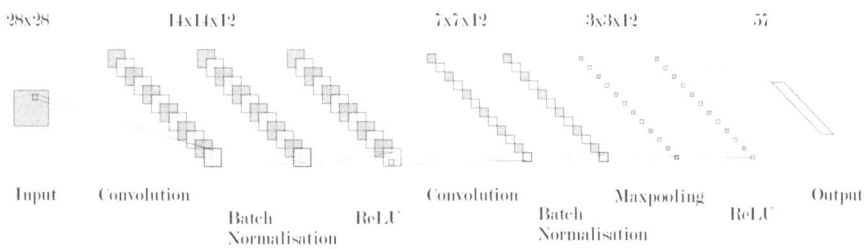

Fig. 1: Neural network for image handling of NNEVCLUS

**Validation of the Implementation** – Initial testing was conducted under favorable conditions, recognising the substantial impact of dissimilarity calculations between images on convergence. Dissimilarities were first computed based on known labels, facilitating the network's convergence to an adjusted rand index (ARI) of 1 without any errors. Even though this method is not supervised in the sense that the labels are not directly incorporated into the loss function, it remains quasi-supervised. As a result, attempts to generalise on 60,000 images after training on 5,000 images achieve high scores: an ACC of 0.97, an NMI of 0.92, and an ARI of 0.93.

Grad-CAM analysis [19] on the test data revealed a logical distribution of areas of interest activated by neurons. Figure 2 depicts each class in a scenario where the network confused digits 6 and 8.

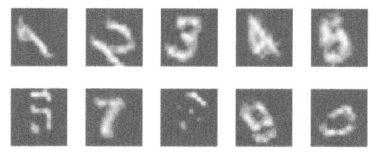

Fig. 2: Grad-CAM for each cluster

**Using Distance on Images** – When dissimilarities were calculated solely through cosine distance between images, without relying on labels, the network's performance decreased to an ARI of 0.25 and a rand index (RI) of 0.8. The disparity between ARI and RI arises from clustering multiple classes, thereby increasing the likelihood of errors across various classes. Errors predominantly occurred with digits that are challenging to distinguish or bear high resemblance, such as digits 1 and 7. This challenge is obvious in the dissimilarity matrix, where certain classes appear closely related. Besides, using distances between images to represent dissimilarity is a risky approach. Indeed, a simple translation between similar images significantly increases the distance. Therefore, other methods of computing dissimilarity can be employed. First, there is the extraction of image-specific features with methods such as Scale Invariant Feature Transform (SIFT) or Oriented FAST and Rotated BRIEF (ORB). Then, there are methods for comparing colour histograms, which are not suitable in our case where the images are black and white. Finally, there are methods that consider the structure of the data, such as Structural Similarity Index Measure (SSIM), or other contour detection methods like Fourier Descriptors, Shape Context Matching, and others. For now, this paper focuses solely on exploiting distances by applying dimensionality reduction to the images.

**Using Distance on Reduced Images** – To enhance the reliability of the dissimilarity matrix, visualisation techniques and dimensional reduction algorithms such as principal components analysis (PCA) and uniform manifold approximation and projection (UMAP) were employed as it can improves clustering performance [14]. Utilising UMAP as an unsupervised method to project the data and derive meaningful dissimilarities based on distance significantly improved clustering performance, resulting in an ARI increase up to 0.65. In addition, using t-distributed Stochastic Neighbor Embedding (t-SNE) reduction increases the ARI up to 0.73.

**Using Pre-clustering on Reduced Images** – As we have seen, the closer the dissimilarity matrix is to the labels, the easier it is for the algorithm to converge. However, there's nothing to rule out the use of an initial unsupervised clustering to obtain the dissimilarity matrix. The interest of this method then lies in the generalisation of this model. Indeed, on 60,000 test images, a conventional method consisting in reducing the images via UMAP and then applying clustering with K-means is relatively time-consuming, taking 25 s on average compared with 0.5 for DEEM. In addition, DEEM's generalisability is much better, as shown in Table 1.

Table 1: Comparison of clustering methods on 60K testing MNIST images

|     | DEEM | K-means | Hierarchical |
|-----|------|---------|--------------|
| ACC | **0.96** | 0.84 | 0.83 |
| NMI | **0.89** | 0.75 | 0.76 |
| ARI | **0.91** | 0.71 | 0.76 |

## 3.2 Effect of Dissimilarity Calculation Method

As specified in Sect. 2.2, the loss function calculates the difference between conflict and dissimilarity of two objects. For this reason, the calculation of the later is crucial for clustering. Different distances calculation methods can be employed such as Euclidean distance or cosine distance among others. Furthermore, extracted features as for original NN-EVCLUS can be used to generate the dissimilarity matrix instead of the set of images. To illustrate the dependence of the network on the dissimilarity, clustering was carried out on a two-class digit clustering for several dissimilarity matrices. To do so, a set of 10 matrices were designed from the perfect binary one to fuzzier ones.

As shown on Fig. 3a, a random Gaussian noise is added to the previous binary matrix for different range of standard deviation ranging from 0 to 0.8. Each of the 10 clustering was performed 30 times to take into account variability in the convergence. Finally, results are expressed thanks to ARI. The mean of results for each matrix is shown on Fig. 3b. The algorithm decently tolerates fuzziness until the seventh point corresponding to 0.53 standard deviation of the Gaussian noise added, which is pretty much encouraging. After that, the algorithm struggles to achieve convergence. Moreover, Gaussian noise was chosen to be the closest from real cases, but conventional random noise also provides similar results. Finally, the first point of the diagram draws our attention to its low value. Indeed, perfect dissimilarity leads to a drop in performance. In such a case, the loss function seems to be highly sensitive to small changes. This leads to instability during training, with small fluctuations in the predicted dissimilarities causing disproportionately large changes. Using ADAM optimiser, a lower learning rate helps to get through the issue and gives a perfect percentage of convergence. The architecture was not optimised which can also explains the sensitivity.

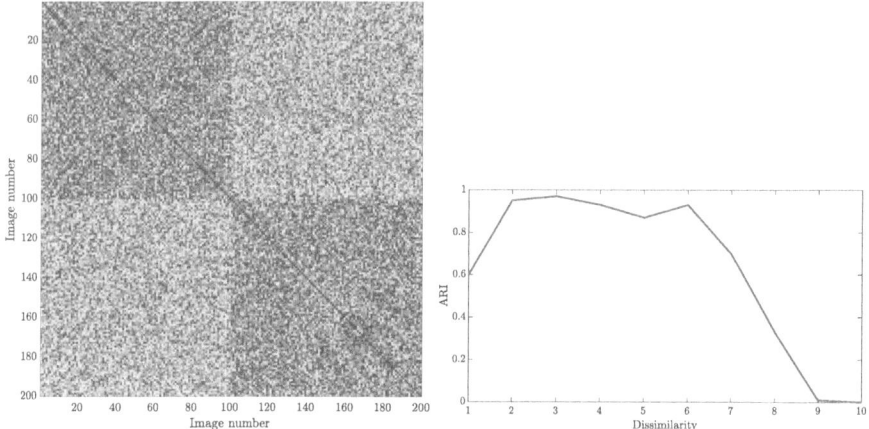

(a) Dissimilarity matrix with added Gaussian noise

(b) ARI variation as a function of fuzziness in dissimilarity

Fig. 3: Noised distance experimentation

## 4 Conclusion

DEEM is an extended version of NN-EVCLUS algorithm to handle images instead of feature vectors. The layers and optimisers were adapted for a highly configurable neural network. We firstly showed performance of the new algorithm on MNIST data in a favorable configuration where distances where determined thanks to the labels. By the desire to have a true unsupervised learning, the next tests were conduced using distance dissimilarity which leads to confusion between close digits such as 1 and 7. In order to improve the performance, dissimilarity where calculated on UMAP reduction, and highly increased the ARI. Finally, further work remains to be done, particularly in try to cluster acoustic emission spectrograms. The results of DEEM were promising on several applications, either on standard benchmarks and from our laboratory. Its generalisation capabilities were also highly encouraging. Current work is on an extensive experimental validation of DEEM and its comparison with other algorithms of the literature on large datasets. Finally, as NN-EVCLUS did, it is possible to add constraints to some of the labels. This feature seems promising, but has yet to be quantified in terms of the number of apriori used.

**Acknowledgments.** This work has been achieved in the frame of the EIPHI Graduate school (contract "ANR-17-EURE-0002"), with the financial support of SAFRAN CERAMICS and Agence de l'innovation de défense (AID).

# References

1. Ayoun, A., Smets, P.: Data association in multi-target detection using the transferable belief model. Int. J. Intell. Syst. **16**(10), 1167–1182 (2001). https://doi.org/10.1002/int.1054
2. Bartholomew-Biggs, M., Brown, S., Christianson, B., Dixon, L.: Automatic differentiation of algorithms. J. Comput. Appl. Math. **124**(1), 171–190 (2000). https://doi.org/10.1016/S0377-0427(00)00422-2
3. Bello, R., Falcón, R., Pedrycz, W., Kacprzyk, J.: Granular Computing: At the Junction of Rough Sets and Fuzzy Sets. Springer, Heidelberg (2008). https://doi.org/10.1007/978-3-540-76973-6
4. Ben Hariz, S., Elouedi, Z., Mellouli, K.: Clustering approach using belief function theory. In: Euzenat, J., Domingue, J. (eds.) AIMSA 2006. LNCS (LNAI), vol. 4183, pp. 162–171. Springer, Heidelberg (2006). https://doi.org/10.1007/11861461_18
5. Bromley, J., Guyon, I., LeCun, Y., Säckinger, E., Shah, R.: Signature verification using a "Siamese" time delay neural network. Adv. Neural Inf. Process. Syst. **6** (1993). https://doi.org/10.1142/S0218001493000339
6. Caron, M., Bojanowski, P., Joulin, A., Douze, M.: Deep clustering for unsupervised learning of visual features. In: Ferrari, V., Hebert, M., Sminchisescu, C., Weiss, Y. (eds.) Computer Vision – ECCV 2018. LNCS, vol. 11218, pp. 139–156. Springer, Cham (2018). https://doi.org/10.1007/978-3-030-01264-9_9
7. Chen, L., Brown, S.D.: Bayesian estimation of membership uncertainty in model-based clustering. J. Chemometr. **28**(5), 358–369 (2014). https://doi.org/10.1002/cem.2511
8. Denœux, T.: NN-EVCLUS: neural network-based evidential clustering. Inf. Sci. **572**, 297–330 (2021). https://doi.org/10.1016/j.ins.2021.05.011
9. Denœux, T., Masson, M.H.: EVCLUS: evidential clustering of proximity data. IEEE Trans. Syst. Man Cybern. Part B Cybern. **34**, 95–109 (2004). https://doi.org/10.1109/TSMCB.2002.806496
10. Gao, B., Liu, T.Y., Qin, T., Zheng, X., Cheng, Q.S., Ma, W.Y.: Web image clustering by consistent utilization of visual features and surrounding texts. In: Proceedings of the 13th Annual ACM International Conference on Multimedia, pp. 112–121. ACM (2005). https://doi.org/10.1145/1101149.1101167
11. Goshtasby, A.A.: Similarity and dissimilarity measures. In: Goshtasby, A.A. (ed.) Image Registration, pp. 7–66. Springer, London (2012). https://doi.org/10.1007/978-1-4471-2458-0_2
12. Li, Z., Liu, F., Yang, W., Peng, S., Zhou, J.: A survey of convolutional neural networks: analysis, applications, and prospects. IEEE Trans. Neural Netw. Learn. Syst. **33**(12), 6999–7019 (2021). https://doi.org/10.1109/TNNLS.2021.3084827
13. Liu, Z., Pan, Q., Dezert, J., Mercier, G.: Credal C-means clustering method based on belief functions. Knowl.-Based Syst. **74**, 119–132 (2015). https://doi.org/10.1016/j.knosys.2014.11.013

14. Allaoui, M., Kherfi, M.L., Cheriet, A.: Considerably improving clustering algorithms using UMAP dimensionality reduction technique: a comparative study. In: El Moataz, A., Mammass, D., Mansouri, A., Nouboud, F. (eds.) ICISP 2020. LNCS, vol. 12119, pp. 317–325. Springer, Cham (2020). https://doi.org/10.1007/978-3-030-51935-3_34
15. Melnykov, V.: Challenges in model-based clustering. WIREs Comput. Stat. **5**(2), 135–148 (2013). https://doi.org/10.1002/wics.1248
16. Ramasso, E.: Contribution of belief functions to hidden Markov models with an application to fault diagnosis. In: 2009 IEEE International Workshop on Machine Learning for Signal Processing, pp. 1–6. IEEE (2009). https://doi.org/10.1109/MLSP.2009.5306209
17. Ren, Y., et al.: Deep clustering: a comprehensive survey (2022)
18. Sadeghian, A., Mendel, J.M., Tahayori, H.: Advances in Type-2 Fuzzy Sets and Systems: Theory and Applications. Springer, New York (2013). https://doi.org/10.1007/978-1-4614-6666-6
19. Selvaraju, R.R., Cogswell, M., Das, A., Vedantam, R., Parikh, D., Batra, D.: Grad-CAM: visual explanations from deep networks via gradient-based localization. Int. J. Comput. Vision **128**(2), 336–359 (2017). https://doi.org/10.1007/s11263-019-01228-7
20. Sentz, K., Ferson, S.: Combination of evidence in Dempster-Shafer theory (SAND2002-0835, 800792). https://doi.org/10.2172/800792
21. Smets, P., Kennes, R.: The transferable belief model. Artif. Intell. **66**(2), 191–234 (1994). https://doi.org/10.1016/0004-3702(94)90026-4
22. Leung, Y., Zhang, J.-S., Xu, Z.-B.: Clustering by scale-space filtering. IEEE Trans. Pattern Anal. Mach. Intell. **22**(12), 1396–1410 (2000). https://doi.org/10.1109/34.895974

# Incremental Belief-Peaks Evidential Clustering

Chaoyu Gong[1(✉)], Sihan Wang[2], and Zhi-gang Su[1]

[1] Southeast University, Nanjing, China
chaoyugong123@gmail.com
[2] National University of Singapore, Singapore, Singapore

**Abstract.** Despite evidential clustering has been widely employed in pattern recognition problems characterized by imprecise and uncertain membership, its application in the realm of big data remains constrained by excessive computational complexity and limited computational resources. To bridge this research gap, this paper introduces an **I**ncremental **E**vidential **C**lustering (IEC) method based on stream data clustering and belief-peaks, a technique that has demonstrated exceptional effectiveness in detecting cluster centers under numerous cases. Commencing with an initial small dataset, IEC progressively scales its operations to accommodate escalating volumes of data. Through judicious employment of scalar transformations, IEC cost-effectively updates the affinity matrix, ensuring incremental adjustments. Targeting solely the altered segments of the affinity matrix, recalibration of cluster centers takes place currently. For voluminous datasets, users can also adjust the update frequency according to their specific requirements. Compared to state-of-the-art (SoTA) stream clustering algorithms, IEC demonstrates better clustering accuracy and comparable runtime across four benchmark datasets. IEC markedly diminishes runtime contrasted with other SoTA evidential clustering algorithms.

**Keywords:** Incremental Clustering · Evidential Clustering · Belief-peak

## 1 Introduction

Clustering is regarded as a fundamental task within the realms of data mining [5]. Its principal objective is the group of a set of $n$ objects into $c$ distinct clusters, where objects within the same cluster demonstrate greater similarity than those in different clusters [6]. Most clustering algorithms tend to produce either hard partitions or fuzzy partitions [7]. Hard partitioning methods forcefully allocate objects to specific clusters, while fuzzy partitioning methods, grounded in fuzzy set theory, assign objects with Bayesian-like memberships [5]. The evidential clustering, first introduced by Thierry in [3], combines the Dempster-Shafer (DS) theory [14] into clustering methodologies. Owing to its enhanced capacity to describe imperfect information, evidential clustering extends beyond not only traditional hard, fuzzy, but also possibilistic, and rough clustering paradigms. Evidential clustering has also found successful applications across various domains, including machine fault diagnosis [8] and image processing [5].

**Related Work.** Two main research lines are distinguished in evidential clustering [3] based on the nature of available data: objective or relational. The first, exemplified by the evidential $c$-means (ECM) algorithm [11], deals with object data using a prototype-based framework and optimizing an objective function incorporating credal partitioning principles. In [5,9,15], belief-peaks evidential clustering and its variations were introduced as an extension of density-peaks clustering. Belief-peaks visualize the possibility of objects serving as cluster centers on a two-dimensional plot, facilitating automatic center selection. Another research line pertains to the relational data. The pioneering algorithm in this domain is EVCLUS [3], which endeavors to construct a credal partition such that the conflict degree between mass functions associated with any two objects aligns with their similarity. CEVCLUS [1] was formulated leveraging must-link and cannot-link constraints, a method subsequently extended to $k$-CEVCLUS [10] to address the challenges posed by datasets of moderate scale. Gong introduced the BPDNEC method in [7], where cluster centers are autonomously determined by solving an equation linked to neighborhood size. The credal partition is formulated by minimizing an objective function based on the relational dissimilarity matrix of objects. Nonetheless, none of these methods explicitly tackle the clustering of data exhibiting large sizes.

**Motivation.** With the progression of technological advancements, the acquisition of extensive datasets featuring millions of objects has become significantly facilitated, denoting the realm of big data [8]. The burden posed by big data issues on evidential clustering algorithms is substantial, given that the time complexity of these algorithms largely resides in $O(n^2)$, where $n$ represents the data size. To the best of our knowledge, the solitary study aimed to scale up evidential clustering algorithms to address the challenges presented by datasets of millions or even greater is [8], which employed multiple computation nodes and a distributed computing framework. However, practical constraints may preclude the sustained online utilization of multiple computation nodes. For instance, conducting evidential clustering analysis of historical operational data from industrial equipment in remote mountainous regions may necessitate the use of a single notebook computer with limited memory. In this case, it may be viable to draw upon the idea of stream data clustering [12] to enhance evidential clustering algorithms.

Motivated by the above challenge, this paper introduces an incremental evidential clustering algorithm to address the gap between stream data clustering and evidential clustering methods. Commencing with a modest dataset, the Incremental Evidential Clustering (IEC) method progressively scales its operations to handle growing data volumes. Through judicious scalar transformations, IEC efficiently updates the affinity matrix, ensuring incremental adjustments. Targeting only modified segments of the affinity matrix, recalibration of cluster centers and memberships occurs concurrently. Compared to the Evidential Evolving Gustafson-Kessel (E2GK) method [13], IEC allows for the possibility of a decrease in the number of clusters ($c$) during data streaming, while E2GK solely allows for an increase in $c$. This feature, more aligned with practical scenarios, acknowledges the possibility of cluster mergers as data streaming. The contribution of this paper is threefold:

- IEC fills the research gap between streaming data clustering and evidential clustering, providing a new approach to tackle computation constraints in big data issues;
- Through basic matrix transformations, IEC dynamically updates belief-peaks and autonomously identifies cluster numbers and centers, eliminating the need for manual setup;
- Compared to SoTA methods in streaming data clustering and evidential clustering, IEC preserves clustering accuracy while significantly reducing computation time.

## 2 Preliminaries

**Evidence Theory.** A variable $\omega$ takes values from a finite set known as the *frame of discernment* $\Omega = \{\omega_1, \omega_2, \ldots, \omega_c\}$. A mass function $m$ is defined as a mapping from $2^\Omega$ to [0,1], where $\sum_{\mathcal{A} \subseteq \Omega} m^\Omega(\mathcal{A}) = 1$. The subsets $\mathcal{A}$ satisfying $m^\Omega(\mathcal{A}) > 0$ are termed the *focal sets* of $m$. The value of $m^\Omega(\mathcal{A})$ denotes a portion of unit mass allocated to the focal set $\mathcal{A}$, which cannot be assigned to any proper subset of $\mathcal{A}$. Other equivalent representations of a mass function $m^\Omega$ are

$$Bel^\Omega(\mathcal{A}) = \sum_{\emptyset \neq \mathcal{B} \subseteq \mathcal{A}} m^\Omega(\mathcal{B}) \qquad (1)$$

and $Pl^\Omega(\mathcal{A}) = \sum_{\mathcal{A} \cap \mathcal{B} \neq \emptyset} m^\Omega(\mathcal{B})$.

**Credal Partition.** Consider a dataset $\mathcal{X}$ consisting of $n$ objects $\{\mathbf{x}_1, \mathbf{x}_2, \cdots, \mathbf{x}_n\}$. Each object $\mathbf{x}_i, i = 1, 2, \cdots, n$, is denoted as a feature vector. The set of $c$ clusters is denoted as the $\Omega = \{\omega_1, \omega_2, \cdots, \omega_c\}$. When $\mathbf{x}_i$ can not be assigned to the clusters with certainty, one can represent ambiguous and uncertain cluster memberships by a mass function $m_i^\Omega$. The degree of support for the proposition that "the true cluster of object $\mathbf{x}_i$ is in $\mathcal{A}$" is interpreted as the mass value $m_i^\Omega(\mathcal{A})$. The n-tuple $\mathcal{M}^\Omega = (m_1^\Omega, m_2^\Omega, \cdots, m_n^\Omega)$ is known as a credal partition [3], which enables objects to be in a composite cluster (defined as the union of several single clusters) in addition to being contained in a single cluster. A credal partition can be transferred to a hard/fuzzy partition. More details can be found in [5].

## 3 Method: Incremental Evidential Clustering

**Basic Idea.** For a dataset $\mathcal{X}$ with $n$ objects, IEC computes the belief-peaks of objects in an initial dataset $\mathcal{X}_0$ ($n_0 \ll n$) and establishes cluster numbers and centers. With each addition of new objects to the processing dataset $\mathcal{X}_p$, updates are made to the belief-peaks. Upon surpassing a predefined size threshold $\epsilon$, objects are selectively removed from the dataset. When all objects have been processed, the cluster centers are ultimately determined, and the corresponding credal partition is also derived.

**Initialization.** Let $\mathcal{N}_K^{\mathcal{X}_0}(\mathbf{x}_i)$ denote the set of $K$ nearest neighbors (KNNs) of object $\mathbf{x}_i$ in $\mathcal{X}_0$. Define two frame of discernments, $\Omega = \{\omega_1, \omega_2, \cdots, \omega_c\}$ denoting the set of clusters and $\mathcal{C} = \{C, \neg C\}$ discerning whether an object is a cluster center ($C$) or not ($\neg C$). Each neighbor $\mathbf{x}_j$ in $\mathcal{N}_K^{\mathcal{X}_0}(\mathbf{x}_i)$ provides a piece of evidence about $\mathbf{x}_i$ being a cluster center, represented by $m_{ij}^{\mathcal{C}}$

$$m_{ij}^{\mathcal{C}}(\mathcal{A}) = \begin{cases} \exp(-\gamma_i^2 d_{ij}^2), & \mathcal{A} = \{C\} \\ 1 - \exp(-\gamma_i^2 d_{ij}^2), & \mathcal{A} = \mathcal{C} \end{cases} \quad (2)$$

where $\gamma_i$ is an object-specific hyperparameter. By combining all the $K$ mass functions $(m_{i1}^{\mathcal{C}}, m_{i2}^{\mathcal{C}}, \cdots, m_{iK}^{\mathcal{C}})$ using Dempster's rule, normalized mass function $m_i^{\mathcal{C}}$ can be obtained $m_i^{\mathcal{C}} = \bigoplus m_{ij}^{\mathcal{C}}$. According to (1), the corresponding belief function $Bel_i^{\mathcal{C}}(\{C\})$, denoting the belief (possibility) of $\mathbf{x}_i$ to be a cluster center, are computed by

$$Bel_i^{\mathcal{C}}(\{C\}) = m_i^{\mathcal{C}}(\{C\}) = 1 - \left[1 - \phi(d_{ij}^2)\right]. \quad (3)$$

Another *delta* metric $\delta_i$ is defined as

$$\delta_i = \min_{\{j: Bel_j^{\mathcal{C}}(\{C\}) > Bel_i^{\mathcal{C}}(\{C\})\}} d_{ij} \quad (4)$$

for objects that do not have the highest degree of belief and $\delta_i = \max_{1 \leq j \leq n_0}\{d_{ij}\}$ for the object with highest degree of belief. In [15], Su and Thierry manually select objects that are noticeably distinct from ordinary samples in the $Bel - delta$ plot as cluster centers. However, manual selection cannot participate in real-time incremental clustering processes. Thus, IEC automatically selects objects as cluster centers by computing the Z-score of $Bel + delta$, whereby objects with a Z-score greater than 3.5 are automatically chosen as cluster centers.

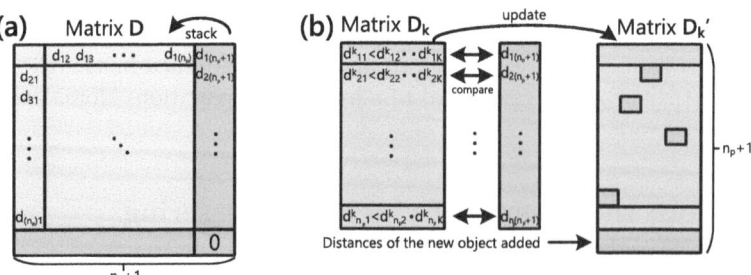

**Fig. 1.** Scalar transformations within matrices $\mathbf{D}$ (a) and $\mathbf{D}_k$ (b). The rectangles ■ and $\mathbf{D}_k'$ in sub-figure (b) represent the scalars updated in $\mathbf{D}_k$ and the updated $\mathbf{D}_k$.

**Incremental Process.** Equations (2)–(4) reveal that the calculations of *Bel* and *delta* (i.e., the determination of cluster centers) rely on $\mathbf{D}_k \in \mathbb{R}^{n \times K}$ (distance matrix between objects and their KNNs, where the scalars in each row are arranged from left to right in ascending order) and affinity matrix $\mathbf{D} \in \mathbb{R}^{n \times n}$. When a new object joins $\mathcal{X}_p$, the corresponding alteration in $\mathbf{D}_k$ and $\mathbf{D}$ is marginal, thereby necessitating only some slight scalar transformations for the swift update of *Bel* and *delta*.

In Fig. 1, we show the scalar transformations in $\mathbf{D}$ and $\mathbf{D}_k$. After the addition of each new object to $\mathcal{X}_p$, its distances to all objects in $\mathcal{X}_p$ are computed. The resulting distance vector $\mathbf{d}_p = (d_{i(n_p+1)}) \in \mathbb{R}^{n_p \times 1}$ is stacked into the distance matrix $\mathbf{D}$, as shown in Fig. 1(a). When updating $\mathbf{D}_k$ shown in Fig. 1(b), all scalars in vector $\mathbf{d}_p$ are compared to the scalars (denoted as $d_{ik}^k$) in the $k$-th column of $\mathbf{D}_k$. If $d_{i(n_p+1)}$ is smaller than $d_{ik}^k$, $d_{i(n_p+1)}$ is inserted into the respective position in the $i$-th row of $\mathbf{D}_k$ based on its value. The distances between the new object added and its KNNs are also stacked into $\mathbf{D}_k$. After the transformations mentioned above, *Bel* **is recalculated only for the objects corresponding to the row indices where changes occurred in $\mathbf{D}_k$ (also for the new object added)**. Subsequently, *delta* is re-computed based on Eq. (4).

When the dataset size of $\mathcal{X}_p$ exceeds threshold $\epsilon$, adding an object entails discarding the one with the smallest Z-score from $\mathcal{X}_p$. Noted that the object with the lowest Z-score is less likely to serve as a cluster center, as well as its *delta*, is very low. Intuitively, its "influence" and "presence" within $\mathcal{X}_p$ is smallest. After ultimate determination of the cluster centers, the cluster memberships for $\mathbf{x}_i$, described by mass functions $m_i^\Omega$, are calculated by

$$m_{iz}^C(\mathcal{A}) = \begin{cases} m_{iz}^\Omega = \dfrac{|A_j|^{-1} d_{iz}^{-2}}{\sum_{l:A_l \neq \phi} |A_l|^{-1} d_{il}^{-2} + 0.1^{-2}}, & A_z \neq \phi \\ m_{i\phi}^\Omega = 1 - \sum_{j:A_j \neq \phi} m_{ij}^\Omega, & \end{cases} \quad (5)$$

where $d_{iz}^2 = \|\mathbf{x}_i - \overline{v}_z\| = (\mathbf{x}_i - \overline{v}_z)'|A_z|^{-1}(\mathbf{x}_i - \overline{v}_z)$, $\overline{v}_z$ is the center of composite cluster $A_z$, $\overline{v}_z = \dfrac{1}{|A_z|} \sum_{k=1}^{c} s_{kz} v_k$ with $s_{kz} = 1$, if $\omega_k \in A_z$; 0, otherwise. The IEC method can be summarized in Algorithm 1.

---

**Algorithm 1.** IEC method

---

**Input:** $\mathcal{X}$, $\mathcal{X}_0$, $K$, $num_n$, $\epsilon$
**Output:** Credal partition $\mathcal{M}$
1: Calculate *Bel* and *delta* for each object in $\mathcal{X}_0$
2: **repeat**
3:     Add $num_n$ object(s) into $\mathcal{X}_p$
4:     Update the *Bel*, *delta* and determine the cluster centers
5:     If $|\mathcal{X}_p|$ exceeds $\epsilon$, remove $num_n$ object(s) with the lowest Z-scores from $\mathcal{X}_p$
6: **until** All the objects have been processed
7: Calculate clustering memberships based on Eq.(5)

---

**Reduction in Time Complexity.** The time complexity of IEC primarily depends on the dataset size $n$. Computing $Bel$ and $delta$ has a complexity of $O(p^2 n_0(n_0 - 1)/2) + O(2^2 K n_0)$ for the initial dataset $\mathcal{X}_0$. The most time-consuming terms are exploring KNNs and computing $delta$ for each object, both with $(n_0)^2$ complexity. However, during the incremental process, recalculating $Bel$ is reduced to $O(p^2 K n_c)$, where $n_c$ is the object number with changes in the KNNs. Since the largest size of $X_p$ is $\epsilon$, the distance matrix $\mathbf{D}$ can be stored in memory. Thus, computing $delta$ during the incremental process only requires scanning the $\mathbf{D}$, reducing complexity to $O(n \log n + n)$. Roughly speaking, incremental processing reduces complexity from $O(n^2)$ to $O(n \log n)$. If users seek to further economize computational time and computer memory, they may increase the number $num_n$ of objects added each time to $\mathcal{X}_p$. For instance, if $num_n = 10$ objects are added to $\mathcal{X}_p$ each time, and if the size of $\mathcal{X}_p$ exceeds the threshold $\epsilon$, then the 10 objects with the lowest Z-scores are removed.

## 4 Experimental Results

We conduct experiments to answer the following three research questions (RQs):

- **RQ1 (Comparison):** does IEC outperform baseline methods on different benchmark datasets? How about the comparison of running time?
- **RQ2 (Performance deterioration):** compared to BPEC, does the incremental process degrade the performance?
- **RQ3 (Hyperparameter sensitivity):** does the size threshold $\epsilon$ of $\mathcal{X}_p$ significantly affect the performance of IEC algorithm?

**Table 1.** Benchmark description. For datasets with sizes smaller than 100,000 and those larger than 100,000, $\epsilon$ is set to 1/5 and 1/500 of their sizes. On letter, kddcup and Poker, the cardinality of focal sets in a credal partition is set to 1. The cluster numbers discovered by IEC are indicated within parentheses.

| Dataset | #Objects | #Dimensions | #Classes | Dataset | #Samples | #Dimensions | #Classes |
|---|---|---|---|---|---|---|---|
| Colon | 62 | 2000 | 2 (2) | Winnipeg | 325,834 | 175 | 7 (7) |
| Vehicle | 846 | 18 | 4 (4) | Kddcup | 494,020 | 40 | 23 (23) |
| Page-block | 5473 | 10 | 5 (5) | Poker | 1,025,010 | 11 | 10 (9) |
| Letter | 20000 | 16 | 26 (26) | Susy | 5,000,000 | 19 | 2 (2) |

**Comparing Baselines.** IEC is compared with 8 baselines, including 4 evidential clustering methods (CBP-EKNN [6], SREC [5], BPDNEC [7], kCEVCLUS [10]) and 4 stream clustering methods (E2GK [13], SDBSCAN [12], Autocloud [2], ORSL [4]). All the hyperparameters are set to values recommended in the corresponding paper, and $(K, num_n)$ is set to (30, 1) in IEC. After transferring the obtained credal partition to a hard partition, the ARI metric is applied.

**RQ1: Comparison.** The comparative results between IEC and 8 baselines are shown in Table 2, where the paired $t$-test is still used to compare IEC with any other baselines statistically. As can be seen, compared to the evidential baselines, IEC is the better one in most cases. It shows that the cluster centers derived through an incremental approach are also reasonably. Statistically, IEC has better performance on 10/16 cases than evidential baselines. IEC also achieves better performance than the stream clustering baselines on Winni, Kddcup, and Susy datasets. On the Poker dataset, the ARI of IEC is lower than that of ORSL. The reason is that IEC only found the 9 out of 10 clusters, as shown in Table 1. In addition, compared to the evidential baselines, IEC significantly reduces running time, and consumes essentially the same amount of runtime with stream clustering methods.

**Table 2.** ARI (mean$_{\text{std.deviation}}$) and running time averaged across five runs, where • indicates whether IEC is statistically superior to a certain comparing algorithm based on a paired $t$-test at a 0.05 significance level.

| ARI | Colon | Vehicle | Page | Letter | ARI | Winni | Kddcup | Poker | Susy |
|---|---|---|---|---|---|---|---|---|---|
| IEC | **.632**$_{.01}$ | .556$_{.01}$ | **.703**$_{.02}$ | **.893**$_{.01}$ | IEC | **.732**$_{.02}$ | **.537**$_{.01}$ | .649$_{.02}$ | **.707**$_{.04}$ |
| CBP-EKNN | .624$_{.02}$ | .568$_{.01}$ | .651$_{.05}$• | .827$_{.06}$• | E2GK | .594$_{.02}$• | .499$_{.01}$• | .612$_{.03}$• | .651$_{.02}$• |
| SREC | **.632**$_{.02}$ | **.579**$_{.01}$ | .678$_{.02}$• | **.893**$_{.01}$ | SDBSCAN | .625$_{.02}$• | .505$_{.03}$ | .639$_{.03}$ | .653$_{.04}$• |
| BPDNEC | .610$_{.02}$• | .558$_{.01}$ | .670$_{.02}$• | .859$_{.01}$• | Autocloud | .707$_{.16}$• | .505$_{.03}$ | .627$_{.03}$ | .681$_{.04}$• |
| kCEVCLUS | .581$_{.02}$• | .550$_{.01}$• | .606$_{.03}$• | .850$_{.02}$• | ORSL | .708$_{.01}$• | .501$_{.03}$• | **.654**$_{.03}$ | .667$_{.04}$ |
| Time (second) | Colon | Vehicle | Page | Letter | Time (second) | Winni | Kddcup | Poker | Susy |
| IEC | 0.32 | 2.31 | 5.61 | 14.51 | IEC | 16.34 | 22.78 | 57.84 | 213.1 |
| CBP-EKNN | 1.21 | 54.31 | 879.14 | 3789.21 | E2GK | 14.54 | 19.57 | 64.32 | 228.93 |
| SREC | 0.96 | 16.87 | 237.14 | 989.32 | SDBSCAN | 22.15 | 42.56 | 79.78 | 267.31 |
| BPDNEC | 1.03 | 24.42 | 423.45 | 1543.21 | Autocloud | 19.35 | 26.95 | 55.61 | 241.34 |
| kCEVCLUS | 0.87 | 3.54 | 75.21 | 149.87 | ORSL | 15.78 | 26.75 | 61.32 | 243.84 |

**RQ2: Performance Deterioration.** In Table 3, we compare IEC and BPEC on four small datasets. Both methods accurately determine the number of clusters, with no degradation of ARI during incremental process. Even on dataset Colon, IEC exhibits a higher ARI. This may be because IEC consistently discards objects with the lowest Z-scores in $\mathcal{X}_p$.

**Table 3.** The cluster number ($\#CluN$) and ARI achieved by IEC and BPEC.

| $\#CluN$ | Colon | Vehicle | Page | Letter | ARI | Colon | Vehicle | Page | Letter |
|---|---|---|---|---|---|---|---|---|---|
| IEC | 2 | 4 | 5 | 26 | IEC | 0.632 | 0.556 | 0.703 | 0.893 |
| BPEC | 2 | 4 | 5 | 26 | BPEC | 0.624 | 0.581 | 0.703 | 0.893 |

**RQ3: Hyperparameter Sensitivity.** In Table 4, we present the performance of IEC under different $\epsilon$ settings. When $\epsilon$ is set to $1/1000$ and $1/2000$ of the dataset size, not all 26 clusters can be identified on the Susy dataset, and there is also a slight degradation in ARI values. However, on the remaining three datasets, IEC demonstrates strong robustness to $\epsilon$.

**Table 4.** The performance achieved by IEC with different hyperparameter settings.

| $\#CluN$ | Winni | Kddcup | Poker | Susy | ARI | Winni | Kddcup | Poker | Susy |
|---|---|---|---|---|---|---|---|---|---|
| $\epsilon = \text{size}/500$ | 2 | 4 | 5 | 26 | | 0.732 | 0.537 | 0.649 | 0.707 |
| $\epsilon = \text{size}/1000$ | 2 | 4 | 5 | (25) | | 0.738 | 0.529 | 0.652 | 0.684 |
| $\epsilon = \text{size}/2000$ | 2 | 4 | 5 | (24) | | 0.721 | 0.545 | 0.642 | 0.677 |

## 5 Conclusion

This paper presents an IEC method, which starts from a small dataset and incrementally increases the size of the processed dataset until it reaches a threshold. Solely through scalar transformations, IEC continuously updates cluster centers automatically until the algorithm terminates. The cluster memberships are finally computed based on the found cluster centers. Experimental results demonstrate that in most cases, IEC significantly reduces runtime and achieves the highest ARI.

## References

1. Antoine, V., Quost, B., Masson, M.H., Denoeux, T.: CEVCLUS: evidential clustering with instance-level constraints for relational data. Soft. Comput. **18**, 1321–1335 (2014)
2. Bezerra, C.G., Costa, B.S.J., Guedes, L.A., Angelov, P.P.: An evolving approach to data streams clustering based on typicality and eccentricity data analytics. Inf. Sci. **518**, 13–28 (2020)
3. Denœux, T., Masson, M.H.: EVCLUS: evidential clustering of proximity data. IEEE Trans. Syst. Man Cybern. Part B (Cybern.) **34**(1), 95–109 (2004)
4. Din, S.U., Shao, J., Kumar, J., Ali, W., Liu, J., Ye, Y.: Online reliable semi-supervised learning on evolving data streams. Inf. Sci. **525**, 153–171 (2020)
5. Gong, C., Li, Y., Fu, D., Liu, Y., Wang, P.H., You, Y.: Self-reconstructive evidential clustering for high-dimensional data. In: 2022 IEEE 38th International Conference on Data Engineering (ICDE), pp. 2099–2112. IEEE (2022)
6. Gong, C., Su, Z.G., Wang, P.H., Wang, Q.: Cumulative belief peaks evidential k-nearest neighbor clustering. Knowl.-Based Syst. **200**, 105982 (2020)
7. Gong, C., Su, Z.G., Wang, P.H., Wang, Q.: An evidential clustering algorithm by finding belief-peaks and disjoint neighborhoods. Pattern Recogn. **113**, 107751 (2021)

8. Gong, C., Su, Z.G., Wang, P.H., You, Y.: Distributed evidential clustering toward time series with big data issue. Expert Syst. Appl. **191**, 116279 (2022)
9. Gong, C., You, Y.: Sparse reconstructive evidential clustering for multi-view data. IEEE/CAA J. Autom. Sinica **11**(2), 459–473 (2024)
10. Li, F., Li, S., Denœux, T.: k-CEVCLUS: constrained evidential clustering of large dissimilarity data. Knowl.-Based Syst. **142**, 29–44 (2018)
11. Masson, M.H., Denoeux, T.: ECM: an evidential version of the fuzzy c-means algorithm. Pattern Recogn. **41**(4), 1384–1397 (2008)
12. Mu, C., Hou, Y., Zhao, J., Wei, S., Wu, Y.: Stream-DBSCAN: a streaming distributed clustering model for water quality monitoring. Appl. Sci. **13**(9), 5408 (2023)
13. Serir, L., Ramasso, E., Zerhouni, N.: Evidential evolving Gustafson-Kessel algorithm for online data streams partitioning using belief function theory. Int. J. Approximate Reasoning **53**(5), 747–768 (2012)
14. Shafer, G.: A Mathematical Theory of Evidence, vol. 42. Princeton University Press, Princeton (1976)
15. Su, Z.G., Denoeux, T.: BPEC: belief-peaks evidential clustering. IEEE Trans. Fuzzy Syst. **27**(1), 111–123 (2018)

# Imprecise Deep Networks for Uncertain Image Classification

Chuanqi Liu[1]($\boxtimes$), Zuowei Zhang[1], Zechao Liu[2], Liangbo Ning[3], and Zhunga Liu[1]

[1] Northwestern Polytechnical University, Xi'an 710000, Shaanxi, China
liuchuanqi@mail.nwpu.edu.cn
[2] Zhejiang Lab, Hangzhou, China
[3] The Hong Kong Polytechnic University, Hong Kong 999077, China

**Abstract.** Deep learning techniques have been successfully applied in image classification tasks. Still, data uncertainty is hindering the demand for higher performance. Existing techniques can only suppress the uncertainty caused by one reason, considered a passive strategy. Here, we introduce an open imprecise deep network (ImpNN) framework for image classification to actively handle data uncertainty. The ImpNN can model and reason data uncertainty-caused imprecision by using meta-class, defined as the union of different specific categories to constrain and improve the network performance. In addition, ImpNN characterizes and exploits imprecision by introducing two new loss functions (Imprecision loss function and Denoising loss function), which separately contribute to exploit the mined imprecision and alleviate the side effect of imprecision. We employ several typical networks to theoretically and experimentally analyze the ImpNN, which presents better performance compared to other methods based on open datasets. Experimental evaluations also show that our ImpNN can characterize imprecise information in results, potentially for cautious decision-making applications.

**Keywords:** Data uncertainty · imprecision · deep learning · meta-class · neural networks

## 1 Introduction

In recent years, deep learning techniques have attracted much attention of researchers and made remarkable achievements in computer vision [1], including image classification tasks [20]. As applications become more specific and complex, however, we increasingly need to deal with imperfect (uncertain) datasets. Here we define that data uncertainty[1] consists of image uncertainty and label uncertainties. Image uncertainty is mainly caused by multiple reasons such as blur, occlusion, angle, darkness, and illumination. And label uncertainty is

---
[1] Uncertainty results from the incompleteness of knowledge.

mainly caused by the reasons such as noisy labels, weak labels, incomplete labels, and multi-labels.

When facing with image uncertainty, Deng et al. [2] used graph Laplace to complete occluded face images before recognizing them. Besides, Haut [3] handled the image uncertainty by extracting robust features and mitigated the overfitting problem by data augmentation technique with random occlusion. For label uncertainty, noisy label is one of the most significant manifestations of imperfect labels, where a training image is occasionally assigned with a wrong label [8]. Data cleaning or noise filtering is considered a common strategy, where the clean data are fed into learners for training after pre-processing the noisy data [15]. Despite the success of the above approaches in dealing with image uncertainty and label uncertainty, each of them is only able to process one single type of data uncertainty.

Different from the above passive defense strategy without mining and reasoning the imprecision[2] induced by uncertainty [17,18]. In this paper, we propose an open imprecise deep network (ImpNN) framework to empower deep neural networks to actively characterize the uncertainty and imprecision in datasets. The contributions of this papre are summarized as follows:

- We propose an open universal framework to characterize the imprecision induced by data uncertainty, which effectively exploits the imprecision to improve the network performance.
- We propose a dynamic network to characterize imprecision. We embed the characterization of imprecision into the training phase, which does not waste additional computing resources.
- A complete system is proposed to not only mine the imprecision but also evaluate the performance of the proposed method. An evaluation metric is introduced to assess the performance numerically.

The rest of this paper is organized as follows. Section 2 present the ImpNN in detail. Extensive experimental results and discussion are available in Sect. 3. We conclude the whole work in Sect. 4.

## 2 Methodology

In this section, we will explain how ImpNN mines and utilizes imprecision to achieve outstanding performance. Here we give some definitions to present ImpNN explicitly. $\{\mathbf{x}, \mathbf{y}\}$ denotes the input sample and its ground truth, where $\mathbf{x} \in \mathbb{R}^d$ and $\mathbf{y}$ is the one-hot label. $d$ represents the dimension of feature space. The neural network is summarized as a function $f$ that maps input image $\mathbf{x}$ to its predicted label $\hat{\mathbf{y}}$, which is defined by

$$\hat{\mathbf{y}} = f(\mathbf{x}), \tag{1}$$

---

[2] Imprecision results from the fuzziness of knowledge.

where $\hat{\mathbf{y}} = \{\hat{y}_1, \ldots, \hat{y}_c\}$ is subject to $\hat{y}_i \geq 0$ and $\sum_{i=1}^{c} \hat{y}_i = 1$. $c$ is the number of categories in the training and test sets. $\{\hat{y}_i\}, i = 1, \cdots, c$ represents the probability of the sample belonging to the $i$-th category. The overview of the ImpNN is shown in Fig. 1.

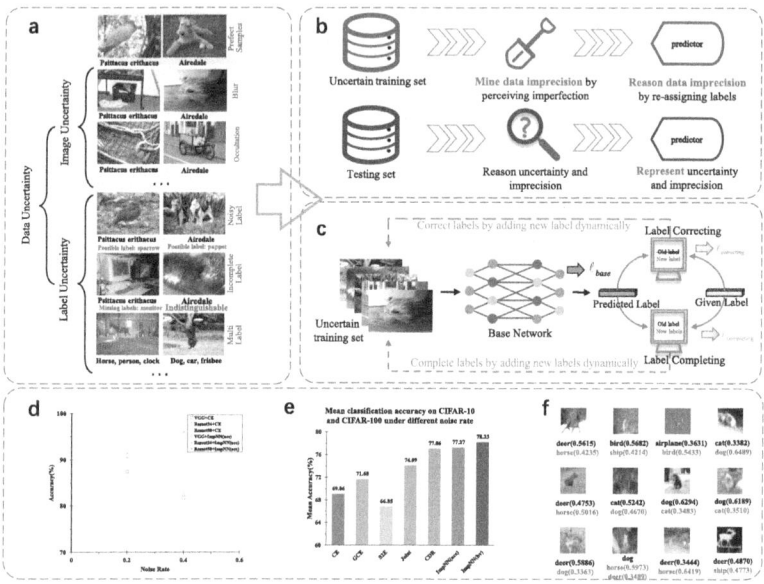

**Fig. 1.** The overview of ImpNN. **a**, **b**, and **c** show the overall framework of ImpNN. **d** and **e** represent some results in image classification tasks. **f** gives some illustrations of the label correcting process. Labels in black are original labels, and those in red are generated dynamically by ImpNN. (Color figure online)

### 2.1 Imprecision Loss Function

We argue that perfect samples contain distinctive features of a specific category. In contrast, imperfect samples may consist of indistinguishable features of a set of categories. However, Only assigning one specific label to these samples may leads to the networks cannot learn the adequate features of each category. To address this problem, we aim to assign potential labels to some samples during iterations and propose an imprecision loss function to exploit the mined imprecision. By doing this, networks are equipped with the ability to learn distinct features from imperfect datasets.

To mine and exploit the potential labels effectively, we propose a label reassignment mechanism and an imprecision loss function. It is assumed that original labels $\mathbf{y}$ are incomplete. $\overline{\mathbf{y}} = \{\overline{y}_1, \ldots, \overline{y}_c\}$ is used to represent the latent label and it could be automatically generated by $\hat{\mathbf{y}}$, which is shown as follows

$$\overline{y}_i = \begin{cases} 0, & \text{if } \hat{y}_i < 0.5 \times \gamma, \\ 0.5, & \text{else.} \end{cases} \quad (2)$$

For convenience, we averagely assign 1 to two latent labels. Original labels mean the current samples belong to a specific category. The new latent label $\overline{\mathbf{y}}$ is used to represent that the current sample belongs to a set of categories. To make full use of the potential labels (i.e., imprecision), an incomplete classification loss function is designed to constrain the network to learn the representation of the latent category, defined by

$$\ell_{weak} = \alpha \sum_i^c -\overline{y}_i \log \hat{y}_i, \quad (3)$$

s.t.

$$\alpha = \begin{cases} 0, & \text{if } |Max\{0, \hat{\mathbf{y}} - 0.5 \times \gamma\}| < 2, \\ 0.5 \times I(\hat{\mathbf{y}}), & \text{else,} \end{cases} \quad (4)$$

and

$$I(\hat{\mathbf{y}}) = \exp\left(\sum_i^c Max\{\hat{y}_i - 0.5 \times \gamma, 0\}\right), \quad (5)$$

where $\alpha$ is a trade-off parameter. If there are massive latent labels, $\ell_{weak}$ will play a significant role in the overall objective function. This enforces the network sufficiently learn the features of the potential category. In Eq. 4, $|\cdot|$ is used to represent the number of non-zero items. To make sure that there are only two latent labels in $\overline{\mathbf{y}}$, $0.5 \times \gamma$ should be bigger than $\frac{1}{3}$. Thus $\gamma \in \left(\frac{2}{3}, 1\right)$ is the default. In practice, the optimal value of $\gamma$ varies among different datasets according to the quality of images.

## 2.2 Denoising Loss Function

Label errors are common and unavoidable in the existing benchmark datasets. For example, there are 2916 label errors (about 6%) in the ImageNet validation set [7]. Since noisy samples guide networks to learn wrong features and bring fuzzy knowledge to networks, label noise is also regarded as imprecision induced by label uncertainty. To alleviate the side effect of this kind of imprecision, we design a denoising loss function to make ImpNN more robust to the wrong labels. If $\arg\max \hat{\mathbf{y}} \neq \arg\max \mathbf{y}$ and $\alpha = 0$, we consider that the current label is wrong since the representation of the input sample mainly belongs to the other category. When this condition is triggered, $\tilde{\mathbf{y}} = \{\tilde{y}_1, \cdots, \tilde{y}_c\}$ is obtained, which denotes the potential correct label. $\tilde{\mathbf{y}}$ is defined by

$$\tilde{y}_i = \begin{cases} 1, & \text{if } \hat{y}_i = \arg\max \hat{\mathbf{y}}, \\ 0, & \text{else.} \end{cases} \quad (6)$$

The denoising loss function is designed to guide the network to learn the representation based on the potential correct label, defined by

$$\ell_{noisy} = \beta \sum_i^c -\tilde{y}_i \log \hat{y}_i, \tag{7}$$

where $\beta$ is a trade-off parameter similar to $\alpha$. It is not equal to zero only if the prediction is complete and the predicted label is not equal to $\mathbf{y}$, defined by

$$\beta = \begin{cases} 1, & \text{if } \arg\max \hat{\mathbf{y}} \neq \arg\max \mathbf{y} \text{ and } \alpha = 0, \\ 0, & \text{else.} \end{cases} \tag{8}$$

### 2.3 Overall Loss Function

To alleviate the effect of noisy samples, the bootstrapped cross entropy [9] is used as the basic loss function, defined by

$$\ell_{base} = (1-\omega) \sum_{i=1}^c -y_i \log \hat{y}_i + \omega \sum_{i=1}^c -\hat{y}_i \log \hat{y}_i. \tag{9}$$

When making decisions, the network assigns probabilities to the current samples according to the potential knowledge. In Eq. 9, if the prediction $\hat{\mathbf{y}}$ is different from the true label $\mathbf{y}$, the second item is exploited to hold its decision by introducing a reasonable trade-off parameter $\omega$.

The randomly initialized network makes random decisions $\hat{\mathbf{y}}$ at the beginning of the training. To alleviate the impact of the uncertain decision at the beginning of the training, a gradually increased parameter $\lambda$ is introduced, shown as follows

$$\lambda = \frac{epoch_{cur}}{epoch_{max}}, \tag{10}$$

where $epoch_{cur}$ is the current number of iterations and $epoch_{max}$ is the maximal number of iterations. By considering all aforementioned components, we define the overall objective function as

$$\ell = -\sum_{i=1}^c [(1-\lambda\omega)y_i + \lambda\omega\hat{y}_i] \log \hat{y}_i + \lambda\ell_{weak} + \lambda\ell_{noisy}. \tag{11}$$

## 3 Experiments

### 3.1 The Evaluation Metric for Imprecise Classification Results

In order to demonstrate the effectiveness and rationality for imprecise classification of the ImpNN adequately, we not only use accuracy [19], but utilise the benefit value [6].

The input sample **x** is fed into the network and we can get the prediction result $\hat{\mathbf{y}}$, $\overline{\mathbf{y}}$ and $\tilde{\mathbf{y}}$ according to aforementioned procedures. The benefit value $\mathcal{B}$ is defined by

$$\mathcal{B} = \begin{cases} 0, & \text{if } \arg\max \overline{\mathbf{y}} \neq \arg\max \mathbf{y}; \\ (\frac{1}{|\overline{\mathbf{y}}|})^{\eta}, & \text{if } |\overline{\mathbf{y}}| > 1, \arg\max \mathbf{y} \in \arg\max \overline{\mathbf{y}}; \\ 1, & \text{if } |\overline{\mathbf{y}}| = 1, \arg\max \overline{\mathbf{y}} = \arg\max \mathbf{y}. \end{cases} \quad (12)$$

If the sample is classified correctly, the benefit value is set to 1. If the true label is not included in $\overline{\mathbf{y}}$, its benefit value is 0. When the prediction consists of two or more labels, we use $(\frac{1}{|\overline{\mathbf{y}}|})^{\eta}$ to compute the benefit value. $\eta$ is a hyper-parameter. We perceive that the imprecise result is better than randomly assigning a specific label to the sample that is hard to classify.

### 3.2 Evaluation Based on Different Benchmark Networks

We use two benchmark datasets (*i.e.*, CIFAR-10, CIFAR-100 [5]) to conduct experiments. To verify the effectiveness of the ImpNN sufficiently, we manually add noisy samples to the pure benchmark datasets [13]. In experiments, each original dataset is corrupted by noisy samples with a noise rate of 0.2 and 0.4 to demonstrate the robustness. Moreover, three base networks are employed, *i.e.*, VGG16 [10], ResNet34 [4] and ResNet50 [4].

Tables 1 and 2 respectively show the accuracy and benefit value results of ImpNN and the evidential deep neural network on the CIFAR-10 and CIFAR-100 dataset. And ImpNN outperforms the baselines trained by the cross entropy loss, which verifies the effectiveness of the proposed imprecision and denoising loss functions.

**Table 1.** The results of different methods based on ResNet50 with diverse noise rates on the CIFAR-10 dataset.

| Network | Method | ACCURACY (%) | | BENEFIT VALUE | |
|---|---|---|---|---|---|
| | | noise rate | | noise rate | |
| | | 0.2 | 0.4 | 0.2 | 0.4 |
| VGG16 | CE | 87.41 ± 0.28 | 81.74 ± 0.55 | 0.8771 ± 0.0024 | 0.7572 ± 0.0042 |
| | ImpNN (acc) | 87.51 ± 0.22 | 82.27 ± 0.33 | 0.8791 ± 0.0023 | 0.7849 ± 0.0028 |
| | ImpNN (bv) | **87.58 ± 0.23** | **82.37 ± 0.44** | **0.8801 ± 0.0019** | **0.7854 ± 0.0053** |
| ResNet34 | CE | 90.20 ± 0.24 | 81.73 ± 1.00 | 0.9029 ± 0.0025 | 0.7621 ± 0.0078 |
| | ImpNN (acc) | 90.50 ± 0.16 | 82.86 ± 0.47 | 0.9081 ± 0.0016 | 0.8005 ± 0.0023 |
| | ImpNN (bv) | **90.75 ± 0.06** | **83.50 ± 0.78** | **0.9102 ± 0.0007** | **0.8008 ± 0.0045** |
| ResNet50 | CE | 86.73 ± 0.36 | 75.33 ± 2.72 | 0.8776 ± 0.0108 | 0.7725 ± 0.0075 |
| | ImpNN (acc) | 91.40 ± 0.15 | 86.02 ± 0.83 | 0.9167 ± 0.0019 | 0.8277 ± 0.0087 |
| | ImpNN (bv) | **91.77 ± 0.11** | **87.15 ± 0.73** | **0.9207 ± 0.0032** | **0.8356 ± 0.0067** |

**Table 2.** The results of different methods based on ResNet50 with diverse noise rates on the CIFAR-100 dataset.

| Network | Method | ACCURACY (%) | | BENEFIT VALUE | |
|---|---|---|---|---|---|
| | | noise rate | | noise rate | |
| | | 0.2 | 0.4 | 0.2 | 0.4 |
| VGG16 (pretrained) | CE | 65.19 ± 0.22 | 55.97 ± 0.69 | 0.6557 ± 0.0024 | 0.5635 ± 0.0063 |
| | ImpNN (acc) | 66.25 ± 0.47 | 58.56 ± 0.43 | 0.6677 ± 0.0042 | 0.5914 ± 0.0033 |
| | ImpNN (bv) | **66.45 ± 0.19** | **58.86 ± 0.22** | **0.6687 ± 0.0021** | **0.5948 ± 0.0022** |
| ResNet34 | CE | 63.56 ± 0.54 | 49.72 ± 0.49 | 0.6411 ± 0.0054 | 0.5104 ± 0.0050 |
| | ImpNN (acc) | 64.44 ± 0.32 | 51.80 ± 0.91 | 0.6496 ± 0.0027 | 0.5306 ± 0.0089 |
| | ImpNN (bv) | **64.44 ± 0.22** | **51.97 ± 0.68** | **0.6501 ± 0.0018** | **0.5313 ± 0.0080** |
| ResNet50 | CE | 63.33 ± 0.29 | 50.84 ± 0.89 | 0.6302 ± 0.0035 | 0.5273 ± 0.0069 |
| | ImpNN (acc) | 70.03 ± 0.15 | 61.63 ± 0.47 | 0.7044 ± 0.0014 | 0.6237 ± 0.0041 |
| | ImpNN (bv) | **71.27 ± 0.37** | **62.71 ± 0.11** | **0.7170 ± 0.0041** | **0.6321 ± 0.0023** |

CE means we only use the cross entropy loss function to train the network. ImpNN(acc) and ImpNN(bv) use accuracy and benefit value to trigger Early Stopping, respectively. By considering the imprecision sufficiently, networks learn representations of different categories precisely. This leads to the significant growth of the accuracy and benefit value. For example, on the CIFAR-10 dataset, the accuracy is improved by 10.49% based on ResNet50. Besides, it is also shown that using the benefit value as the evaluation metric for model selection in Early Stopping is better than using accuracy. Since the benefit value is more suitable for evaluating the imprecise classification results, we can choose a more robust model that generalizes well according to the benefit value.

### 3.3 Comparison with Different Methods

To demonstrate the superiority of ImpNN, we compare it with the state-of-the-art classification methods (*i.e.*, GCE [16], S2E [14], Joint [11], CDR [12]) dedicated to solving one single type of imprecision.

**Table 3.** The results of different methods based on ResNet50 with diverse noise rates.

| Datasets | Noise rate | ACCURACY (%) | | | | | | |
|---|---|---|---|---|---|---|---|---|
| | | method | | | | | | |
| | | CE | GCE | S2E | Joint | CDR | ImpNN (acc) | ImpNN (bv) |
| CIFAR10 | 0.2 | 86.73 | 88.02 | 90.32 | 90.43 | 91.14 | 91.40 | **91.77** |
| | 0.4 | 75.33 | 76.89 | 68.93 | 85.23 | 86.25 | 86.02 | **87.15** |
| CIFAR100 | 0.2 | 63.33 | 66.67 | 61.08 | 65.91 | 69.82 | 70.03 | **71.27** |
| | 0.4 | 50.84 | 55.14 | 47.06 | 55.09 | 61.03 | 61.63 | **62.71** |

We conduct extensive experiments on CIFAR-10 and CIFAR-100. ResNet50 is chosen as the benchmark. The results are displayed in Table 3. It is shown that ImpNN achieves the best performance in all cases. It can not only distinguish between noisy and clean samples but also exploit the mined information to constrain the network training.

## 4 Conclusion

In this paper, we proposed an open framework to improve the performance of deep neural networks, which characterized and exploited the imprecision induced by data uncertainty. We only modified the loss functions to mine imprecision effectively and dynamically. The imprecision loss function was beneficial to find the latent labels, which made networks learn the generalized representation adequately. During this, we characterized the imprecision by introducing potential labels. The denoising loss function was proposed to correct noisy labels and use the samples with new labels to guide the network training. This method could be readily embedded into the existing networks to yield better performance. Sufficient experiments were conducted to verify the effectiveness of ImpNN. In the future, we aim to provide a comprehensive system to mine, utilize, measure and reduce imprecision.

**Acknowledgments.** This work was supported in part by the National Natural Science Foundation of China under Grant U20B2067, and in part by the Fundamental Research Funds for the Central Universities under Grant G2023KY05102.

# References

1. Chen, X., Wang, X., Zhou, J., Qiao, Y., Dong, C.: Activating more pixels in image super-resolution transformer. In: Proceedings of the IEEE/CVF Conference on Computer Vision and Pattern Recognition, pp. 22367–22377 (2023)
2. Deng, Y., Dai, Q., Zhang, Z.: Graph laplace for occluded face completion and recognition. IEEE Trans. Image Process. **20**(8), 2329–2338 (2011)
3. Haut, J.M., Paoletti, M.E., Plaza, J., Plaza, A., Li, J.: Hyperspectral image classification using random occlusion data augmentation. IEEE Geosci. Remote Sens. Lett. **16**(11), 1751–1755 (2019)
4. He, K., Zhang, X., Ren, S., Sun, J.: Deep residual learning for image recognition. In: Proceedings of the IEEE Conference on Computer Vision and Pattern Recognition, pp. 770–778 (2016)
5. Krizhevsky, A., Hinton, G., et al.: Learning multiple layers of features from tiny images (2009)
6. Liu, Z.G., Pan, Q., Dezert, J., Martin, A.: Combination of classifiers with optimal weight based on evidential reasoning. IEEE Trans. Fuzzy Syst. **26**(3), 1217–1230 (2017)
7. Northcutt, C.G., Athalye, A., Mueller, J.: Pervasive label errors in test sets destabilize machine learning benchmarks. arXiv preprint arXiv:2103.14749 (2021)
8. Quinlan, J.R.: The effect of noise on concept. Mach. Learn.: Artif. Intell. Approach **2**, 149 (1983)

9. Reed, S., Lee, H., Anguelov, D., Szegedy, C., Erhan, D., Rabinovich, A.: Training deep neural networks on noisy labels with bootstrapping. arXiv preprint arXiv:1412.6596 (2014)
10. Simonyan, K., Zisserman, A.: Very deep convolutional networks for large-scale image recognition. arXiv preprint arXiv:1409.1556 (2014)
11. Tanaka, D., Ikami, D., Yamasaki, T., Aizawa, K.: Joint optimization framework for learning with noisy labels. In: Proceedings of the IEEE Conference on Computer Vision and Pattern Recognition, pp. 5552–5560 (2018)
12. Xia, X., Liu, T., Han, B., Gong, C., Wang, N., Ge, Z., Chang, Y.: Robust early-learning: hindering the memorization of noisy labels. In: International Conference on Learning Representations (2020)
13. Xia, X., et al.: Part-dependent label noise: towards instance-dependent label noise. Adv. Neural. Inf. Process. Syst. **33**, 7597–7610 (2020)
14. Yao, Q., Yang, H., Han, B., Niu, G., Kwok, J.T.Y.: Searching to exploit memorization effect in learning with noisy labels. In: International Conference on Machine Learning, pp. 10789–10798. PMLR (2020)
15. Zhang, Y., Zheng, S., Wu, P., Goswami, M., Chen, C.: Learning with feature-dependent label noise: a progressive approach. arXiv preprint arXiv:2103.07756 (2021)
16. Zhang, Z., Sabuncu, M.: Generalized cross entropy loss for training deep neural networks with noisy labels. Adv. Neural Inf. Process. Syst. **31** (2018)
17. Zhang, Z.W., Liu, Z., Martin, A., Liu, Z.G., Zhou, K.: Dynamic evidential clustering algorithm. Knowl.-Based Syst. **213**, 106643 (2021)
18. Zhang, Z.W., Liu, Z.G., Martin, A., Zhou, K.: BSC: belief shift clustering. IEEE Trans. Syst. Man Cybern.: Syst. **53**(3), 1748–1760 (2022)
19. Zhang, Z., Liu, Z., Ning, L., Martin, A., Xiong, J.: Representation of imprecision in deep neural networks for image classification. IEEE Trans. Neural Netw. Learn. Syst. (2023)
20. Zhang, Z., Ning, L., Liu, Z., Yang, Q., Ding, W.: Mining and reasoning of data uncertainty-induced imprecision in deep image classification. Inf. Fusion **96**, 202–213 (2023)

# Dempster-Shafer Credal Probabilistic Circuits

David Ricardo Montalván Hernández[✉][iD], Thomas Krak[iD], and Cassio de Campos[iD]

Eindhoven University of Technology, Eindhoven, Netherlands
{d.r.montalvan.hernandez,t.e.krak,c.decampos}@tue.nl

**Abstract.** Probabilistic circuits are deep, tractable generative models capable of computing various types of exact inferences. However, their traditional specifications do not fully account for epistemic uncertainty. To address this, credal probabilistic circuits were introduced, incorporating a way to manage such uncertainty. We propose a novel framework for learning the structure and parameters of credal probabilistic circuits, leveraging the Dempster-Shafer theory of evidence. Unlike previous credal approaches, the framework handles both discrete and continuous data and allows for the use of multiple classification criteria. We conclude by presenting some preliminary experimental results, demonstrating the performance of the proposed models compared to commonly used probabilistic circuits across a range of classification tasks.

**Keywords:** Probabilistic Circuits · Credal Probabilistic Circuits · Dempster-Shafer Theory · Imprecise Probabilities · Evidential Clustering · Generative Models

## 1 Introduction

Probabilistic circuits (PCs) are a class of tractable deep generative models that have shown promising results in several machine learning tasks. Mauá *et al.* [6] have previously proposed the credal probabilistic circuit (CPC), a generalization of PCs more capable of handling epistemic uncertainty.

The graph structure and parameters of PCs are typically learned from data, e.g. using the commonly employed *LearnSPN* framework [3,8]. This method recursively employs independence testing and data clustering to obtain a tree-structured PC that describes the distribution of the training data. Less work has been done on learning CPCs from data. One example is the work of Levray and Belle [4], who presented an approach that accounts for missing values. Their method is limited to discrete domains.

We propose a novel framework for learning CPCs using the Dempster-Shafer theory of evidence. Our approach is based on modifying the clustering step of the *LearnSPN* framework using evidential clustering [2], which allows us to work with both discrete and continuous data. Furthermore, we propose some alternatives for estimating the (set-valued) numerical parameters from data.

## 2 Preliminaries

### 2.1 Probabilistic Circuits

Probabilistic Circuits (PCs) [8] are a class of probabilistic graphical models whose particular graph structure encodes a probability distribution over some domain, in such a way that it can be directly used for computational purposes, that is, to perform probabilistic inference. More specifically, PCs are represented with a singly-rooted Directed Acyclic Graph (DAG) containing three types of nodes: sum, product, and leaf (distribution) nodes.

Let $\boldsymbol{X} = \{X_1, \ldots, X_d\}$ be a set of real-valued random variables. A PC $\mathcal{S}$ encodes a joint probability distribution defined over $\boldsymbol{X}$. Every node $N \in \mathcal{S}$ encodes a distribution over a subset of variables $\boldsymbol{Y} \subseteq \boldsymbol{X}$, and the types of these nodes and their position in the graph determine the relation of these distributions. This subset of variables $\boldsymbol{Y}$ is called the scope of node $N$ and we denote it as $\operatorname{sc}(N)$. This scope is determined inductively with respect to the graph structure: let $D_{\boldsymbol{Y}}$ be a leaf node encoding a distribution over the variables $\boldsymbol{Y} \subseteq \boldsymbol{X}$; then the scope of $D_{\boldsymbol{Y}}$ is defined as $\operatorname{sc}(D_{\boldsymbol{Y}}) = \boldsymbol{Y}$. Conversely, for product or sum nodes $N$, their scope is defined as the union of the scope of their children, that is, $\operatorname{sc}(N) = \cup_{C \in \operatorname{ch} N} \operatorname{sc}(C)$, where $\operatorname{ch} N$ are the children of $N$. By convention, we assume that the scope of the root node is $\boldsymbol{X}$ (that is, for every $X_i \in \boldsymbol{X}$ there is at least one leaf node $D_{\boldsymbol{Y}}$ for which $X_i \in \boldsymbol{Y}$).

We assume that the DAG satisfies the following: a) **Completeness:** For every sum node $S$, its children have the same scope: $\operatorname{sc}(C) = \operatorname{sc}(C')$ for all $C, C' \in \operatorname{ch} S$. b) **Decomposability:** For every product node $P$, its children have disjoint scope: $\operatorname{sc}(C) \cap \operatorname{sc}(C') = \emptyset$ for all $C, C' \in \operatorname{ch} P$.

In addition to this graphical structure, the specification of a PC requires some further parameters. First, for each leaf node $D_{\boldsymbol{Y}}$, we need to specify some distribution over $\boldsymbol{Y}$; this depends on the type of these variables, and we leave this implicit in the remainder. However, we assume that some tractable description is given, for example that it is an independent product of exponential family distributions over individual variables in $\boldsymbol{Y}$. Finally, for each sum node $S$, we require the specification of a probability mass function $\boldsymbol{w}_S$ over its children $\operatorname{ch} S$.

**Inference with Probabilistic Circuits.** For our current purposes, performing inference with a PC amounts to computing the expected value of a function $f$ over $\boldsymbol{X}$, with respect to the distribution encoded by the PC. In particular, we focus attention on *factorising functions*, which are of the form $f = \prod_{i=1}^{d} f_i$, where each $f_i$ is a function of the variable $X_i$ [7]. Note that this covers, for instance, the evaluation of likelihoods as a special case; simply take the indicator function $\mathbb{I}_{x_i}$ for feature $x_i$ of data point $\mathbf{x}$ as the choice for each $f_i$. The graph structure of a PC is well-suited for such computations, in that they can be performed efficiently with a recursive scheme that associates a value $\mathcal{V}^f(N)$ to each node $N$. For a leaf node $D_{\boldsymbol{Y}}$, this value is defined as $\mathcal{V}^f(D_{\boldsymbol{Y}}) := \mathbb{E}_{D_{\boldsymbol{Y}}}[f(\boldsymbol{Y})]$, where, with a slight abuse of notation, $f(\boldsymbol{Y})$ is the product of factors $f_i$ such that $X_i \in \boldsymbol{Y}$, and where the expectation is taken with respect to the distribution

**Algorithm 1.** *LearnSPN*($\mathcal{D}_X$, $N$)

**Require:** Dataset $\mathcal{D}_X$ with variables $X$, Parent node $N$
   **if** $|X| = 1$ **then**
      Add a leaf node $D_X$ as child of $N$ and estimate its parameters from $\mathcal{D}_X$
   **else**
      Try to partition $X$ into approximately independent subsets $Y_j$
      **if** success **then**
         Add a product node $P$ as child of $N$
         For each subset $Y_j$ call *LearnSPN*($\mathcal{D}_{Y_j}$, $P$)
      **else**
         Cluster the instances in $\mathcal{D}_X$ into subsets $\mathcal{D}_X^j$       ▷ Clustering step
         Add a sum node $S$ as child of $N$
         Set mass $\boldsymbol{w}_S(C_j) = |\mathcal{D}_X^j|/|\mathcal{D}_X|$ for child $C_j$ as relative size of cluster $j$
         For each $\mathcal{D}_X^j$ call *LearnSPN*($\mathcal{D}_X^j$, $S$)
      **end if**
   **end if**

---

associated with this node. Conversely, the value of sum and product nodes is defined recursively as a function of the values of their children. For a product node $P$, we simply let $\mathcal{V}^f(P) := \prod_{C \in \mathrm{ch}\, P} \mathcal{V}^f(C)$. For a sum node $S$, we take the associated mass function $\boldsymbol{w}_S$ and then let $\mathcal{V}^f(S) := \sum_{C \in \mathrm{ch}\, S} \boldsymbol{w}_S(C) \mathcal{V}^f(C)$. The expectation $\mathbb{E}_\mathcal{S}[f(X)]$ of $f$ with respect to the PC $\mathcal{S}$ is then understood to be the value $\mathcal{V}^f(R)$ of the root node $R$ of the DAG of $\mathcal{S}$.

**Learning Probabilistic Circuits.** The learnable parameters of a PC are divided into three groups: the DAG structure itself; the collection $\boldsymbol{W}$ of mass functions of sum nodes; and the collection $\boldsymbol{\Theta}$ of parameters of leaf nodes.

The popular *LearnSPN* framework [3,8] estimates all of these parameters from a given dataset $\mathcal{D}$. It works by recursively partitioning $\mathcal{D}$, at each step splitting the (remaining) dataset either column- or row-wise. Partitioning column-wise corresponds to finding independent subsets of features, and involves adding a product node to the DAG; the specific independence test used is a meta-parameter. Conversely, partitioning row-wise corresponds to grouping similar instances, and involves adding a sum node to the DAG; this uses a clustering method, the specifics of which the framework is agnostic to. Once the partitioning terminates—for example because a maximum depth is reached or only a single feature variable remains—a distribution is fit to the remaining dataset, and represented by adding an associated leaf node to the DAG. The graphical structure of the DAG obtained in this way is always a tree. Algorithm 1 sketches the pseudo-code for the *LearnSPN* framework.

## 2.2 Credal Probabilistic Circuits

To better handle epistemic uncertainty, Mauá *et al.* [6] proposed a generalization of PCs called credal probabilistic circuits (CPCs), which are based on the theory of imprecise probabilities [12]. A CPC can be understood as a *set* of normal ('precise') PCs, all with the same DAG structure and the same parameters $\Theta$ for the leaf nodes, but with different parameters $\mathbf{W}$ for the sum nodes.[1] The specification of a CPC thus requires (in addition to the DAG and the parameters $\Theta$), a *set $\mathcal{C}$* of (collections of) parameters for sum nodes. Let us denote, for some fixed DAG structure and parameters $\Theta$, and any choice of $\mathbf{W} \in \mathcal{C}$, the corresponding (precise) PC as $\mathcal{S}_\mathbf{W}$. Then the associated CPC can be characterized as the set $\{\mathcal{S}_\mathbf{W} : \mathbf{W} \in \mathcal{C}\}$. From an inference perspective, for CPCs we are interested in the corresponding *lower-* and *upper expectations*, which are defined as

$$\underline{\mathbb{E}}_\mathcal{S}[f(\boldsymbol{X})] := \inf_{\mathbf{W} \in \mathcal{C}} \mathbb{E}_{\mathcal{S}_\mathbf{W}}[f(\boldsymbol{X})], \quad \overline{\mathbb{E}}_\mathcal{S}[f(\boldsymbol{X})] := \sup_{\mathbf{W} \in \mathcal{C}} \mathbb{E}_{\mathcal{S}_\mathbf{W}}[f(\boldsymbol{X})]. \quad (1)$$

The efficiency with which these inferences can be computed depends on a number of factors involving, for example, the underlying DAG structure, the structure of $\mathcal{C}$, and the structure of $f$. Under some basic regularity conditions[2] on $\mathcal{C}$, and when the DAG is a tree, efficient algorithms are known for computing lower- and upper expectations of arbitrary factorising functions [7, Theorem 2].

### 2.3 Dempster-Shafer Theory & Evidential Clustering

The theory of evidence, introduced by Dempster and Shafer [1,9], is a formalism for representing epistemic knowledge. Instead of using probabilities for representing chances, Dempster and Shafer propose the use of belief functions. Here we briefly present the definitions from this theory that are useful for our work. Consider a finite set $\Omega = \{\omega_1, \ldots, \omega_k\}$ of $k$ classes. A **mass function** defined on $\Omega$ is a function $m : 2^\Omega \to [0,1]$ such that $\sum_{A \subseteq \Omega} m(A) = 1$. If $m(\emptyset) = 0$, then we say that $m$ is normalized. Given a mass function $m$, we can obtain the **belief** and **plausibility** functions, defined respectively as

$$Bel_m(A) := \sum_{\emptyset \neq B \subseteq A} m(B) \quad \text{and} \quad Pl_m(A) := \sum_{B \cap A \neq \emptyset} m(B) \text{ for all } A \subseteq \Omega. \quad (2)$$

It may sometimes be useful to derive from a (normalized) mass function $m$ a probability mass function on $\Omega$. For our present purposes, we use the **pignistic probability** [10] defined as the function $BetP_m : \Omega \to [0,1]$ given by

$$BetP_m(\omega) := \sum_{\omega \in A \subseteq \Omega} \frac{m(A)}{|A|} \quad \text{for all } \omega \in \Omega. \quad (3)$$

**Evidential Clustering.** As discussed in Sect. 2.1, the clustering step plays a fundamental role in the original *LearnSPN* framework. We replace this by a

---
[1] The leaf distributions parameterised by $\Theta$ may also be relaxed to imprecise models in an analogous manner, but for simplicity we will not consider this here.
[2] Essentially, that $\mathcal{C}$ is 'separately specified' using 'local models' for each sum node, and that each local model is sufficiently nice, i.e. non-empty, closed, and convex.

method based on the Dempster-Shafer theory of evidence, which is known as *evidential clustering* [2,5]. Let us briefly describe the basic idea.

Consider first the traditional clustering problem of a dataset $\mathcal{D} = \{\mathbf{x}_1, \ldots, \mathbf{x}_n\}$ among $k$ potential clusters, say $\Omega = \{\omega_1, \ldots, \omega_k\}$. The task is to associate with each observation $\mathbf{x} \in \mathcal{D}$ some (potentially degenerate) probability mass function $p_\mathbf{x}$ over $\Omega$, such that $p_\mathbf{x}(\omega)$ describes the probability that we attribute to observation $\mathbf{x}$ belonging to cluster $\omega$. Conversely, in evidential clustering, each observation $\mathbf{x}$ has associated to it a (belief) mass function $m_\mathbf{x}$ over the powerset $2^\Omega$. This mass function represents the partial knowledge that we have regarding the cluster membership of $\mathbf{x}$. For example, the case where $m_\mathbf{x}(\Omega) = 1$ represents a scenario of complete ignorance about to which cluster $\mathbf{x}$ belongs. If $\emptyset = \arg\max_{A \subseteq \Omega} m_\mathbf{x}(A)$, then $\mathbf{x}$ can be interpreted as an outlier. Through the associated belief and plausibility functions, further assertions can be derived about cluster membership. Evidential clustering provides a more expressive approach compared to traditional methods.

An assignment of such mass functions to the entire dataset $\mathcal{D}$, say $\mathcal{M} = \{m_{\mathbf{x}_1}, \ldots, m_{\mathbf{x}_n}\}$, is called a **credal partition** of $\mathcal{D}$.

## 3 Learning Credal Evidential Circuits

We propose a novel framework for learning credal circuits by means of evidential clustering. In what follows, $\mathcal{D} = \{\mathbf{x}_1, \ldots, \mathbf{x}_n\}$, $\mathbf{x}_i \in \mathbb{R}^d$ is the dataset at hand after performing evidential clustering and removing outliers, and $\Omega = \{\omega_1, \ldots, \omega_k\}$ is the set of possible cluster assignments. Every mass function, $m_\mathbf{x}$, is assumed to be normalized.

The weight vector of a credal circuit belongs to a subset of the probability simplex. This subset is usually defined through a set of inequalities, namely, defining lower and upper bounds for each weight component. We can use the credal partition $\mathcal{M}$ to obtain these bounds and hence to obtain a credal circuit. For that, we propose a few approaches.

The first approach uses the lower and upper approximations for a cluster $\omega$ as defined in [5]. The lower and upper approximations for cluster $\omega$ are given respectively by

$$\omega^L := \{\mathbf{x} \in \mathcal{D} : \{\omega\} = \arg\max_{A \subseteq \Omega} m_\mathbf{x}(A)\}; \qquad (4)$$

$$\omega^U := \{\mathbf{x} \in \mathcal{D} : \omega \in A \text{ and } A = \arg\max_{B \subseteq \Omega} m_\mathbf{x}(B)\}. \qquad (5)$$

We call this approach *lu-max* (lower and upper approximations via maximum mass). The second approach follows [2], where for each $m_\mathbf{x}$ we define the set of non-dominated clusters as

$$\widetilde{\Omega}(m_\mathbf{x}) := \{\omega \in \Omega : \forall \omega' \in \Omega, Pl_{m_\mathbf{x}}(\omega) \geq m_\mathbf{x}(\omega')\}. \qquad (6)$$

Using the non-dominated clusters, we define the lower and upper approximations for $\omega$, respectively, as

$$\omega^L := \{\mathbf{x} \in \mathcal{D} : \{\omega\} = \widetilde{\Omega}(m_\mathbf{x})\}, \quad \text{and} \quad \omega^U := \{\mathbf{x} \in \mathcal{D} : \omega \in \widetilde{\Omega}(m_\mathbf{x})\}. \qquad (7)$$

We refer to this approach as *lu-int* (lower and upper approximations via interval dominance). For these two approaches, using their corresponding approximations, we constrain each weight component $\boldsymbol{w}_i$ to lie in the interval $\left[\underline{\boldsymbol{w}_i}, \overline{\boldsymbol{w}_i}\right] := \left[\frac{|\omega_i^L|}{n}, \frac{|\omega_i^U|}{n}\right]$ for all $i = 1, \ldots, k$. Furthermore, the observations $\mathcal{D}_X^j$ passed to the recursive call in the *LearnSPN* algorithm (after the clustering step) are the observations contained in the upper approximations $\omega^U$.

Our third approach uses the argmax of the pignistic probability of Equation (3) for assigning the observations to the corresponding cluster. We call this approach *pignistic-belpla*, and we determine the lower and upper weight bounds using the average belief and plausibility, namely,

$$\underline{\boldsymbol{w}_i} := \frac{1}{|\mathcal{D}|} \sum_{\mathbf{x} \in \mathcal{D}} Bel_{m_\mathbf{x}}(\omega_i) \text{ and, } \overline{\boldsymbol{w}_i} := \frac{1}{|\mathcal{D}|} \sum_{\mathbf{x} \in \mathcal{D}} Pl_{m_\mathbf{x}}(\omega_i). \tag{8}$$

## 3.1 Decision Criteria

We assess the performance of each model by measuring their accuracy under different classification tasks. More specifically, let $X_q$ and $\boldsymbol{X}_e$ be, respectively, the class and the observed variables; define $J_c^{\boldsymbol{x}_e} = \mathbb{1}_c(X_q) \prod_{X \in \boldsymbol{X}_e} \mathbb{1}_{\boldsymbol{x}_e|X}(X)$, where $\boldsymbol{x}_e|X$ is the projection of $\boldsymbol{x}_e$ onto $X$ (we are not assuming that variables are independent. Rather, $J_c^{\boldsymbol{x}_e}$ simply serves as a test function for some class $c$ and evidence $\mathbf{x}_e$ such that $\mathbb{E}_\mathcal{S}[J_c^{\boldsymbol{x}_e}] = P_\mathcal{S}(X_c = c, \boldsymbol{X}_e = \mathbf{x}_e)$, where $P_\mathcal{S}$ is the distribution encoded by $\mathcal{S}$). According to the type of circuit we follow distinct classification criteria to obtain the predicted class $c^*$ for the class variable $X_q$.

**Classification with Precise Circuits:** In the case of precise circuits, we have

$$c^* = \text{argmax}_{c \in \text{val } X_q} \mathbb{E}_\mathcal{S}[J_c^{\boldsymbol{x}_e}]. \tag{9}$$

**Classification with Credal Circuits:** Conversely, for credal circuits, we follow three of the criteria presented in [11].

$\Gamma$**-maximin:** $\quad c^* = \text{argmax}_{c \in \text{val } X_q} \underline{\mathbb{E}}_\mathcal{S}[J_c^{\boldsymbol{x}_e}].$ (10)

$\Gamma$**-maximax:** $\quad c^* = \text{argmax}_{c \in \text{val } X_q} \overline{\mathbb{E}}_\mathcal{S}[J_c^{\boldsymbol{x}_e}].$ (11)

**(Determinate) Maximality:** For $a, b \in \text{val } X_q$, we define the strict partial order $a \succ_M b$ if and only if $\underline{\mathbb{E}}_\mathcal{S}[J_a^{\boldsymbol{x}_e} - J_b^{\boldsymbol{x}_e}] > 0$. The set of $\succ_M$-non-dominated classes is given by

$$N_{\succ_M} := \{c \in \text{val } X_q : c' \not\succ_M c \text{ for all } c' \in \text{val } X_q\}. \tag{12}$$

Then, the predicted class is given by

$$c^* = \text{argmax}_{c \in N_{\succ_M}} \frac{1}{|N_{\succ_M}|} \sum_{c' \in N_{\succ_M}} \underline{\mathbb{E}}_\mathcal{S}[J_c^{\boldsymbol{x}_e} - J_{c'}^{\boldsymbol{x}_e}]. \tag{13}$$

Note that this is a variation of the well-known maximality criterion where we force the choice of a single prediction by looking at Eq. (13). This choice is for computational convenience (lower expectations in the equation are immediately available after computing maximal elements) and to allow us to easily compare credal classifiers with non-credal counterparts.

## 4 Experiments

We present some preliminary results obtained while performing classification for several binary datasets.[3,4] For the sake of speed and to deal with most challenging classification tasks, each training data is down-sampled to contain at most 4000 observations. We randomly select 5 query variables and for each query variable we compute the test accuracy obtained by each model over 1000 instances in the test set; we report the average of these accuracies across 10 different seeds.

In the following tables, the column *most freq* presents the results obtained by predicting the most frequent class, *kmeans* refers to a precise circuit learned using K-Means algorithm in the clustering step. Similarly, *gmm* is a precise circuit using a Gaussian Mixture Model in the clustering step and using hard-cluster assignment. These models serve as our benchmarks.

To obtain our credal evidential circuits, we use the IRQP-EVCLUS[5] algorithm proposed in [2]. The number of clusters used in all the experiments is set equal to 3 and, as is commonly done in evidential clustering, we restrict to subsets $A \subseteq \Omega$ with $1 \leq |A| \leq 2$ plus the set $\Omega$. The rest of the parameters for the evidential clustering algorithm are the ones proposed in [2].

Table 1 shows the results obtained with the benchmark models. Tables 2, 3 and 4 contain the accuracies obtained with the proposed credal models. As we can see from these tables, classifying with credal models via maximality criterion, often gives us higher accuracies than when we use $\Gamma$-maximin and $\Gamma$-maximax criteria. We can also observe that credal models learned using *lu-max* approach have slightly higher accuracies. Conversely, the model *pig-belpla* achieves slightly lower accuracies compared with the remaining credal models. Lastly, we can see that credal models perform on par when compared to precise PCs.

---

[3] https://github.com/UCLA-StarAI/Density-Estimation-Datasets.
[4] It is worth noting that the criteria from Sect. 3.1 also applies to the general case of multi-class classification.
[5] Iterative Row-wise Quadratic Programming Evidential Clustering.

**Table 1.** Benchmark models

| Dataset | most freq. | kmeans | gmm |
|---|---|---|---|
| accidents | 0.605 | 0.565 | 0.536 |
| adult | 0.893 | 0.853 | 0.837 |
| baudio | 0.787 | 0.777 | 0.775 |
| bnetflix | 0.486 | 0.529 | 0.518 |
| dna | 0.740 | 0.732 | 0.731 |
| jester | 0.561 | 0.651 | 0.656 |
| mushrooms | 0.802 | 0.631 | 0.638 |
| nips | 0.590 | 0.574 | 0.596 |
| nltcs | 0.603 | 0.714 | 0.700 |
| plants | 0.818 | 0.829 | 0.831 |

**Table 2.** Maximality criterion

| Dataset | lu-int | lu-max | pig-belpla |
|---|---|---|---|
| accidents | 0.726 | 0.728 | 0.681 |
| adult | 0.891 | 0.891 | 0.891 |
| baudio | 0.781 | 0.777 | 0.761 |
| bnetflix | 0.463 | 0.464 | 0.465 |
| dna | 0.747 | 0.747 | 0.709 |
| jester | 0.406 | 0.406 | 0.415 |
| mushrooms | 0.792 | 0.799 | 0.775 |
| nips | 0.626 | 0.632 | 0.625 |
| nltcs | 0.661 | 0.672 | 0.652 |
| plants | 0.816 | 0.816 | 0.815 |

**Table 3.** $\Gamma$-maximin criterion

| Dataset | lu-int | lu-max | pig-belpla |
|---|---|---|---|
| accidents | 0.551 | 0.532 | 0.538 |
| adult | 0.839 | 0.840 | 0.841 |
| baudio | 0.782 | 0.772 | 0.749 |
| bnetflix | 0.529 | 0.533 | 0.530 |
| dna | 0.734 | 0.736 | 0.700 |
| jester | 0.658 | 0.651 | 0.653 |
| mushrooms | 0.735 | 0.695 | 0.678 |
| nips | 0.571 | 0.582 | 0.570 |
| nltcs | 0.734 | 0.706 | 0.703 |
| plants | 0.816 | 0.816 | 0.813 |

**Table 4.** $\Gamma$-maximax criterion

| Dataset | lu-int | lu-max | pig-belpla |
|---|---|---|---|
| accidents | 0.548 | 0.533 | 0.538 |
| adult | 0.839 | 0.840 | 0.841 |
| baudio | 0.781 | 0.773 | 0.759 |
| bnetflix | 0.532 | 0.534 | 0.529 |
| dna | 0.739 | 0.738 | 0.710 |
| jester | 0.658 | 0.654 | 0.656 |
| mushrooms | 0.733 | 0.694 | 0.682 |
| nips | 0.584 | 0.584 | 0.584 |
| nltcs | 0.741 | 0.717 | 0.720 |
| plants | 0.816 | 0.816 | 0.813 |

## 5 Conclusions

In this work we propose a novel framework for learning credal probabilistic circuits directly from data. This framework is rooted in the Dempster-Shafer theory of evidence via the use of evidential clustering algorithms. Unlike previous approaches for learning credal PCs, our methodology allow us to handle discrete and continuous variables. We propose three classification criteria to be used with credal probabilistic circuits, and empirically investigate whether models obtained under the proposed framework perform on par with precise PCs.

This work introduces several future research directions. A natural next step is to combine the individual mass functions through the use of combination rules, such as Dempster's or Yager's rule. Using the combined mass function we can propose new ways for defining the bounds for the weight vectors of sum nodes. One of the main drawbacks of the IRQP-EVCLUS algorithm (employed here) is that for every observation in the dataset we have to solve a quadratic program.

Since learning the structure of a credal circuit involves several clustering steps, the proposed framework does not scale well for large datasets; therefore a possible area to explore is the use of alternative evidential clustering algorithms that scale better with respect to the size of the dataset. Furthermore, having faster clustering algorithms allow us to use cross-validation in order to determine the number of clusters to use. Finally, a deeper empirical evaluation of all these ideas is warranted, as this work only presents some preliminary analyses.

**Acknowledgements.** The authors thank the support from the Eindhoven Artificial Intelligence Systems Institute and the Department of Mathematics and Computer Science of TU Eindhoven. The authors thank Erik Quaeghebeur for their valuable insights and discussions on credal PCs. Cassio de Campos thanks the support of EU European Defence Fund Project KOIOS (EDF-2021-DIGIT-R-FL-KOIOS).

# References

1. Dempster, A.P.: A generalization of Bayesian inference. J. Roy. Stat. Soc. Ser. B (Methodol.) **30**(2), 205–247 (1968)
2. Denœux, T., Sriboonchitta, S., Kanjanatarakul, O.: Evidential clustering of large dissimilarity data. Knowl.-Based Syst. **106**, 179–195 (2016)
3. Gens, R., Domingos, P.: Learning the structure of sum-product networks. In: Dasgupta, S., McAllester, D. (eds.) Proceedings of the 30th International Conference on Machine Learning. Proceedings of Machine Learning Research, vol. 28, pp. 873–880. PMLR, Atlanta (2013)
4. Levray, A., Belle, V.: Learning credal sum product networks. In: Automated Knowledge Base Construction (2020)
5. Masson, M.H., Denœux, T.: ECM: an evidential version of the fuzzy C-means algorithm. Pattern Recogn. **41**(4), 1384–1397 (2008)
6. Mauá, D.D., Conaty, D., Cozman, F.G., Poppenhaeger, K., De Campos, C.P.: Robustifying sum-product networks. Int. J. Approximate Reasoning **101**, 163–180 (2018)
7. Montalván Hernández, D.R., Centen, T., Krak, T., Quaeghebeur, E., de Campos, C.: Beyond tree-shaped credal probabilistic circuits. Int. J. Approximate Reasoning 109047 (2023)
8. Poon, H., Domingos, P.: Sum-product networks: a new deep architecture. In: 2011 IEEE International Conference on Computer Vision Workshops (ICCV Workshops), pp. 689–690 (2011)
9. Shafer, G.: A Mathematical Theory of Evidence. Princeton University Press, Princeton (1976)
10. Smets, P.: Decision making in the TBM: the necessity of the pignistic transformation. Int. J. Approximate Reasoning **38**(2), 133–147 (2005)
11. Troffaes, M.C.: Decision making under uncertainty using imprecise probabilities. Int. J. Approximate Reasoning **45**(1), 17–29 (2007)
12. Walley, P.: Statistical Reasoning with Imprecise Probabilities. Chapman and Hall (1991)

# Uncertainty Quantification in Regression Neural Networks Using Likelihood-Based Belief Functions

Thierry Denœux[1,2]([✉])

[1] Université de technologie de Compiègne, CNRS, Heudiasyc, Compiègne, France
[2] Institut universitaire de France, Paris, France
tdenoeux@utc.fr

**Abstract.** We introduce a new method for quantifying prediction uncertainty in regression neural networks using evidential likelihood-based inference. The method is based on the Gaussian approximation of the likelihood function and the linearization of the network output with respect to the weights. Prediction uncertainty is described by a random fuzzy set inducing a predictive belief function. Preliminary experiments suggest that the approximations are very accurate and that the method allows for conservative uncertainty-aware predictions.

**Keywords:** Evidence theory · Dempster-Shafer theory · machine learning · deep learning · random fuzzy sets

## 1 Introduction

In recent years, research in machine learning (ML) has been increasingly focused on developing models that not only have good prediction performance, but also provide some measure of prediction uncertainty [1,10]. The mainstream Bayesian approach is computationally intensive and it requires the existence of prior knowledge about the model parameters, an unrealistic assumption in the case of neural networks with thousands of weights. The Bayesian approach also does not clearly separate aleatory uncertainty (due to variability of the response given the predictors) from epistemic uncertainty (due to lack of knowledge of the true data distribution). In this paper, continuing previous work, I propose to explore another direction referred to as *evidential machine learning* (EML), in which uncertainty is quantified using belief functions. In particular, a belief function induced by a random set [13] or a random fuzzy set [7] has a probabilistic component, suitable for representing aleatory uncertainty, and a set-based component that can express epistemic uncertainty.

At least two main approaches have been proposed for supervised learning in the evidential ML framework. The *distance-based* approach consists in computing a predictive belief function by assessing the similarity between the input vector and training instances or prototypes. This idea was first proposed for classification [3,4,9] and was only recently applied to regression [5,6]; it does not assume any parametric statistical model. In contrast, the *likelihood-based* approach to statistical prediction, first introduced in [11] and revisited in [7] using the new concept of epistemic random fuzzy set, starts with a parametric model and treats the relative likelihood function as a possibility distribution. By expressing the response variable as a function of the parameter and a random variable with known probability distribution, one obtains a random fuzzy set modeling prediction uncertainty. Noticeably, this approach boils down to Bayesian inference when a prior probability distribution is assumed.

Likelihood-based evidential prediction was applied to linear regression in [12] and to logistic regression in [8]. Applying it to nonlinear models with a large number of parameters while keeping computations tractable is particularly challenging. First results in this direction are reported in this paper, with a focus on regression neural networks. The rest of this paper is organized as follows. Necessary notions about random fuzzy sets and likelihood-based evidential inference will first be recalled in Sect. 2. Our new approach is then described in Sect. 3 and experimental results are reported in Sect. 4.

## 2 Background

Background notions about possibility theory and epistemic random fuzzy sets will first be recalled in Sect. 2.1. Evidential likelihood-based inference will then be summarized in Sect. 2.2.

### 2.1 Possibility Theory and Random Fuzzy Sets

*Possibility and Necessity Measures.* Let $\boldsymbol{\theta}$ be a variable taking values in $\Theta$. Assume that we receive a piece of evidence telling us that "$\boldsymbol{\theta}$ is $\widetilde{F}$", where $\widetilde{F}$ is a normal fuzzy subset of $\Theta$ (i.e., a map $\widetilde{F} : \Theta \to [0,1]$ such that $\sup_{\theta \in \Theta} \widetilde{F}(\theta) = 1$). This evidence induces a *possibility measure* $\Pi_{\widetilde{F}}$ from $2^\Theta$ to $[0,1]$ defined by $\Pi_{\widetilde{F}}(B) = \sup_{\theta \in B} \widetilde{F}(\theta)$, for all $B \subseteq \Theta$. The number $\Pi_{\widetilde{F}}(B)$ is interpreted as the degree of possibility that $\boldsymbol{\theta} \in B$, given that $\boldsymbol{\theta}$ is $\widetilde{F}$ [16]. The corresponding *possibility distribution* is the mapping $\pi_{\widetilde{F}} : \Theta \to [0,1]$ defined by $\pi_{\widetilde{F}}(\theta) = \Pi_{\widetilde{F}}(\{\theta\}) = \widetilde{F}(\theta)$. It is identical to $\widetilde{F}$: the degree of possibility that $\boldsymbol{\theta} = \theta$ given the flexible constraint "$\boldsymbol{\theta}$ is $\widetilde{F}$" is equal to the degree of membership of $\theta$ to fuzzy set $\widetilde{F}$. The dual *necessity measure* is defined as $N_{\widetilde{F}}(B) = 1 - \Pi_{\widetilde{F}}(B^c)$, where $B^c$ denotes the complement of $B$ in $\Theta$.

*Gaussian Fuzzy Vectors.* A *Gaussian fuzzy vector (GFV)* is a normal fuzzy subset $\widetilde{F}$ of $\Theta = \mathbb{R}^p$ (with $p \geq 1$) such that $\widetilde{F}(\theta) = \exp\left(-\frac{1}{2}(\theta - m)^T \boldsymbol{H} (\theta - m)\right)$, where $m \in \mathbb{R}^p$ is the mode of $\widetilde{F}$, and $\boldsymbol{H} \in \mathbb{R}^{p \times p}$ is a symmetric and positive semidefinite precision matrix. We write $\widetilde{F} \sim \text{GFV}(m, \boldsymbol{H})$. When $p = 1$, we say that $\widetilde{F}$ is a Gaussian fuzzy number (GFN). The following proposition (proved in [8]) states that the image of a GFV by a linear mapping is still a GFV.

**Proposition 1.** *Let $\boldsymbol{\theta} \in \mathbb{R}^p$ be a p-dimensional variable constrained by a possibility distribution $\pi_\theta \sim \text{GFV}(m, \boldsymbol{H})$ with mode $m \in \mathbb{R}^p$ and positive definite precision matrix $\boldsymbol{H} \in \mathbb{R}^{p \times p}$. Let $\boldsymbol{U} \in \mathbb{R}^{q \times p}$ be a real matrix of rank*

$q \leq p$, $v \in \mathbb{R}^q$ and $z = U\theta + v \in \mathbb{R}^q$. Variable $z$ is constrained by $\pi_z \sim GFV\left(Um + v, (UH^{-1}U^T)^{-1}\right)$.

*Random Fuzzy Sets.* Let $(\Omega, \Sigma_\Omega, P)$ denote a probability space, $(\Theta, \Sigma_\Theta)$ a measurable space, and $\widetilde{X}$ a mapping from $\Omega$ to the set $[0,1]^\Theta$ of fuzzy subsets of $\Theta$. For any $\alpha \in [0,1]$, let $^\alpha\widetilde{X}$ be the mapping from $\Omega$ to $2^\Theta$ such that $\omega \mapsto \{\theta \in \Theta : \widetilde{X}(\omega)(\theta) \geq \alpha\}$. If, for any $\alpha \in [0,1]$, $^\alpha\widetilde{X}$ is $\Sigma_\Omega - \Sigma_\Theta$ strongly measurable [13], the tuple $(\Omega, \Sigma_\Omega, P, \Theta, \Sigma_\Theta, \widetilde{X})$ is said to be a *random fuzzy set* (RFS) [2]. In Epistemic Random Fuzzy Set theory, a RFS represents a piece of evidence about a variable $\theta$ taking values in $\Theta$. The set $\Omega$ is seen as a *set of interpretations* of this piece of evidence, which may be unreliable, vague (fuzzy), or both. If interpretation $\omega \in \Omega$ holds, we only know that $\theta$ is constrained by the possibility distribution defined by fuzzy set $\widetilde{X}(\omega)$. To any RFS verifying normalization conditions [7], we can associate a belief function representing one's beliefs based on the available evidence. For any $\omega \in \Omega$, a conditional possibility measure $\Pi_{\widetilde{X}(\omega)}$ on $\Theta$ can be defined as follows: for any $B \subseteq \Theta, \Pi_{\widetilde{X}(\omega)}(B) = \sup_{\theta \in B} \widetilde{X}(\omega)(\theta)$. For any $B \in \Sigma_\Theta$, let $Bel_{\widetilde{X}}(B)$ and $Pl_{\widetilde{X}}(B)$ denote, respectively, the *expected necessity* and the *expected possibility* of $B$ wrt $P$. The corresponding mappings $Bel_{\widetilde{X}} : \Sigma_\Theta \to [0,1]$ and $Pl_{\widetilde{X}} : \Sigma_\Theta \to [0,1]$, are, respectively, belief and plausibility functions [2].

## 2.2 Evidential Likelihood-Based Inference

We consider an observed random vector $Y$ with probability density function (pdf) $f_{Y|\theta}$, where $\theta \in \Theta$ is the unknown parameter. The likelihood of any value $\theta$ of the parameter after observing $Y = y$ is $L(\theta; y) = \eta f_{Y|\theta}(y)$, where $\eta$ is an arbitrary positive constant. Assuming that $\sup_\theta L(\theta; y) < +\infty$, we can define the relative likelihood of $\theta$ as

$$\pi_{\theta|y}(\theta) = \frac{L(\theta; y)}{\sup_{\theta' \in \Theta} L(\theta'; y)}. \tag{1}$$

As proposed in [7], we interpret mapping $\pi_{\theta|y} : \Theta \to [0,1]$ as a *possibility distribution* over $\Theta$ or, equivalently, as the *fuzzy set* of likely values of $\theta$ after observing $Y = y$. It is, thus, a representation of the information about $\theta$ provided by observation $y$.

Assuming $\ln \pi_{\theta|y}(\theta)$ to be twice differentiable, a tractable approximation of function $\pi_{\theta|y}(\theta)$ can often be obtained by computing a Taylor expansion of its logarithm about a solution $\widehat{\theta}$ of the score equation $\frac{\partial \ln \pi_{\theta|y}}{\partial \theta} = 0$ up to the second order [14]. We then obtain

$$\pi_{\theta|y}(\theta) \approx \exp\left[-\frac{1}{2}(\theta - \widehat{\theta})^T \mathcal{I}(\widehat{\theta})(\theta - \widehat{\theta})\right], \tag{2}$$

where $\mathcal{I}(\widehat{\theta})$ is the *observed information matrix* defined as

$$\mathcal{I}(\widehat{\theta}) = -\left.\frac{\partial^2 \ln \pi_{\theta|y}}{\partial \theta \partial \theta^T}\right|_{\theta=\widehat{\theta}}.$$

As noted in [14], (2) is usually a good approximation when $\boldsymbol{Y} = (Y_1, \ldots, Y_n)$ is an independent sample and $n$ is large.

Let us now consider a prediction problem, where we want to predict the value of a new $Y_0$ with sample space $\mathcal{Y}$, whose distribution also depends on $\boldsymbol{\theta}$. We can always write $Y_0 = \varphi(\boldsymbol{\theta}, U)$, where $U$ is a pivotal random variable with known distribution and sample space $\mathcal{U}$, and $\varphi$ is a mapping from $\Theta \times \mathcal{U}$ to $\mathcal{Y}$ [12]. After observing the data $\boldsymbol{y}$, our knowledge about $\boldsymbol{\theta}$ is represented by the possibility distribution $\pi_{\boldsymbol{\theta}|\boldsymbol{y}}$. By Zadeh's extension principle [15], our knowledge of $Y_0$ conditionally on $U = u$ is, thus, represented by the possibility distribution $\pi_{Y_0|\boldsymbol{y},u} = \varphi(\pi_{\boldsymbol{\theta}|\boldsymbol{y}}, u)$ defined as

$$\pi_{Y_0|\boldsymbol{y},u}(y) = \sup_{\{\theta \in \Theta : \varphi(\theta, u) = y\}} \pi_{\boldsymbol{\theta}|\boldsymbol{y}}(\theta) \qquad (3)$$

for all $y \in \mathcal{Y}$. The mapping $\widetilde{Y} : [0,1] \to [0,1]^{\mathcal{Y}}$ such that $u \mapsto \pi_{Y_0|\boldsymbol{y},u}$ is, then, a RFS representing statistical evidence about $Y_0$.

## 3 Application to Regression Neural Networks

We consider a neural network for regression with weight vector $\boldsymbol{w} \in \mathbb{R}^N$. The output for input $x$ is denoted by $f(x; \boldsymbol{w})$. We assume that the response variable for an input vector $x$ can be written as $Y = f(x; \boldsymbol{w}) + \sigma U$, where $\sigma$ is the error standard deviation and $U \sim N(0,1)$ is a random variable with standard normal distribution. Given iid data $\{(x_i, y_i)\}_{i=1}^n$, the network is trained by maximizing the penalized log-likelihood

$$\ell_\lambda(\boldsymbol{\theta}) = -n \log \sigma - \frac{1}{2} \log(2\pi) - \frac{1}{2\sigma^2} \sum_{i=1}^n (y_i - f(x_i; \boldsymbol{w}))^2 - \sum_{j=1}^N \lambda_j w_j^2, \qquad (4)$$

where $\boldsymbol{\theta} = (\boldsymbol{w}^T, \sigma)^T$ is the vector of all parameters in the model and $\boldsymbol{\lambda} = (\lambda_1, \ldots, \lambda_N)$ is a vector of $N$ regularization coefficients. The general form of the regularizer in (4) allows us to specify a distinct regularization coefficient for each weight; typically $\lambda_j$ is set to 0 if $w_j$ is a bias term. Our approach is based on a second-order approximation of the penalized log-likelihood (4) and a linear approximation of the map $\boldsymbol{w} \mapsto f(x; \boldsymbol{w})$ for a given input $x$. These two approximations are detailed below.

*Possibility Distribution of $\boldsymbol{\theta}$.* Let $\widehat{\boldsymbol{\theta}} = (\widehat{\boldsymbol{w}}^T, \widehat{\sigma})^T$ be a global maximizer of $\ell_\lambda(\boldsymbol{\theta})$. We define the joint possibility distribution of $\boldsymbol{\theta}$ as $\pi_{\boldsymbol{\theta}|\boldsymbol{y}}(\boldsymbol{\theta}) = \exp\left[\ell_\lambda(\boldsymbol{\theta}) - \ell_\lambda(\widehat{\boldsymbol{\theta}})\right]$. We note that $\pi_{\boldsymbol{\theta}|\boldsymbol{y}}$ is proportional to the product of the relative likelihood (1) and a GFV $\pi_0 \sim \text{GFV}(\boldsymbol{0}, 2 \operatorname{diag}(\boldsymbol{\lambda}))$, which can be seen as encoding prior information. Using the normal approximation (2), $\pi_{\boldsymbol{\theta}|\boldsymbol{y}}$ can be approximated by a GFV with mode $\widehat{\boldsymbol{\theta}}$ and precision matrix $\mathcal{I}_\lambda(\widehat{\boldsymbol{\theta}}) = -\left.\frac{\partial^2 \ell_\lambda}{\partial \boldsymbol{\theta} \partial \boldsymbol{\theta}^T}\right|_{\boldsymbol{\theta}=\widehat{\boldsymbol{\theta}}}$. Using simple calculations, it can be shown that

$$\mathcal{I}_\lambda(\widehat{\boldsymbol{\theta}}) = \begin{pmatrix} \boldsymbol{H} & \boldsymbol{v} \\ \boldsymbol{v}^T & a \end{pmatrix} \qquad (5)$$

with

$$H = -\left.\frac{\partial^2 \ell_\lambda}{\partial w \partial w^T}\right|_{\theta=\widehat{\theta}}, \quad v = -\left.\frac{\partial^2 \ell_\lambda}{\partial w \partial \sigma}\right|_{\theta=\widehat{\theta}} = (4/\widehat{\sigma})\lambda \odot \widehat{w},$$

and $a = -\left.\frac{\partial^2 \ell_\lambda}{\partial \sigma^2}\right|_{\theta=\widehat{\theta}} = 2n/\widehat{\sigma}^2$, where $\odot$ denotes pointwise multiplication.

*Prediction.* We now wish to predict a new outcome $Y_0$ of the response for $x = x_0$; it can be written as $Y_0 = f(x_0, w) + \sigma U = \varphi(x_0, \theta, U)$, with $U \sim N(0,1)$. Given $U = u$, the uncertainty $Y_0$ is constrained by the possibility distribution $\pi_{Y_0|y,u} = \varphi(x_0, \pi_{\theta|y}, u)$. This possibility distribution can be approximated by linearizing $f(x_0; w)$ around $\widehat{w}$, which gives

$$f(x_0; w) \approx f(x_0; \widehat{w}) + g(x_0)^T(w - \widehat{w}),$$

with $g(x_0) = \left.\frac{\partial f(x_0;w)}{\partial w}\right|_{w=\widehat{w}}$. With this approximation, we have

$$Y_0 \approx (g(x_0)^T, U)\theta + f(x_0, \widehat{w}) - g(x_0)^T \widehat{w}.$$

From Proposition 1, assuming matrix $\mathcal{I}_\lambda(\widehat{\theta})$ to be positive definite, possibility distribution $\pi_{Y_0|y,u}$ can then be approximated by a GFN with mode $f(x_0, \widehat{w}) + \widehat{\sigma}u$ and precision

$$h(x_0, u) = \left[(g(x_0)^T \ u)\,\mathcal{I}_\lambda(\widehat{\theta})^{-1}\begin{pmatrix} g(x_0) \\ u \end{pmatrix}\right]^{-1}. \tag{6}$$

The inverse of the precision matrix (5) can be written as

$$\mathcal{I}_\lambda(\widehat{\theta})^{-1} = \frac{1}{c}\begin{pmatrix} cC^{-1} & -H^{-1}v \\ -v^T H^{-1} & 1 \end{pmatrix},$$

where $C = H - vv^T/a = H - 8(\lambda\lambda^T) \odot (\widehat{w}\widehat{w}^T)/n$, and

$$c = a - v^T H^{-1} v = \frac{2n}{\widehat{\sigma}^2}\left(1 - \frac{8}{n}(\lambda\lambda^T) \odot (\widehat{w}^T H^{-1} \widehat{w})\right).$$

Hence, (6) can be written as

$$h(x_0, u) = c\left[\{cg(x_0)^T C^{-1} - v^T H^{-1} u\} g(x_0) - [g(x_0)^T H^{-1} v + u] u\right]^{-1}$$
$$= \frac{1}{\alpha + \gamma u + u^2/c} \tag{7}$$

with $\alpha = g(x_0)^T C^{-1} g(x_0)$ and $\gamma = -2c^{-1}g(x_0)^T H^{-1} v$. The predictive RFS is, thus, $\widetilde{Y}(x_0) : U \mapsto \text{GFN}(f(x_0;\widehat{w}) + U\widehat{\sigma}, h(x_0, U))$. We can observe that both the mode and the precision of $\widetilde{Y}(x_0)(U)$ depend on $U$: $\widetilde{Y}(x_0)$ is, thus, not a Gaussian random fuzzy number (GRFN) as defined in [7]. The degrees of belief and plausibility for any real interval can easily be computed by Monte Carlo simulation. Alternatively, we can observe that, for large $n$, the terms $\gamma u$ and $c^{-1}u^2$

become negligible compared to $\alpha$ in the denominator on the right-hand side of (7); replacing $u$ and $u^2$ by their expectations, $h(x_0, u)$ can be approximated by $1/(\alpha + 1/c)$. RFS $\widetilde{Y}(x_0)$ is then, approximately, a GRFN with mean $f(x_0; \widehat{\boldsymbol{w}})$, variance $\widehat{\sigma}^2$ and precision $h(x_0) = 1/(\alpha + 1/c)$.

**Remark 1.** *In the above derivations, we have assumed that (i) $\widehat{\boldsymbol{\theta}}$ is a global maximizer of $\ell_\lambda(\boldsymbol{\theta})$, and (ii) precision matrix (5) is positive definite. The first assumption is necessary to ensure that the possibility distribution $\pi_{\boldsymbol{\theta}|\boldsymbol{y}}$ does not take values greater than one. It is very difficult, if not impossible, to guarantee that this assumption is verified, but we can ensure that we have reached a high enough maximum by running the optimization algorithm a large number of times. Assumption (ii) ensures that the inverse of the precision matrix exists and the precisions (6) are positive. As we will see in Sect. 4, this assumption is usually not verified exactly as matrix $\mathcal{I}_\lambda(\widehat{\boldsymbol{\theta}})$ typically has a small number of negative eigenvalues. If necessary, we may add a small quantity to the diagonal elements of $\mathcal{I}_\lambda(\widehat{\boldsymbol{\theta}})$ to make it nonsingular and well conditioned.*

## 4 Simulation Results

To evaluate the quality of the approximations performed in Sect. 3 and study some properties of the corresponding predictive belief functions, we considered the `Boston` dataset included in the R package `MASS`. We considered only three of the most informative predictors: `crim`, `zn` and `lstat`, which were normalized with zero mean and unit standard deviation. The data were split into a training set of size 300 and a test set of size 206. A network with one layer of 50 hidden units with Exponential Linear Unit (ELU) activation functions was fit to the data. The regularization coefficients had the same value $\lambda_j = \lambda$ for non-bias weights and $\lambda_j = 0$ for bias weights. Coefficient $\lambda$ was determined by five-fold cross-validation, yielding $\lambda = 0.1$.

Figure 1 shows three examples of exact and approximated possibility distributions $\widetilde{Y}(x)(u)$ for different test input vectors $x$ and random numbers $u$. The "exact" possibilities $\pi_{Y_0|y,u}(y)$ were computed by maximizing the log-likelihood $\ell_\lambda(\boldsymbol{\theta})$ subject to the constraint $f(x, \boldsymbol{w}) = y$. As we can see, the exact possibility distributions are almost undistinguishable from their Gaussian approximations, which are themselves very well approximated by GRFNs.

The lower and upper predictive cdfs computed using the Gaussian approximation are shown in Fig. 2, together with the GRFN approximation with fixed precision. Again, we can see that this latter approximation is excellent: the predictive RFSs are very well approximated by GRFNs. Figure 2 also displays the upper and lower predictive cdfs obtained by the ENNreg model [6].

Figure 3 shows calibration curves for the likelihood-belief functions introduced in this paper and for those computed by ENNreg. In [6], we defined calibration curves as plots of coverage probabilities of intervals centered on $\widehat{\mu}(x_0)$, with degree of belief $\alpha$, for different values of $\alpha \in [0, 1]$; the predictive belief functions are calibrated if the curve is above the diagonal. In Fig. 3, we display more detailed information in that we consider not only two-sided belief

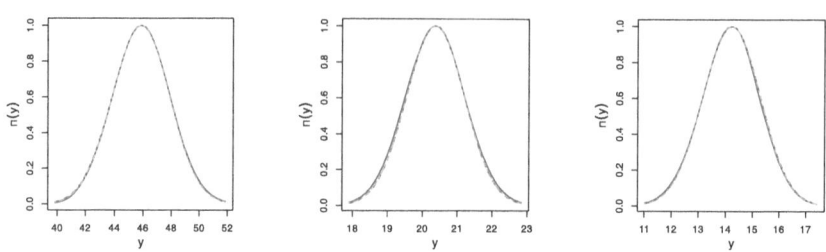

**Fig. 1.** Exact possibility distributions $\widetilde{Y}(x)(u)$ (solid blue lines), Gaussian approximations (red dashed lines) and Gaussian approximations with fixed precision (green dash-dotted lines) for three values of $x$ and $u$. (Color figure online)

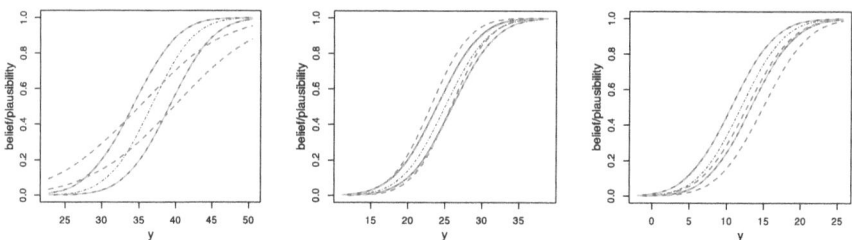

**Fig. 2.** Lower and upper cdfs of predictive RFSs $\widetilde{Y}(x)$ for the same three input vectors as those of Fig. 1 (solid blue lines), Gaussian approximations with fixed precision (red dashed lines), and cdf of probabilistic prediction (blue dotted line). The predictive cdfs obtained by ENNreg are shown as cyan dash-dotted lines. (Color figure online)

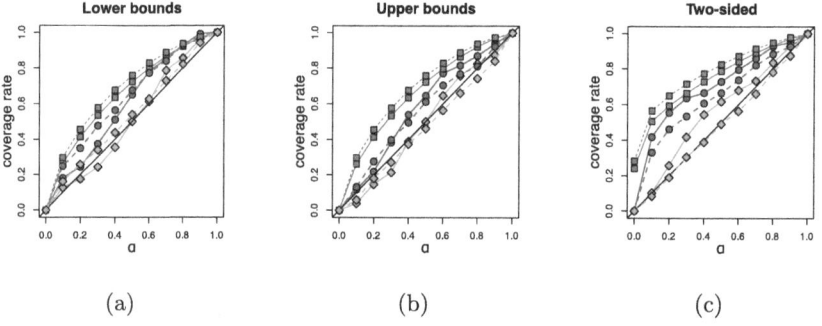

**Fig. 3.** Calibration curves for belief lower bounds (a), belief upper bounds (b) and two-sided belief intervals (c) for likelihood-based belief functions (solid lines) and ENNreg (dashed lines). The lower green curves, middle blue curve and upper red curves correspond, respectively, to the coverage rates of probabilistic predictive intervals, coverage rates of belief intervals, and average plausibilities of belief intervals. (Color figure online)

intervals centered at $\widehat{\mu}(x_0)$ (Fig. 3c), but also one-sided intervals defined by a lower bound (Fig. 3a) or an upper bound (Fig. 3b). Furthermore, we display not only the coverage probabilities of different belief intervals, but also their average plausibilities.

We can see that, for both methods, the belief intervals are conservative (i.e., their coverage rates are greater than their belief degrees), and the coverage rates are bounded above by their average plausibilities, which corresponds to a stronger notion of calibration than that introduced in [6]. For this dataset, there appears to be little difference between the calibration graphs of the predictions obtained by two methods. As noted in [6], predictions can be adjusted using a validation sets to be as precise as possible, while remaining calibrated. A more extensive comparison between the two methods remains to be done.

## 5 Conclusions

We have shown how to apply likelihood-based inference to regression neural network. The method is based on two approximations: the Gaussian approximation of the likelihood function, and the linearization of the network output with respect to the weights. These approximation make it possible to quantify prediction uncertainty by a RFS, which can itself be approximated by a Gaussian random fuzzy number as introduced in [7]. Experimental results with a real dataset suggest that these approximations are very accurate and that they allow us to compute calibrated predictive belief functions with low complexity (the most computationally expensive step being the calculation and inversion of the Hessian matrix, which need to be done only once). These preliminary results will need to be confirmed by much more extensive experiments. Also, in future work, our approach will be applied to a more realistic heteroscedastic model in which the conditional variance is a function of the inputs.

## References

1. Abdar, M., et al.: A review of uncertainty quantification in deep learning: techniques, applications and challenges. Inf. Fusion **76**, 243–297 (2021)
2. Couso, I., Sánchez, L.: Upper and lower probabilities induced by a fuzzy random variable. Fuzzy Sets Syst. **165**(1), 1–23 (2011)
3. Denœux, T.: A $k$-nearest neighbor classification rule based on Dempster-Shafer theory. IEEE Trans. Syst. Man Cybern. **25**(05), 804–813 (1995)
4. Denœux, T.: A neural network classifier based on Dempster-Shafer theory. IEEE Trans. Syst. Man Cybern. A **30**(2), 131–150 (2000)
5. Denœux, T.: An evidential neural network model for regression based on random fuzzy numbers. In: Le Hégarat-Mascle, S., Bloch, I., Aldea, E. (eds.) BELIEF 2022. LNCS, vol. 13506, pp. 57–66. Springer, Cham (2022). https://doi.org/10.1007/978-3-031-17801-6_6
6. Denœux, T.: Quantifying prediction uncertainty in regression using random fuzzy sets: the ENNreg model. IEEE Trans. Fuzzy Syst. **31**, 3690–3699 (2023)
7. Denœux, T.: Reasoning with fuzzy and uncertain evidence using epistemic random fuzzy sets: general framework and practical models. Fuzzy Sets Syst. **453**, 1–36 (2023)

8. Denœux, T.: Uncertainty quantification in logistic regression using random fuzzy sets and belief functions. Int. J. Approximate Reasoning **168**, 109159 (2024)
9. Denœux, T., Kanjanatarakul, O., Sriboonchitta, S.: A new evidential k-nearest neighbor rule based on contextual discounting with partially supervised learning. Int. J. Approximate Reasoning **113**, 287–302 (2019)
10. Hüllermeier, E., Waegeman, W.: Aleatoric and epistemic uncertainty in machine learning: an introduction to concepts and methods. Mach. Learn. **110**(3), 457–506 (2021)
11. Kanjanatarakul, O., Sriboonchitta, S., Denœux, T.: Forecasting using belief functions: an application to marketing econometrics. Int. J. Approximate Reasoning **55**(5), 1113–1128 (2014)
12. Kanjanatarakul, O., Sriboonchitta, S., Denœux, T.: Prediction of future observations using belief functions: a likelihood-based approach. Int. J. Approximate Reasoning **72**, 71–94 (2016)
13. Nguyen, H.T.: On random sets and belief functions. J. Math. Anal. Appl. **65**, 531–542 (1978)
14. Sprott, D.A.: Statistical Inference in Science. Springer, Berlin (2000). https://doi.org/10.1007/b98955
15. Zadeh, L.A.: The concept of a linguistic variable and its application to approximate reasoning -I. Inf. Sci. **8**, 199–249 (1975)
16. Zadeh, L.A.: Fuzzy sets as a basis for a theory of possibility. Fuzzy Sets Syst. **1**, 3–28 (1978)

# An Evidential Time-to-Event Prediction Model Based on Gaussian Random Fuzzy Numbers

Ling Huang[1](✉), Yucheng Xing[1], Thierry Denœux[3,4], and Mengling Feng[1,2]

[1] Saw Swee Hock School of Public Health, National University of Singapore, Singapore, Singapore
huang.l@nus.edu.sg
[2] Institute of Data Science, National University of Singapore, Singapore, Singapore
[3] Université de technologie de Compiègne, CNRS, Heudiasyc, Compiègne, France
[4] Institut Universitaire de France, Paris, France

**Abstract.** We introduce an evidential model for time-to-event prediction with censored data. In this model, uncertainty on event time is quantified by Gaussian random fuzzy numbers, a newly introduced family of random fuzzy subsets of the real line with associated belief functions, generalizing both Gaussian random variables and Gaussian possibility distributions. Our approach makes minimal assumptions about the underlying time-to-event distribution. The model is fit by minimizing a generalized negative log-likelihood function that accounts for both normal and censored data. Comparative experiments on two real-world datasets demonstrate the very good performance of our model as compared to the state-of-the-art.

**Keywords:** Survival analysis · belief functions · Dempster-Shafer theory · random fuzzy sets · uncertainty · machine learning

## 1 Introduction

Time-to-event analysis, also known as survival analysis, focuses on analyzing the time it takes for an event of interest to occur, such as time to death or machine failure. The main challenge of time-to-event prediction is that the observed outcomes are typically censored, meaning that the exact event time is unknown for some data due to early-end experiments or a lack of follow-up, making the estimation problem challenging. Conventional statistical machine learning techniques focus on the estimation of the hazard function, mathematically defined as the ratio of the probability density to the time-to-event function, representing the conditional probability density that a single nonrepeatable event will occur in a particular time interval, given that the item did not experience the event before that time.

The Cox proportional hazards model (Cox model) [1], proposed by Cox in 1972, offers a straightforward approach to handling censoring by assuming proportional hazards across covariates while leaving the baseline hazard function unspecified. Faraggi and Simon [8] introduced an extension of the Cox model by replacing its linear predictor with a one-hidden layer multilayer perceptron (MLP). Recent advancements of the Cox model with deep neural networks, e.g., DeepSurve [10] and Cox-CC [11], show promising performance. However, the proportional hazards assumption may not hold in complex scenarios, potentially leading to biased estimates and inaccurate predictions. To address this limitation, Kvamme et al. [11] proposed a time-dependent Cox model to account for time-varying covariates. Furthermore, the Cox-based model estimates the baseline hazard function solely based on observed event times, which can introduce extra biases or information loss when data is limited. Random Survival Forests (RSF) [9], a non-parametric model that builds upon the random forest algorithm and ensemble learning, shows advantages where traditional parametric or semi-parametric methods may not be suitable or when the underlying survival distribution is complex. In addition to estimating the time-to-event distribution, recent deep-learning advanced approaches also focus on improving prediction performance with new optimization strategies. For instance, DeepHit [12], a probability mass function-based discrete-time model, shows promising concordance index results with a loss function designed to improve event time ranking while disregarding the calibration of the predictions.

In this paper, we propose an evidence-based time-to-event prediction model that does not rely on specific forms of data distribution assumptions. Instead, we calculate the evidence of a time interval directly under the framework of belief functions [2,13] and random fuzzy sets [3,7]. The proposed approach modifies the ENNreg model introduced in [4,6] to account for censored data. Prediction uncertainty is quantified using Gaussian random fuzzy numbers (GRFNs) [7], a newly introduced family of random fuzzy subsets of the real line. In addition to providing the most plausible event time, our model outputs two additional quantities: standard deviation and precision measuring, respectively, aleatory and epistemic prediction uncertainties. The model is fitted by minimizing a generalized negative log-likelihood loss function.

The rest of this paper is organized as follows. Background notions are first recalled in Sect. 2. The proposed model is then introduced in Sect. 3, and experimental results are reported in Sect. 4. Finally, Sect. 5 concludes the paper.

## 2 Background

The theory of epistemic random fuzzy sets (RFSs) was introduced in [3,7] as an extension of Dempster-Shafer theory allowing us to represent both partially reliable and vague evidence. In short, an RFS is a mapping from a probability space to the fuzzy powerset of another space, verifying some measurability property. The reader is referred to the cited references for a general exposition of this theory. Hereafter, we briefly recall the notions of Gaussian and lognormal random fuzzy numbers in Sects. 2.1 and 2.2, respectively.

## 2.1 Gaussian Random Fuzzy Numbers

A Gaussian fuzzy number (GFN) is a fuzzy subset of the real line with membership function $x \mapsto \exp(-0.5h(x-m)^2)$, where $m \in \mathbb{R}$ and $h \geq 0$ are the mode and precision parameters. A Gaussian random fuzzy number (GRFN) $\widetilde{T}$ is an RFS defined as a GFN whose mode $M$ is a Gaussian random variable with mean $\mu$ and variance $\sigma^2$ [7]. It is then defined by three parameters $\mu$, $\sigma^2$ and $h$ and we write $\widetilde{T} \sim \widetilde{N}(\mu, \sigma^2, h)$. The family of GRFNs is closed under the product-intersection rule, a combination operator generalizing Dempter's rule [7]. A GRFN defines a belief function of the real line. Formulas for the degrees of belief and plausibility of any real interval are given in [7].

## 2.2 Lognormal Random Fuzzy Numbers

A GRFN is a model of uncertainty about a variable taking values in the whole real line. In contrast, in time-to-event analysis, the response variable is positive. Uncertainty about such a variable is better represented by a lognormal random fuzzy number as introduced in [5].

In general, let $\psi$ be a one-to-one mapping from $\mathbb{R}$ to a subset $\Lambda \subseteq \mathbb{R}$. Its extension $\widetilde{\psi}$ maps each fuzzy subset $\widetilde{F}$ of $\mathbb{R}$ to a fuzzy subset $\widetilde{\psi}(\widetilde{F})$ of $\Lambda$ with membership function $\lambda \mapsto \widetilde{F}[\psi^{-1}(\lambda)]$. Let $[0,1]^{\mathbb{R}}$ denote the set of all fuzzy subsets of $\mathbb{R}$, and $\widetilde{Y}: \Omega \to [0,1]^{\mathbb{R}}$ be a RFS. By composing $\widetilde{\psi}$ with $\widetilde{Y}$, we obtain a new RFS $\widetilde{\psi} \circ \widetilde{Y}: \Omega \to [0,1]^{\Lambda}$. For any event $C \subseteq \Lambda$, we have

$$Bel_{\widetilde{\psi} \circ \widetilde{Y}}(C) = Bel_{\widetilde{Y}}(\psi^{-1}(C)) \quad \text{and} \quad Pl_{\widetilde{\psi} \circ \widetilde{X}}(C) = Pl_{\widetilde{Y}}(\psi^{-1}(C)). \quad (1)$$

Taking $\widetilde{Y} \sim \widetilde{N}(\mu, \sigma^2, h)$, $\Lambda = [0, +\infty)$ and $\psi = \exp$, we obtain a *lognormal random fuzzy number* $\widetilde{T}: \widetilde{\exp} \circ \widetilde{Y}$ denoted by $\widetilde{T} \sim T\widetilde{N}(\mu, \sigma^2, h, \log)$. We can remark that $\widetilde{\log} \circ \widetilde{Y} \sim \widetilde{N}(\mu, \sigma^2, h)$. Degrees of belief $Bel_{\widetilde{T}}(I)$ and $Pl_{\widetilde{T}}(I)$ for any interval $I \subseteq [0, +\infty)$ can easily be computed from (1) and formulas given in [7] for GRFNs.

## 3 Model

Our approach is based on the ENNreg model introduced in [6]. This model will be recalled in Sect. 3.1. The loss function adapted to censored data will then be described in Sect. 3.2.

### 3.1 Evidential Time-to-Event Prediction Network

In time-to-event analysis, the response of an event time is always positive, while a GRFN is a model about a variable taking values in the whole real line. Following the idea of *Lognormal random fuzzy numbers* as we introduced in Sect. 2.2, we construct a transformed GRFN-based evidential neural network to map predictions into the positive real timeline with $Y = \log(T)$, where $T$ is the time to event. Here the network is composed of three layers: the distance layer, the evidence mapping layer, and the fusion layer. The distance layer computes the distances between the input vector $x$ and each prototype $p_k$ with a positive scale parameter $\gamma_k$: $s_k(x) = \exp(-\gamma_k^2 \|x - p_k\|^2)$. For each prototype $p_k$, the evidence mapping computes a GRFN $\tilde{N}(\mu_k(x), \sigma_k^2, s_k(x)h_k)$, where $\sigma_k^2$ and $h_k$ are variance and precision parameters, and $\mu_k(x)$ is given by $\mu_k(x) = \beta_k^T x + \beta_{k0}$, where $\beta_k$ is a $p$-dimensional vector of coefficients and $\beta_{k0}$ is a scalar parameter. The evidence fusion layer combines evidence from the $K$ prototypes using the unnormalized product-intersection combination rule $\boxplus$ [7] and outputs a final GRFN $\tilde{Y}(x) \sim \tilde{N}(\mu(x), \sigma^2(x), h(x))$ given by

$$\mu(x) = \frac{\sum_{k=1}^K s_k(x)h_k\mu_k(x)}{\sum_{k=1}^K s_k(x)h_k}, \quad \sigma^2(x) = \frac{\sum_{k=1}^K s_k^2(x)h_k^2\sigma_k^2}{(\sum_{k=1}^K s_k(x)h_k)^2},$$

and $h(x) = \sum_{k=1}^K s_k(x)h_k$. Output $\mu(x)$ denotes the most plausible time-to-event prediction, $\sigma^2(x)$ denotes the variance around $\mu(x)$ (aleatory uncertainty), and $h(x)$ denotes the precision of the prediction (epistemic uncertainty). Uncertainty about $T$ is then described by the lognormal RFN $\tilde{T} \sim T\tilde{N}(\mu(x), \sigma^2(x), h(x), \log)$.

### 3.2 Loss Function

To optimize the proposed framework, we use negative generalized log-likelihood loss defined in [6], and adapt it to account for both uncensored and censored data. If the data is not censored, the continuous event time $\tilde{Y}$ is always observed with finite precision. Therefore, instead of observing an exact value, we actually observe an interval $[y]_\epsilon = [y - \epsilon, y + \epsilon]$ centered at $y$. Our prediction evidence can, therefore, be characterized by either the degree of belief $Bel([y]_\epsilon)$ or the plausibility $Pl([y]_\epsilon)$. Conversely, if the data is censored, the event time $\tilde{Y}$ will be observed in interval $[y, \infty)$. Our prediction evidence can now be represented as the degree of belief $Bel([y, \infty))$ or plausibility functions $Pl([y, \infty))$ in the time interval $[y, \infty)$. We can optimize the time-to-event function based either on $L_{Bel}$ or $L_{Pl}$. While none of these two functions adequately measures the quality of the imprecise predictions as mentioned in [6]. Let $\tilde{Y}$ be the output GRFN, $y = \log(t)$ the observation, and $D$ a binary censoring variable such that $D = 1$ if $Y = y$, and $D = 0$ if it is only known that $Y \geq y$. We, therefore, consider the following weighted sum of $L_{Bel}$ or $L_{Pl}$

$$\mathcal{L}_{\lambda,\epsilon}(\tilde{Y}, y, D) = \lambda \overline{\mathcal{L}}(\tilde{Y}, y, D) + (1 - \lambda)\underline{\mathcal{L}}(\tilde{Y}, y, D),$$

with

$$\overline{\mathcal{L}}(\tilde{Y}, y, D) = -D \ln Bel_{\tilde{Y}}([y - \epsilon, y + \epsilon]) - (1 - D) \ln Bel_{\tilde{Y}}([y, \infty)),$$

and

$$\underline{\mathcal{L}}(\tilde{Y}, y, D) = -D \ln Pl_{\tilde{Y}}([y - \epsilon, y + \epsilon]) - (1 - D) \ln Pl_{\tilde{Y}}([y, \infty)),$$

where $\lambda$ is the hyperparameter that controls the cautiousness of the prediction (the smaller, the more cautious). We set $\lambda = 0.1$ to enable the model to focus more on plausibility optimization. The hyperparameter $\epsilon$ was set to $10^{-6}$ to present an infinitesimal time interval.

The network is trained by minimizing the regularized average loss

$$C^{(R)}_{\lambda,\epsilon,\xi,\rho}(\Psi) = \frac{1}{n}\sum_{i=1}^{n}\mathcal{L}_{\lambda,\epsilon}(\widetilde{Y}(x_i;\Psi), y_i, D_i) + \frac{\xi}{K}\sum_{k=1}^{K}h_k + \frac{\rho}{K}\sum_{k=1}^{K}\gamma_k^2,$$

where $\Psi$ is the vector of all parameters (prototypes are included as well) in the network, $\widetilde{Y}(x_i;\Psi)$ is the network output for input $x_i$, and $\xi$, $\rho$ are two regularization parameters. The first regularizer term has the effect of reducing the number of prototypes used for the prediction (e.g., setting $h_k = 0$ to discarding prototype $k$), while the second regularizer term shrinks the solution towards a linear model (e.g., setting $\gamma_k = 0$ for all $k$ yields a linear model) In [6], $\xi$ and $\rho$ are tuned by cross-validation. In the experiments reported in Sect. 4, we kept them fixed at $\xi = \rho = 0.1$ for simplicity.

## 4 Experimental Results

We will now show some qualitative results of our method applied to a simulated dataset with various data censoring scenarios (Sect. 4.1), and compare its performance to state-of-the-art time-to-event prediction methods on two real-world datasets (Sect. 4.2).

### 4.1 Illustrative Example on Simulated Dataset

We first consider artificial data with the following distribution: the input $X$ has a uniform distribution in the interval $[-2, 2]$, and the response is

$$T = \exp\left[1.5X + 2\cos(3X)^3 + \frac{X+5}{3\sqrt{5}}V\right], \quad (2)$$

where $V \sim N(0,1)$ is a standard normal random variable. To simulate data censoring scenarios, two elements were incorporated: the event censoring state indicator $D$ and a random censoring value $C$. The event indicator $D$ has a Bernoulli distribution, denoted as $D \sim B(p)$, where $1 - p$ represents the censoring rate (set as 0.1 and 0.7). For events flagged with a censoring indicator $D = 1$, a negative value $C$ is added to $T$ to emulate right censoring. Here, the value $C$ follows a uniform distribution, with $C \sim U(-1,0)$ and $C \sim U(-2,0)$ representing different degrees of censoring severity. Learning and validation sets of size $n = 4000$ and $n = 1000$ were generated.

The model was initialized with $K = 40$ prototypes. The targets $y = \log(t)$, the network outputs $\mu(x)$, along with belief prediction intervals (BPIs) at levels $\alpha \in \{0.5, 0.9, 0.99\}$ are shown in Fig. 1. BPIs, as defined in [6], are intervals centered at $\mu(x)$ with the degree of belief $\alpha$ to contain the true value of the response variable. When only 10% of the data are censored, our model predicts a time-to-event function (red line) that closely aligns with the ground truth function (blue broken line), and the predicted BPIs effectively encompass the majority of data points, as shown in Fig. 1a. With 70% of the data censored (Fig. 1b), the predicted time-to-event function becomes smoother with fewer details and exhibits an upward bias relative to the true sample distribution, as expected. Nevertheless, the BPIs still effectively encompass the majority of data points, though they are wider. When the censoring interval increases, for example, from $[0,1]$ (Fig. 1a) to $[0,2]$ (Fig. 1c), our model continues to perform well with even wider BPIs.

We can conclude that for different data censoring scenarios, the predicted time-to-event functions closely model the actual regression function, even when the data is highly censored. Furthermore, the BPIs effectively encompass the majority of data points. These observations illustrate the robustness of our approach to varying data censoring conditions.

### 4.2 Comparative Results on Real-World Datasets

We further evaluated our approach using two real-world time-to-event datasets provided by [10]. These are the Molecular Taxonomy of Breast Cancer International Consortium (METABRIC) dataset, comprising 1904 samples with a censoring rate of 42%, and the Rotterdam Tumor Bank and German Breast Cancer Study Group (GBSG) dataset, containing 2232 samples with a censoring rate of 43%. Following [11], we used the concordance index ($C_{idx}$) to evaluate the prediction performance, as well as the integrated Brier score (IBS) and the integrated binomial log-likelihood (IBLL) to evaluate the calibration of the estimates. We used five-fold cross-validation and repeated it five times. We compared our methods to the baseline Cox method, RSF [9], Deepsurv [10], Cox-CC [11], Cox-Time [11] and DeepHit [12]. Hyperparameter values for the compared methods are given in the documentation of the Pycox package[1].

Results for the two datasets are reported in Tables 1 and 2. We can remark that all the compared methods are based on specific assumptions, and it is not surprising that some of them perform quite well for specific data distribution. The Cox model performs rather poorly, which was expected as it is based on very restrictive model assumptions. The methods that assume proportional hazards without linearity assumption, i.e., Cox-CC and DeepSurv perform worse, in general, than the less restrictive methods, namely, RFS and Deephit.

Our ENNreg method achieved the best performance according to the $C_{idx}$ and IBLL criteria, and the second best. Notably, our proposal outperforms the continuous Cox-time model by a large amount and performs slightly better than the discrete DeepHit model. This result is interesting considering that we did not develop a time-dependent prediction model like Cox-time, nor did we use concordance for hyperparameter tuning as in Deephit. As we can see from the IBS and

---

[1] https://github.com/havakv/pycox.

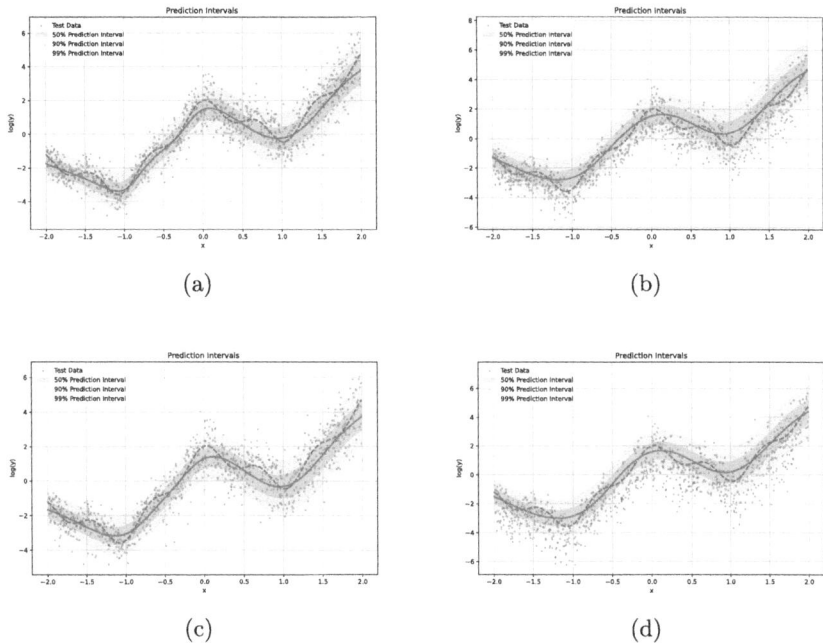

**Fig. 1.** Simulated data, actual regression function (blue broken lines), and predictions obtained from the trained model. Predicted values $\log(y)$ are depicted by red solid lines, while belief prediction intervals (BPIs) at levels $\alpha \in \{0.5, 0.9, 0.99\}$ are represented by shaded areas in blue, green, and orange. The first and second rows are data with censoring intervals $[-1, 0]$ and $[-2, 0]$, respectively. The first and second columns are data with 10% and 70% censoring rates, respectively. (Color figure online)

IBLL results, the promising concordance performance of Deephit comes at the cost of poorly calibrated survival estimates. In contrast, our proposal exhibits good calibration properties, with statistically significant differences observed in calibrated survival estimates for both datasets. We can, therefore, conclude that our evidence-based time-to-event prediction model, based on minimal assumptions, demonstrates greater flexibility and robustness compared to state-of-the-art models that rely on restrictive hypotheses such as the proportional hazard assumption.

**Table 1.** Means and standard errors of $C_{idx}$, IBS and IBLL scores on the Metabric database for our method (ENNreg) and alternative methods. The best and second best results are, resp., in bold and underlined.

| Methods | $C_{idx}$ ↑ | IBS↓ | IBLL ↓ |
|---|---|---|---|
| Cox | $0.633 \pm 9.3 \times 10^{-3}$ | $0.164 \pm 3.3 \times 10^{-3}$ | $0.497 \pm 1.1 \times 10^{-2}$ |
| RFS | $0.644 \pm 1.2 \times 10^{-3}$ | $0.173 \pm 0.9 \times 10^{-3}$ | $0.510 \pm 2.0 \times 10^{-3}$ |
| Deepsurv | $0.646 \pm 7.4 \times 10^{-3}$ | $\mathbf{0.162 \pm 3.6 \times 10^{-3}}$ | $\underline{0.493} \pm 1.2 \times 10^{-2}$ |
| Cox-cc | $0.641 \pm 2.1 \times 10^{-3}$ | $\underline{0.163} \pm 3.3 \times 10^{-3}$ | $\mathbf{0.490 \pm 8.6 \times 10^{-3}}$ |
| Cox-time | $\underline{0.663} \pm 1.0 \times 10^{-2}$ | $0.164 \pm 4.6 \times 10^{-3}$ | $0.488 \pm 1.1 \times 10^{-2}$ |
| DeepHit | $\mathbf{0.672 \pm 1.0 \times 10^{-2}}$ | $0.173 \pm 2.6 \times 10^{-3}$ | $0.516 \pm 6.5 \times 10^{-3}$ |
| ENNreg | $\mathbf{0.672 \pm 9.4 \times 10^{-3}}$ | $\underline{0.163} \pm 2.1 \times 10^{-3}$ | $\mathbf{0.490 \pm 5.0 \times 10^{-3}}$ |

**Table 2.** Means and standard errors of $C_{idx}$, IBS and IBLL scores on the GBSG database for our method (ENNreg) and alternative methods. The best and second best results are, resp., in bold and underlined.

| Methods | $C_{idx}$ ↑ | IBS↓ | IBLL ↓ |
|---|---|---|---|
| Cox | $0.669 \pm 2.5 \times 10^{-3}$ | $\mathbf{0.174 \pm 3.3 \times 10^{-3}}$ | $\underline{0.519} \pm 1.7 \times 10^{-3}$ |
| RFS | $0.655 \pm 0.3 \times 10^{-3}$ | $0.190 \pm 0.5 \times 10^{-3}$ | $0.539 \pm 1.0 \times 10^{-3}$ |
| Deepsurv | $0.666 \pm 8.4 \times 10^{-3}$ | $0.180 \pm 1.9 \times 10^{-3}$ | $0.531 \pm 5.1 \times 10^{-3}$ |
| Cox-cc | $0.672 \pm 3.3 \times 10^{-3}$ | $\mathbf{0.174 \pm 0.5 \times 10^{-3}}$ | $0.529 \pm 3.4 \times 10^{-3}$ |
| Cox-time | $\underline{0.678} \pm 4.7 \times 10^{-3}$ | $\underline{0.177} \pm 1.5 \times 10^{-3}$ | $0.523 \pm 3.7 \times 10^{-3}$ |
| DeepHit | $\underline{0.678} \pm 4.5 \times 10^{-3}$ | $0.195 \pm 1.0 \times 10^{-3}$ | $0.565 \pm 2.6 \times 10^{-3}$ |
| ENNreg | $\mathbf{0.681 \pm 2.2 \times 10^{-3}}$ | $\mathbf{0.174 \pm 1.1 \times 10^{-3}}$ | $\mathbf{0.518 \pm 2.8 \times 10^{-3}}$ |

## 5 Conclusion

In time-to-event analysis, some proportion of the data is usually censored. In this paper, we have adapted the ENNreg model introduced in [6] to account for censored data, and applied it to time-to-event prediction. The model is trained using the logarithm of the response variable $T$ as the target variable and outputs a GRFN. Prediction uncertainty is, thus, quantified by a lognormal random fuzzy number, from which degrees of belief and plausibility of various events can be straightforwardly computed. In this paper, we have focused on prediction accuracy and calibration (assessed using standard performance criteria) and showed the good performance of our model on two datasets as compared to the state-of-the-art. In the future, we will further explore the advantages of uncertainty quantification in time-to-event tasks using belief functions, e.g., studying the standard deviation and precision of the prediction. We will also extend the comparison with state-of-the-art to a wider range of clinical medical datasets for different time-to-event tasks. Moreover, the study of mixtures of GFRNs to fit applications, e.g., finance analysis, should also be interesting.

**Acknowledgment.** This research is supported by A*STAR, CISCO Systems (USA) Pte. Ltd, and National University of Singapore under its Cisco-NUS Accelerated Digital Economy Corporate Laboratory (Award I21001E0002) and the National Research Foundation Singapore under AI Singapore Programme (Award AISG-GC-2019-001-2B).

# References

1. Cox, D.R.: Regression models and life-tables. J. Roy. Stat. Soc.: Ser. B (Methodol.) **34**(2), 187–202 (1972)
2. Dempster, A.P.: Upper and lower probabilities induced by a multivalued mapping. Ann. Math. Stat. **38**, 325–339 (1967)
3. Denœux, T.: Belief functions induced by random fuzzy sets: a general framework for representing uncertain and fuzzy evidence. Fuzzy Sets Syst. **424**, 63–91 (2021)
4. Denœux, T.: An evidential neural network model for regression based on random fuzzy numbers. In: Le Hégarat-Mascle, S., Bloch, I., Aldea, E. (eds.) Belief Functions: Theory and Applications, pp. 57–66. Springer, Cham (2022). https://doi.org/10.1007/978-3-031-17801-6_6
5. Denœux, T.: Parametric families of continuous belief functions based on generalized gaussian random fuzzy numbers. Fuzzy Sets Syst. **471**, 108679 (2023)
6. Denœux, T.: Quantifying prediction uncertainty in regression using random fuzzy sets: the ENNreg model. IEEE Trans. Fuzzy Syst. **31**, 3690–3699 (2023)
7. Denœux, T.: Reasoning with fuzzy and uncertain evidence using epistemic random fuzzy sets: general framework and practical models. Fuzzy Sets Syst. **453**, 1–36 (2023)
8. Faraggi, D., Simon, R.: A neural network model for survival data. Stat. Med. **14**(1), 73–82 (1995)
9. Ishwaran, H., Kogalur, U.B., Blackstone, E.H., Lauer, M.S.: Random survival forests. Ann. Appl. Stat. **2**(2), 841–860 (2008)
10. Katzman, J.L., Shaham, U., Cloninger, A., Bates, J., Jiang, T., Kluger, Y.: DeepSurv: personalized treatment recommender system using a cox proportional hazards deep neural network. BMC Med. Res. Methodol. **18**(1), 1–12 (2018)
11. Kvamme, H., Borgan, Ø., Scheel, I.: Time-to-event prediction with neural networks and cox regression. J. Mach. Learn. Res. **20**(129), 1–30 (2019)
12. Lee, C., Zame, W., Yoon, J., Van Der Schaar, M.: DeepHit: a deep learning approach to survival analysis with competing risks. In: Proceedings of the AAAI Conference on Artificial Intelligence, vol. 32 (2018)
13. Shafer, G.: A Mathematical Theory of Evidence, vol. 42. Princeton University Press (1976)

# Object Hallucination Detection in Large Vision Language Models via Evidential Conflict

Zhekun Liu[1,2], Tao Huang[1,2], Rui Wang[1,2](✉), and Liping Jing[1,2]

[1] School of Computer Science and Technology, Beijing Jiaotong University, Beijing, China
rui.wang@bjtu.edu.cn
[2] Beijing Key Lab of Traffic Data Analysis and Mining, Beijing, China

**Abstract.** Despite their remarkable ability to understand both textual and visual data, large vision-language models (LVLMs) still face issues with hallucination. This is particularly presented as the object hallucination, where the models inaccurately describe objects in the images. Current efforts mainly focus on detecting such erroneous behaviors through the semantic consistency of outputs via multiple inferences or by evaluating the entropy-based uncertainty of predictions. However, the former is resource-intensive, while the latter is often considered a less precise measure due to generally recognized overconfident predictions. To address the issue, we propose an object hallucination detection method based on evidential conflict. To be specific, we view the features in the last layer of the transformer decoder as evidence. Then, we combine the evidence based on Dempster's rule, following the approach presented in the work [6]. Hence, this enables us to detect hallucinations by evaluating the conflict among evidence. Preliminary experiments were conducted on a state-of-the-art LVLM, mPLUG-Owl2. Results show that our approach exhibits an enhancement over baseline methods, particularly in cases with highly uncertain inputs.

**Keywords:** LVLM · object hallucination · uncertainty quantification · Dempster-Shafer theory

## 1 Introduction

Since their introduction, large language models (LLMs) have attracted significant attention [3,21], with their capabilities now being extended to multimodal tasks [4,18]. Even though they present excel performance in tasks dealing with both visual and textual data, such as image captioning and visual question answering (VQA), the LVLMs still suffer from hallucinations, similar to LLMs. Particularly, LVLMs present the inaccuracies in describing the object within target images, dubbed as object hallucinations, as shown in Fig. 1. Such unexpected behaviors severely limit their application in safety-critical domains [2].

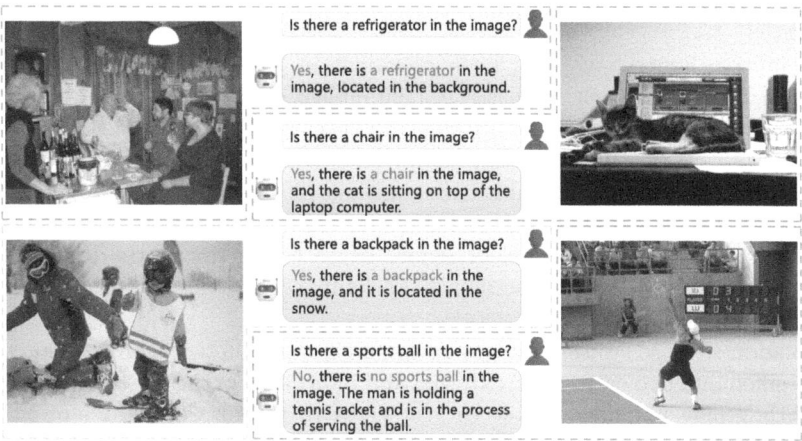

**Fig. 1.** Examples of object hallucination in LVLMs. Red fonts indicate hallucinations. (Color figure online)

Researchers [19] observed that when confident, a model consistently generates similar responses across multiple inferences. Based on this observation, prior studies [14,19,22] focused on evaluating the semantic consistency of responses through repeated testing for hallucination detection. However, such methods are computationally expensive due to their reliance on repeated model inference. Other approaches assess predictive uncertainty with various metrics for this misbehavior detection. Nevertheless, the overconfidence in misclassification of neural network classifier has been widely recognized [9,11].

To mitigate these issues, we resort to the evidential conflict based on the Dempster-Shafer (DS) theory [5,25] to detect object hallucinations, treating the features in the last layer of the transformer decoder in LVLMs as evidence. Following the approach presented in the work [6], we fuse this evidence from the transformer decoder using Dempster's rule. Finally, we evaluate the evidential conflict among the combined results to quantify the uncertainty. We use a state-of-the-art (SOTA) LVLM named mPLUG-Owl2, conducting experiments on three in-distribution datasets and two out-of-distribution datasets. Experimental results demonstrate the effectiveness of our method in object hallucination detection, especially in the conditions with high uncertainty.

## 2 Prerequisites

In this section, we have organized the current methods for hallucination detection and analyzed their limitations. Additionally, to pave the way for our methodology, we have presented some fundamental concepts from DS theory.

## 2.1 Hallucination Detection

Hallucination issues have long been prevalent in generative models [24] for a wide range of tasks, such as abstractive summarization [8], dialogue generation [13], question answering [15], and machine translation [20]. LLMs represented by GPT series [1,3], demonstrating significant performance improvements in the aforementioned tasks, still exhibit hallucinations [29]. For LVLMs, hallucinations are mainly in the form of object hallucinations [23]. They tend to generate depictions of nonexistent items or ignore those that exist in the image. This phenomenon is particularly common in tasks like image description and image captioning.

To address these issues, previous research aiming to detect hallucinations can be classified into two categories, external and internal approaches. For model API users, they may not always able to get access to the internal states of LLMs. Thus, researchers propose external approaches. By introducing semantic entropy, Kuhn et al. [14] assessed the semantic similarity among the multiple outputs generated by the model. While Raj et al. [22] and Manakul et al. [19] detected hallucinations by analyzing the semantic consistency of the large language model's responses to multiple questions with the same meaning, or multiple responses to the same question. With regard to model developers, they prefer internal methods, which enable the direct uncertainty evaluation based on the outputs of the softmax layer for detecting or even mitigating hallucinations. Hence, existing methods primarily focus on the metrics such as log-probability [10] and entropy [27]. However, these quantified uncertainties are less accurate due to the widely recognized overconfidence issue [9,11] of the softmax function.

## 2.2 Basics for DS Theory

The DS theory is a formal framework for processing and combining uncertain information, especially when the evidence is incomplete. In the DS theory, a mass function $m$ is a mapping from the power set of the frame of discernment $\Omega$, which represents the set of all possible hypotheses, to the closed interval $[0,1]$. It is used to represent the degree of belief in a certain hypothesis, and verify the following property:

$$\sum_{A \subseteq \Omega} m^{\Omega}(A) = 1, \tag{1}$$

where $A$ is a subset of $\Omega$.

For each subset $A \subseteq \Omega$, if $m(A) > 0$, $A$ is called a *focal set* of $m$. A mass function $m$ is *simple* if it can be expressed as:

$$m(A) = s, \quad m(\Omega) = 1 - s, \tag{2}$$

where $A \subseteq \Omega$ and $A \neq \emptyset$, $s \in [0,1]$ is the *degree of support* in $A$. Additionally, we adopt the *weight of evidence* associated to $m$,

$$w := -\ln(1-s), \tag{3}$$

to facilitate the combination of evidence. Thus, this *simple* mass function can be denoted as $A^w$.

Given two mass functions $m_1$ and $m_2$, which represent the evidence from two distinct sources (e.g., different agents). The combined mass function for all $A \subseteq \Omega$, $A \neq \emptyset$ through Dempster's rule is:

$$(m_1 \oplus m_2)(A) := \frac{1}{1-\kappa} \sum_{B \cap C = A} m_1(B) m_2(C), \tag{4}$$

where $(m_1 \oplus m_2)(\emptyset) := 0$, and the conflict $\kappa$ is an important metric that used to measure the degree of conflict between mass functions,

$$\kappa := \sum_{B \cap C = \emptyset} m_1(B) m_2(C). \tag{5}$$

For two simple mass functions $m_1$ and $m_2$, assume that they have the same focal set $A$ and the degrees of support are $s_1$ and $s_2$, we have:

$$(m_1 \oplus m_2)(A) = 1 - (1-s_1)(1-s_2), \tag{6}$$

$$(m_1 \oplus m_2)(\Omega) = (1-s_1)(1-s_2). \tag{7}$$

The corresponding weight of evidence associated to $m_1 + m_2$ is:

$$w = -\ln\left[(1-s_1)(1-s_2)\right] = -\ln(1-s_1) - \ln(1-s_2) = w_1 + w_2. \tag{8}$$

Hence, when fusing *simple* mass functions that have identical focal sets using Dempster's rule, the resulting mass function remains a *simple* one, with their weights summed. This property is represented as follows:

$$A^{w_1} \oplus A^{w_2} = A^{w_1+w_2}. \tag{9}$$

## 3 Method

Inspired by the capability of DS theory to handle uncertain inferences, in this section, we introduce our novel method for quantifying predictive uncertainty at the token level within LVLMs. This approach is designed for monitoring the occurrence of object hallucination.

### 3.1 Evidential Conflict of Neural Network Classifier

Consider a multi-classification problem with a number of categories $K > 2$. Following the work [6], for each category $\theta_k$, we consider the features $\phi_j(x)$ from layer before logits as evidence for output predictions, and thus assign belief either to the singleton $\{\theta_k\}$ or to its complement $\overline{\{\theta_k\}}$, depending on the sign of its weight,

$$w_{jk} := \beta_{jk} \phi_j(x) + \alpha_{jk}, \tag{10}$$

where $\beta_{jk}$ and $\alpha_{jk}$ are parameters to be identified. The weights of evidence for $\{\theta_k\}$ and $\overline{\{\theta_k\}}$ are equivalent to the positive and negative components of $w_{jk}$, which are denoted by $w_{jk}^+$ and $w_{jk}^-$, respectively. Thus, we have

$$m_{jk}^+ := \{\theta_k\}^{w_{jk}^+}, \quad m_{jk}^- := \overline{\{\theta_k\}}^{w_{jk}^-}. \tag{11}$$

Combining separately the positive and the negative evidence to each category $\theta_k$ through Dempster's rule, we have

$$m_k^+ := \bigoplus_{j=1}^{J} m_{jk}^+ = \{\theta_k\}^{w_k^+}, \tag{12}$$

$$m_k^- := \bigoplus_{j=1}^{J} m_{jk}^- = \overline{\{\theta_k\}}^{w_k^-}. \tag{13}$$

where $w_k^+ := \sum_{j=1}^{J} w_{jk}^+$ and $w_k^- := \sum_{j=1}^{J} w_{jk}^-$. Due to the limited space, refer to [6] for the detailed inference process. We directly provide the formula to calculate the degree of conflict between positive mass functions $m^+$ and negative ones $m^-$ as:

$$\begin{aligned}
\kappa &= \sum_{k=1}^{K} \left\{ m^+ (\{\theta_k\}) \sum_{\theta_k \notin A} m^-(A) \right\} \\
&= \sum_{k=1}^{K} \left\{ \eta^+ \left( \exp\left(w_k^+\right) - 1 \right) \left[ 1 - \eta^- \exp\left(-w_k^-\right) \right] \right\}.
\end{aligned} \tag{14}$$

where $\eta^+ = (\sum_{l=1}^{K} exp(w_l^+) - K + 1)^{-1}$, and $\eta^- = (1 - \prod_{l=1}^{K}[1 - exp(-w_l^-)])^{-1}$ are normalization factors.

### 3.2 Object Hallucination Detection via Evidential Conflict

As shown in Fig. 2, LVLMs commonly feature the architecture with encoders for input data from distinct sources and multiple compiled transformer decoders. Therefore, they can take in both text and image inputs with their respective encoders for processing. Following this, the encoded information is merged in a modality fusion layer, designed to synthesize the multimodal data. The integrated representation is passed to the sequence of N transformer decoders for advanced processing. Finally, after the last decoder has completed its work, the model refines the outcomes to produce logits. Logits are model's raw prediction scores for different categories, they are then processed by the softmax layer, converting them into a normalized probability distribution to generate the final outputs. The output prediction based on these transformer decoder characterizes a sequential token-by-token process, and the output of each token is essentially a multi-classification problem. As illustrated in Fig. 2, we view the outputs $\phi_j(x)$ of the last layer in the transformer decoders for one token as evidence, where $j$

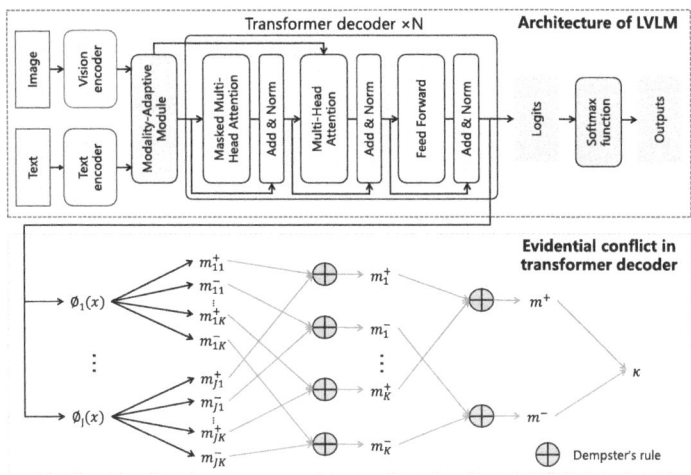

**Fig. 2.** Architecture of LVLMs and the calculation process of $\kappa$.

is the dimension of the outputs. Then, we obtain the mass functions of each evidence $\phi_j(x)$ for each category $\{\theta_k\}$ and its complement $\overline{\{\theta_k\}}$, presented as $m_{jk}^+$ and $m_{jk}^-$, respectively. For each token, $\kappa$ can be calculated through Eq. (14).

For each response generated by LVLMs, we calculate the sum of $\kappa$,

$$\kappa_{sum} = \sum_{i=1}^{L} \kappa_i, \qquad (15)$$

where $L$ is the length of the output tokens and $\kappa_i$ is the $\kappa$ of the $i$-th token in the response. Similarly, we can use other statistical measures, such as the maximum and average values of $\kappa$ in each response.

## 4 Preliminary Experiments

In this section, we show the experiment configurations in Sect. 4.1 and 4.2, as well as the result analysis in Sect. 4.3.

### 4.1 Data Preparation

Our dataset consists of two parts: image set and question set. We chose MSCOCO [17] as our image set. It is a large image dataset developed by Microsoft with 80 labels and more than 160,000 images that can be utilized for a variety of tasks, including segmentation, image recognition, etc.

**In Distribution (ID) Question Set.** We applied the method in POPE [16] when building the ID question set. In details, we selected 500 images featuring over three ground-truth objects through a random process and constructed six

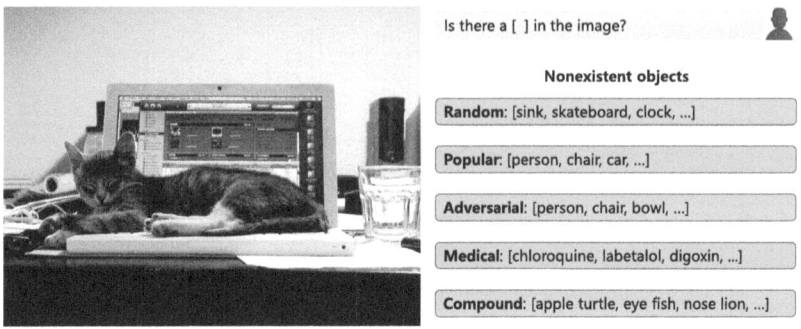

**Fig. 3.** Examples of questions for the five question sets.

questions for each image. Three of them contained existing objects, while the other three involved nonexistent objects. For nonexistent objects: in the random set, objects were chosen in a stochastic manner. In the popular set, we targeted the top-3 most common objects in MSCOCO that are absent from the image. In the adversarial set, we picked the top-3 most frequently co-occurring, yet absent objects based on their relation with present objects (Fig. 3).

**Out of Distribution (OOD) Question Set.** To validate our conjecture that $\kappa$ may performs better for a highly uncertain inputs, we developed two OOD question sets. We manually screened 227 biomedical names from the Therapeutics Data Commons (TDC) [12], with which we randomly replaced the negative sampling objects as the first OOD question set. Meanwhile, we combined common nouns in pairs to form uncommon word compounds like 'glass tiger' and 'sand monkey', which were used to constitute the second OOD question set.

### 4.2 Experimental Setup

We chose a SOTA LVLM mPLUG-Owl2 [28] for the experiment. It is recently released by the Alibaba DAMO Academy, which uses ViT-L (0.3B) [7] as the vision encoder and LLaMA (7B) [26] as the text encoder. We directly used the pre-trained weights on HuggingFace[1] without any fine-tuning. Since our uncertainty measurement method based on evidential conflict is an internal approach, we selected two most commonly used internal uncertainty metrics, self information $I$ based on log probability and entropy $H$ as our baseline methods. Additionally, based on an intuitive hypothesis that models tend to generate longer responses when they are highly uncertainty and making up answers, we also consider the length of the model's response tokens as an impact factor for the occurrence of object hallucinations. To assess the effectiveness of these methods, we used the area under the receiver operating characteristic curve (AUROC). It is a commonly used measure for evaluating the quality of detection metrics.

---

[1] https://huggingface.co/MAGAer13/mplug-owl2-llama2-7b.

## 4.3 Result Analysis

We conducted experiments under five question sets. For each metric, we calculated the maximum, average, sum values of all tokens in the generated responses and assessed their AUROC scores. Results in Table 1 show that the proposed evidential conflict has the best performance on the adversarial and two OOD question sets, particularly outstanding on the two OOD question sets. On the random and popular question sets, the conflict has comparable performance to other metrics. Furthermore, the average of each metric is better than the other two statistic measures. Results also confirm our hypothesis that $\kappa$ is effective for hallucination detection, particularly in highly uncertain situation. It's worth noting that, despite the moderate hallucination occurrence rate on the second OOD question set, $\kappa$ remains consistently effective.

**Table 1.** AUROC scores for hallucination detection under ID and OOD question sets on the validation set of MSCOCO, respectively. The top two results of each setting are indicated bold and underlined. HR stands for actual hallucination occurrence rate.

(a) ID question sets.

| Question set | Random | | | | Popular | | | | Adversarial | | | |
|---|---|---|---|---|---|---|---|---|---|---|---|---|
| Metric | $H$ | $I$ | $L$ | $\kappa$ | $H$ | $I$ | $L$ | $\kappa$ | $H$ | $I$ | $L$ | $\kappa$ |
| Max | 0.60 | 0.60 | 0.47 | 0.62 | 0.56 | 0.57 | 0.49 | 0.61 | 0.57 | 0.57 | 0.53 | 0.62 |
| Avg | **0.66** | **0.66** | 0.47 | <u>0.65</u> | **0.65** | <u>0.64</u> | 0.49 | <u>0.64</u> | <u>0.63</u> | 0.62 | 0.53 | **0.64** |
| Sum | 0.61 | 0.61 | 0.47 | 0.62 | 0.60 | 0.60 | 0.49 | 0.61 | 0.61 | 0.61 | 0.53 | <u>0.63</u> |
| HR | 15.4% | | | | 20.8% | | | | 26.7% | | | |

(b) OOD question sets.

| Question set | Medical | | | | Compound | | | |
|---|---|---|---|---|---|---|---|---|
| Metric | $H$ | $I$ | $L$ | $\kappa$ | $H$ | $I$ | $L$ | $\kappa$ |
| Max | 0.55 | 0.53 | 0.40 | <u>0.81</u> | 0.58 | 0.57 | 0.40 | <u>0.72</u> |
| Avg | 0.59 | 0.58 | 0.40 | **0.82** | 0.64 | 0.63 | 0.40 | **0.78** |
| Sum | 0.51 | 0.51 | 0.40 | 0.77 | 0.56 | 0.56 | 0.40 | 0.70 |
| HR | 30.3% | | | | 13.5% | | | |

## 5 Conclusion

We introduce a method to detect object hallucinations occurring in LVLMs based on evidential conflict. This method utilizes the outputs from the last layer of the transformer decoder as evidence and combines them for token-wise uncertainty quantification. We detect hallucinations with the aggregated the conflicts for all tokens in the generated content. Our preliminary experiments demonstrate the competitive effectiveness of this method in detecting object hallucinations. In particular, they show the great potential for identifying highly uncertain inputs. We hope that our work will inspire future research to further explore the application of the DS theory in mitigate hallucination issues for LVLMs.

## References

1. Achiam, J., et al.: GPT-4 technical report. arXiv preprint arXiv:2303.08774 (2023)
2. Anderljung, M., et al.: Frontier AI regulation: managing emerging risks to public safety. arXiv preprint arXiv:2307.03718 (2023)
3. Brown, T., et al.: Language models are few-shot learners. In: Advances in Neural Information Processing Systems, vol. 33, pp. 1877–1901 (2020)
4. Dai, W., et al.: InstructBLIP: towards general-purpose vision-language models with instruction tuning. In: Advances in Neural Information Processing Systems, vol. 36 (2024)
5. Dempster, A.P.: Upper and lower probabilities induced by a multivalued mapping. In: Yager, R.R., Liu, L. (eds.) Classic Works of the Dempster-Shafer Theory of Belief Functions. STUDFUZZ, vol. 219, pp. 57–72. Springer, Heidelberg (2008). https://doi.org/10.1007/978-3-540-44792-4_3
6. Denœux, T.: Logistic regression, neural networks and Dempster-Shafer theory: a new perspective. Knowl.-Based Syst. **176**, 54–67 (2019)
7. Dosovitskiy, A., et al.: An image is worth $16 \times 16$ words: transformers for image recognition at scale (2021)
8. Falke, T., Ribeiro, L.F., Utama, P.A., Dagan, I., Gurevych, I.: Ranking generated summaries by correctness: an interesting but challenging application for natural language inference. In: Proceedings of the 57th Annual Meeting of the Association for Computational Linguistics, pp. 2214–2220 (2019)
9. Gal, Y., Ghahramani, Z.: Dropout as a Bayesian approximation: representing model uncertainty in deep learning. In: International Conference on Machine Learning, pp. 1050–1059. PMLR (2016)
10. Guerreiro, N.M., Voita, E., Martins, A.F.: Looking for a needle in a haystack: a comprehensive study of hallucinations in neural machine translation. In: Proceedings of the 17th Conference of the European Chapter of the Association for Computational Linguistics, pp. 1059–1075 (2023)
11. Guo, C., Pleiss, G., Sun, Y., Weinberger, K.Q.: On calibration of modern neural networks. In: International Conference on Machine Learning, pp. 1321–1330. PMLR (2017)
12. Huang, K., et al.: Therapeutics data commons: machine learning datasets and tasks for drug discovery and development. In: Proceedings of Neural Information Processing Systems, NeurIPS Datasets and Benchmarks (2021)
13. Jha, S., Jha, S.K., Lincoln, P., Bastian, N.D., Velasquez, A., Neema, S.: Dehallucinating large language models using formal methods guided iterative prompting. In: 2023 IEEE International Conference on Assured Autonomy (ICAA), pp. 149–152. IEEE (2023)
14. Kuhn, L., Gal, Y., Farquhar, S.: Semantic uncertainty: linguistic invariances for uncertainty estimation in natural language generation. In: The Eleventh International Conference on Learning Representations (2022)
15. Li, C., Bi, B., Yan, M., Wang, W., Huang, S.: Addressing semantic drift in generative question answering with auxiliary extraction. In: Proceedings of the 59th Annual Meeting of the Association for Computational Linguistics and the 11th International Joint Conference on Natural Language Processing (Volume 2: Short Papers), pp. 942–947 (2021)
16. Li, Y., Du, Y., Zhou, K., Wang, J., Zhao, W.X., Wen, J.R.: Evaluating object hallucination in large vision-language models. In: Proceedings of the 2023 Conference on Empirical Methods in Natural Language Processing, pp. 292–305 (2023)

17. Lin, T.Y., et al.: Microsoft COCO: common objects in context. In: Fleet, D., Pajdla, T., Schiele, B., Tuytelaars, T. (eds.) ECCV 2014. LNCS, vol. 8693, pp. 740–755. Springer, Cham (2014). https://doi.org/10.1007/978-3-319-10602-1_48
18. Liu, H., Li, C., Wu, Q., Lee, Y.J.: Visual instruction tuning. In: Advances in Neural Information Processing Systems, vol. 36 (2024)
19. Manakul, P., Liusie, A., Gales, M.: SelfCheckGPT: zero-resource black-box hallucination detection for generative large language models. In: The 2023 Conference on Empirical Methods in Natural Language Processing (2023)
20. Martindale, M., Carpuat, M., Duh, K., McNamee, P.: Identifying fluently inadequate output in neural and statistical machine translation. In: Proceedings of Machine Translation Summit XVII: Research Track, pp. 233–243 (2019)
21. Ouyang, L., et al.: Training language models to follow instructions with human feedback. In: Advances in Neural Information Processing Systems, vol. 35, pp. 27730–27744 (2022)
22. Raj, H., Gupta, V., Rosati, D., Majumdar, S.: Semantic consistency for assuring reliability of large language models. arXiv preprint arXiv:2308.09138 (2023)
23. Rohrbach, A., Hendricks, L.A., Burns, K., Darrell, T., Saenko, K.: Object hallucination in image captioning. In: Proceedings of the 2018 Conference on Empirical Methods in Natural Language Processing, pp. 4035–4045 (2018)
24. Semeniuta, S., Severyn, A., Gelly, S.: On accurate evaluation of GANs for language generation (2018). arXiv Preprint (1806)
25. Shafer, G.: A Mathematical Theory of Evidence, vol. 42. Princeton University Press (1976)
26. Touvron, H., et al.: LLaMA: open and efficient foundation language models. arXiv preprint arXiv:2302.13971 (2023)
27. Van Der Poel, L., Cotterell, R., Meister, C.: Mutual information alleviates hallucinations in abstractive summarization. In: Proceedings of the 2022 Conference on Empirical Methods in Natural Language Processing, pp. 5956–5965 (2022)
28. Ye, Q., et al.: mPLUG-Owl2: revolutionizing multi-modal large language model with modality collaboration. In: Proceedings of the IEEE/CVF International Conference on Computer Vision (2024, accepted)
29. Zhuang, Y., Yu, Y., Wang, K., Sun, H., Zhang, C.: ToolQA: a dataset for LLM question answering with external tools. In: Advances in Neural Information Processing Systems, vol. 36 (2024)

# Multi-oversampling with Evidence Fusion for Imbalanced Data Classification

Hongpeng Tian[1], Zuowei Zhang[2(✉)], Zhunga Liu[2], and Jingwei Zuo[3]

[1] School of Electrical and Information Engineering, Zhengzhou University, Zhengzhou, China
[2] School of Automation, Northwestern Polytechnical University, Xi'an, China
{zhangzuowei,liuzhunga}@nwpu.edu.cn
[3] Technology Innovation Institute, Abu Dhabi, UAE
Jingwei.Zuo@tii.ae

**Abstract.** Oversampling methods concentrate on creating a balanced dataset by generating samples, widely utilized in classifying imbalanced data. However, current oversampling methods overlook the uncertainty in the samples produced, potentially shifting the data's distribution and adversely affecting the classification outcomes. To address this problem, we introduce a multi-oversampling with evidence fusion (MOEF) method for imbalanced data classification based on Dempster-Shafer theory. We first design a multi-oversampling strategy to produce various balanced datasets, characterizing the uncertainty of generated samples. Then, we develop a discounting fusion rule based on the inconsistency of data distribution post-oversampling, thereby mitigating the adverse effects of data distribution alterations on classification. Extensive testing on various imbalanced datasets indicates that the proposed MOEF method exhibits more satisfactory performance than other related methods.

**Keywords:** Oversampling · Evidence fusion · Imbalanced data · Dempster-Shafer theory

## 1 Introduction

Imbalanced data classification is a pervasive challenge in machine learning, occurring when datasets exhibit an unequal distribution of samples between classes [1]. This can significantly impair the performance of basic classifiers, which often prioritize overall accuracy, leading to a pronounced bias towards the majority class and neglect of minority class samples. Correctly identifying minority class samples is crucial in many applications [2], such as disease diagnosis, fraud detection, and anomaly identification, where the misclassification of such samples can have severe consequences.

Recently, a large number of methods for classifying imbalanced data have been developed. Among them, sampling methods [3], one of the most popular and effective methods, focus on preprocessing the input data to balance the classes, leading to more equitable model learning and improved performance on minority class prediction compared to other methods. They can be broadly classified into undersampling and oversampling methods. Undersampling methods [4,5], aiming to counteract class imbalance by downsizing the majority class. Random undersampling (RUS) [4], for instance, alleviates class imbalance by randomly eliminating samples from the majority class until a desired balance is achieved. However, RUS may lead to information loss, as valuable samples from the majority class could be discarded. Unlike undersampling, oversampling methods produce artificial data aimed at augmenting minority sample numbers, thereby preserving crucial details about minority samples and enhancing the conceptual portrayal of the minority group. Random oversampling (ROS) [6], for instance, addresses imbalance by duplicating random samples from the minority class, thereby increasing their representation in the dataset. Chawla *et al.* [7] introduce a synthetic minority oversampling technique (SMOTE), synthesizing minority class samples by interpolating between existing samples and their nearest neighbors. This method effectively enhances minority class representation while preserving data distribution characteristics. Han *et al.* [8] present a Borderline-SMOTE, an extension of SMOTE that focuses on oversampling minority samples near the decision boundary. Chao *et al.* [9] propose a novel data augmentation method called H-SMOTE, combining the notion of neighbors and Manhattan distance to produce new samples.

The aforementioned oversampling methods have proven effective in classifying imbalanced data. However, each of these methods yields unique synthetic specimens for the minority class but fails to accurately represent the uncertainty in the produced samples. Inevitably, there is some deviation between the produced sample and the actual sample. Furthermore, this generational divergence could alter the original dataset's distribution, potentially leading to poor classification outcomes.

Dempster-Shafer theory [10,11], also known as the theory of belief functions or evidence reasoning, has been appealing for reasoning uncertain information and widely used in data classification [12–15]. To overcome the above limitations of existing oversampling methods, we propose a multi-oversampling with evidence fusion (MOEF) method for imbalanced data classification with Dempster-Shafer theory in this paper. The contributions of MOEF can be summarized in three aspects. 1) We design a multi-oversampling strategy to generate multiple versions of synthetic samples, characterizing the uncertainty of generated samples. 2) We develop an evidence fusion rule according to the inconsistency of data distribution post-oversampling, weakening the negative impact of changes in data distribution on classification. 3) We apply MOEF to several real imbalanced datasets to demonstrate its superiority over other related methods.

The rest of this paper is arranged as follows. The proposed method is presented in detail in Sect. 2. Then, it is tested in Sect. 3 and compared with several other typical methods, followed by conclusions.

## 2 Multi-oversampling with Evidence Fusion for Imbalanced Data Classification

This part introduces the proposed MOEF method in detail. Assume that a test set $Y = \{\mathbf{y}_1, ..., \mathbf{y}_M\}$ is classified under the discernment framework $\Omega = \{\omega_{min}, \omega_{maj}\}$, based on a training set $X = \{\mathbf{x}_1, ..., \mathbf{x}_N\}$ across $H$ distinct attribute spaces. $X_{min}$ and $X_{maj}$ denote the minority and majority class, respectively.

### 2.1 Multi-oversampling for the Minority Class

In this subsection, we introduce a multi-oversampling technology for the minority class. In this way, various balanced datasets can be acquired, thereby training a basic classifier.

A training sample $\mathbf{x}_i$ is randomly selected from the minority class $X_{min}$. For $\mathbf{x}_i$, we search neighbors from $X_{min}$, denoted as $\mathbf{x}_k$ ($k = 1, ..., K$). Then, we can generate synthetic samples between $\mathbf{y}_i$ and neighbors $\mathbf{x}_k$ ($k = 1, ..., K$) like SMOTE. Then, other training samples from the minority class also generate synthetic samples in this way, and this process is terminated until the number of generated samples is equal to the number ($i.e., |X_{maj}| - |X_{min}|$) of synthetic samples that need to be expanded. However, the generated samples are random and there exists uncertainty in this process. It cannot characterize the uncertainty if we just generate a specific synthetic sample between $\mathbf{x}_i$ and a neighbor $\mathbf{x}_k$. Thus, we introduce a multi-oversampling technology for $\mathbf{x}_i$ to generate multiple synthetic samples, thereby characterizing the uncertainty of synthetic samples. The generated sample $\mathbf{x}_i^t$ ($t = 1, ..., T$) is given by:

$$\mathbf{x}_i^t = \mathbf{x}_i + \alpha_t(\mathbf{x}_k - \mathbf{x}_i) \qquad (1)$$

where $\alpha_t$ is a random number, such that $\alpha_t \in [0, 1]$. By generating $T$ different values of $\alpha_t$, $\mathbf{x}_i$ can generate $T$ synthetic samples in this way. Each sample used for oversampling can generate $T$ synthetic samples, which are combined with training samples to balance the training set. By doing this, we obtain a total of $T$ balanced training sets, denoted as $X^t$ ($t = 1, ..., T$).

A basic classifier is trained using $T$ balanced training sets to classify test samples. With a given test sample $\mathbf{y}_j$, $T$ classification outcomes $P_j^t$ ($t = 1, ..., T$) are achievable. Decision-making for $\mathbf{y}_j$ can be based on the Dempster-Shafer (DS) rule. Nonetheless, oversampling might alter the distribution of the minority class. The outcomes yielded by these classifiers vary in dependability and could lead to discrepancies, resulting in implausible results according to the DS rule in this case. Consequently, the subsequent part will outline methods for efficiently assessing the reliability of various classification outcomes and their combinations.

## 2.2 Evidence Fusion with Discounting Factors

This part assesses the reliability of various classifiers based on the inconsistency in the distribution of data post-oversampling. Subsequently, we implement evidence fusion with different discounting factors using the DS rule [10].

The unavoidable discrepancy between generated synthetic and actual samples leads to a variance in the distribution of the initial minority class compared to the class post-oversampling, potentially adversely impacting the classification outcomes. The maximum mean discrepancy (MMD) [16] is utilized to measure the variance in the distribution between the initial minority class and the class post-oversampling, indicated as:

$$MMD(X_{min}, X_{min}^t) = \left\| \frac{1}{|X_{min}|} \sum_{x_i \in X_{min}} \phi(\mathbf{x}_i) - \frac{1}{|X_{min}^t|} \sum_{x_j \in X_{min}^t} \phi(\mathbf{x}_j) \right\| \quad (2)$$

where $X_{min}^t$ symbolizes the minority class oversampling at the $t$-th level. The symbol $|.|$ represents the number of elements. $\phi(.)$ denotes a mapping function used to align $X_{min}$ and $X_{min}^t$ into a singular space, thus determining their variance.

For the training set $X^t$ after oversampling, the lower value of inconsistency of the minority class, the more reliability of the classification result $P_j^t$ corresponding to $X^t$. Thus, the reliability $\delta^t$ of $P_j^t$ is denoted as:

$$\delta^t = e^{-MMD(X_{min}, X_{min}^t)} / \sum_{t=1}^{T} e^{-MMD(X_{min}, X_{min}^t)} \quad (3)$$

The discounted masses of belief are obtained by:

$$\begin{cases} m_j^t(A) = \delta^t p_j^t(A), A \subset \Omega; \\ m_j^t(\Omega) = 1 - \delta^t + \delta^t p_j^t(\Omega). \end{cases} \quad (4)$$

Through this process, information that has been devalued can be submitted to the total unknown class $(\Omega)$, thereby reducing the degree of discord among pieces of evidence. This permits the application of the DS rule for fusing discounted results. The combined masses of belief $m_j(A)$ for the sample $\mathbf{y}_j$ can be transformed into pignistic probability for the ultimate decision. Defining the pignistic probability is as follows:

$$BetP(\omega_c) = \sum_{A \in 2^\Omega, \omega_c \in A} \frac{1}{|A|} m_j(A) \quad (5)$$

where $|A|$ represents the number of elements in $A$. Subsequently, the sample $\mathbf{y}_j$ can be assigned to the class with the highest probability.

## 3 Experiment Applications

This part evaluates the efficacy of the proposed MOEF method against a range of other commonly used methods. A pair of prevalent indexes [17], named F-measure (FM) and G-mean (GM), commonly applied in classifying imbalanced data, are utilized to assess various methods.

### 3.1 Benchmark Datasets

To assess the effectiveness of diverse methods in classifying imbalanced data, we employ twelve realistic imbalanced datasets extracted from two reputable sources: the KEEL dataset repository (available at http://www.keel.es/) and the UCI Machine Learning Repository (available at http://archive.ics.uci.edu/ml). A summary of the pivotal characteristics of these datasets utilized in the experimental phase is presented in Table 1, detailing the total number of attributes (#Attr), samples (#Size), the majority class samples (#Maj), the minority class samples (#Min), and the imbalance ratio (#IR). To guarantee a sturdy and comprehensive analysis, a rigorous five-fold stratified cross-validation scheme is applied to each dataset, thereby ensuring the reliability and generalizability of the evaluation results.

Table 1. Basic information of the keel datasets

| Data | #Attr. | #Size. | #Min. | #Maj. | #IR. |
|---|---|---|---|---|---|
| Immunotherapy | 7 | 90 | 19 | 71 | 3.74 |
| Climate | 18 | 540 | 46 | 494 | 10.74 |
| Ecoli3 | 7 | 336 | 35 | 301 | 8.60 |
| Ecoli4 | 7 | 336 | 20 | 316 | 15.80 |
| Page-blocks0 | 10 | 5472 | 559 | 4913 | 8.79 |
| Glass | 19 | 2308 | 329 | 1979 | 6.02 |
| Shuttlec2vsc4 | 9 | 129 | 6 | 123 | 20.50 |
| Statlog | 13 | 270 | 120 | 150 | 1.25 |
| Wdbc | 7 | 90 | 19 | 71 | 3.74 |
| Yeast1vs7 | 7 | 457 | 30 | 427 | 14.23 |
| Yeast1 | 8 | 1484 | 429 | 1055 | 2.46 |
| Vehicle2 | 18 | 846 | 218 | 628 | 2.88 |

### 3.2 Comparison Methods

The performance of the proposed method is systematically evaluated in comparison with other related methods. Specifically, RUS [4] alleviates class imbalance by randomly removing majority samples, while ROS [6] duplicates minority samples. SMOTE [7] generates synthetic minority samples along linear interpolations between a sample and its nearest neighbors. Borderline-SMOTE [8] focuses on oversampling minority samples near the decision boundary. H-SMOTE [9] combines nearest neighbors and Manhattan distance to create synthetic samples directed toward the minority class center.

### 3.3 Performance Evaluation

This study utilizes twelve imbalanced datasets to investigate MOEF's efficacy by contrasting it with alternative comparison methods in real-world datasets. Table 2 and Table 3 detail the FM and GM values for various classification methods. In the last row of these tables, we report the number of wins/ties/losses (W/T/L) for each method compared to the one with the highest rank. It's evident that in the majority of datasets, MOEF outperforms other comparison methods. This is due to MOEF producing various iterations of synthetic samples, thereby characterizing the uncertainty in the oversampling process. Furthermore, MOEF assesses the inconsistency in data distribution post-oversampling, thereby mitigating the adverse effects of data distribution alterations on classification. Consequently, MOEF is capable of achieving more robust and satisfactory performance compared to other comparison methods.

**Table 2.** FM of imbalanced datasets by different methods (IN %)

| Datasets | RUS | ROS | SMOTE | Borderline-SMOTE | H-SMOTE | MOEF |
|---|---|---|---|---|---|---|
| Immunotherapy | 49.22 | 47.08 | 52.69 | 45.32 | 37.98 | **54.95** |
| Climate | 54.44 | 42.49 | 62.19 | 60.91 | **62.85** | 62.29 |
| Ecoli3 | 58.39 | 50.37 | 59.38 | 56.89 | 59.74 | **62.97** |
| Ecoli4 | 73.55 | 62.33 | 76.55 | 73.09 | 27.45 | **78.00** |
| Page-blocks0 | 60.27 | 58.70 | 59.62 | 46.03 | **68.57** | 59.54 |
| Glass | 98.48 | 89.76 | 98.48 | 78.92 | 98.33 | **98.48** |
| Shuttlec2vsc4 | 81.33 | 71.82 | **93.33** | **93.33** | **93.33** | **93.33** |
| Statlog | 81.52 | 82.56 | 81.70 | 80.26 | 81.51 | **82.95** |
| Wdbc | 51.48 | 45.33 | 53.48 | 44.46 | 43.33 | **54.71** |
| Yeast1vs7 | 40.03 | 28.98 | 36.87 | 34.03 | 29.26 | **40.33** |
| Yeast1 | 57.58 | 57.37 | 57.57 | 55.57 | 47.07 | **57.81** |
| Vehicle2 | 75.69 | 72.19 | 75.58 | 73.50 | 69.86 | **76.01** |
| W/T/L | 0/0/12 | 0/0/12 | 0/1/11 | 0/1/11 | 2/1/9 | **9/1/2** |

### 3.4 Influence of $T$

The parameter $T$ plays a pivotal role in the proposed MOEF method and could substantially influence the efficiency of MOEF. $T$ denotes the count of different forms of generated synthetic samples. This experiment utilizes a range of imbalanced datasets to examine the efficacy of MOEF across different $T$ values,

Table 3. GM of imbalanced datasets by different methods (IN %)

| Datasets | RUS | ROS | SMOTE | Borderline-SMOTE | H-SMOTE | MOEF |
|---|---|---|---|---|---|---|
| Immunotherapy | 67.47 | 67.13 | 72.19 | 64.83 | 50.55 | **75.05** |
| Climate | 85.27 | 80.15 | 86.91 | 85.85 | 68.41 | **86.92** |
| Ecoli3 | 82.46 | 86.02 | 85.05 | 86.60 | 79.79 | **87.07** |
| Ecoli4 | 95.20 | 92.91 | 95.50 | 94.91 | 72.96 | **95.66** |
| Page-blocks0 | 85.41 | 87.03 | 87.27 | 83.19 | 77.79 | **87.47** |
| Glass | **99.11** | 97.46 | **99.11** | 95.42 | 99.09 | **99.11** |
| Shuttlec2vsc4 | 92.85 | 87.35 | 94.14 | **94.14** | **94.14** | **94.14** |
| Statlog | 82.98 | 83.81 | 82.77 | 81.55 | 82.96 | **84.22** |
| Wdbc | 67.67 | 63.62 | 68.44 | 58.97 | 52.58 | **70.20** |
| Yeast1vs7 | 75.73 | 75.69 | 74.26 | 71.29 | 46.89 | **78.10** |
| Yeast1 | 66.97 | 66.51 | 66.79 | 62.24 | 59.29 | **67.05** |
| Vehicle2 | 86.33 | 83.71 | 86.31 | 84.28 | 79.42 | **86.71** |
| W/T/L | 0/1/11 | 0/0/12 | 0/1/11 | 0/1/11 | 0/1/11 | 10/2/0 |

aiding in the application of $T$. The illustrations displayed in Fig. 1 symbolize the classification outcomes of MOEF, each with varying parameter values. In this context, the x-coordinate signifies $T$ values between 2 and 8, while the y-coordinate indicates the values of FM and GM. The findings clearly show that MOEF remains stable regardless of $T$'s value, with minimal fluctuations noted as $T$ increases. Furthermore, the value of $T$ should not be too small, as it may fail to characterize the uncertainty of generated samples. However, if the $T$ value is set too large, it will bring a lot of computational burden. Therefore, our suggestion is to set $T \in [3,5]$ as the default value in applications.

## 4 Complexity Analysis and Runtime Comparison

MOEF's computational complexity primarily hinges on calculating the distances between between samples of minority class samples during the process of over-sampling. In the minority class $\mathcal{X}_{min}$, the sample $\mathbf{x}_i$ calculates the distance from $\mathbf{x}_i$ to every minority sample in $\mathcal{X}_{min}$, with the computational complexity denoted as $\mathcal{O}(|\mathcal{X}_{min}|)$, where $|\mathcal{X}_{min}|$ signifies the sample count in $\mathcal{X}_{min}$. It's presumed that $N'(N' \leq |\mathcal{X}_{min}|)$ samples require neighbors searching upon the conclusion of the over-sampling procedure. Consequently, MOEF's overall computational complexity is $\mathcal{O}(N'|\mathcal{X}_{min}|)$.

Table 4 displays the execution time in seconds for MOEF and various other methods. It's evident that MOEF's execution duration isn't the briefest, as it requires computing numerous distances among samples to acquire neighboring samples. Within practical scenarios, MOEF proves more apt for situations demanding high precision, in contrast to situations where efficient computing isn't essential.

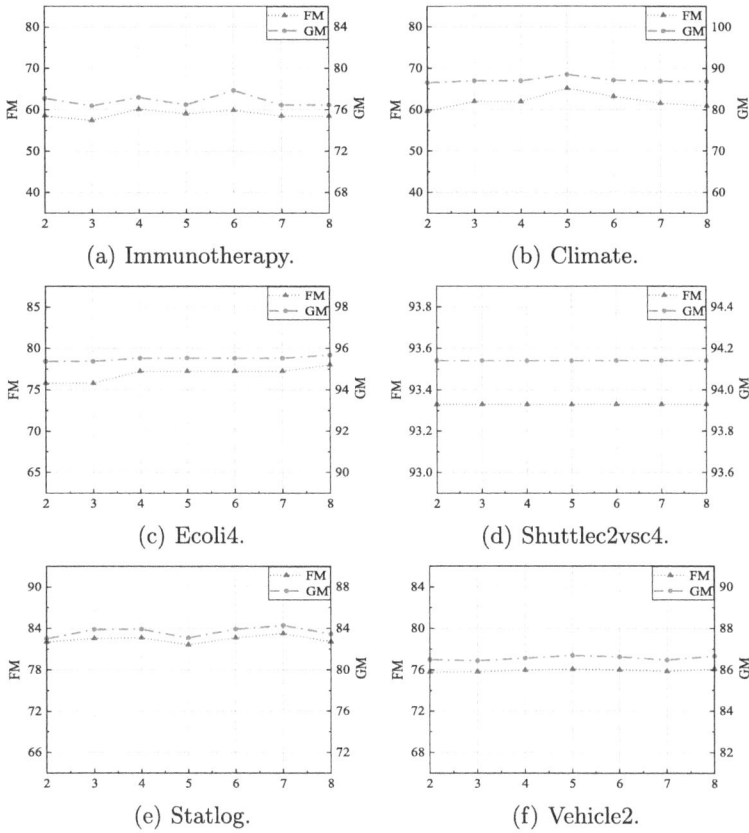

**Fig. 1.** Classification results of MOEF with various values of $T$.

**Table 4.** Execution time of different methods (In seconds)

| Datasets | RUS | ROS | SMOTE | Borderline-SMOTE | H-SMOTE | MOEF |
|---|---|---|---|---|---|---|
| Immunotherapy | 0.17 | 1.28 | 0.26 | 0.11 | 0.23 | 0.53 |
| Climate | 0.15 | 0.78 | 0.38 | 0.31 | 0.30 | 0.81 |
| Ecoli3 | 0.15 | 0.76 | 0.29 | 0.19 | 0.20 | 0.48 |
| Ecoli4 | 0.15 | 0.77 | 0.30 | 0.18 | 0.20 | 0.51 |
| Page-blocks0 | 0.18 | 1.52 | 4.99 | 6.39 | 115.96 | 16.05 |
| Glass | 0.17 | 0.92 | 1.40 | 2.31 | 8.88 | 3.41 |
| Shuttlec2vsc4 | 0.15 | 0.78 | 0.24 | 0.12 | 0.19 | 0.29 |
| Statlog | 0.15 | 0.76 | 0.21 | 0.10 | 0.19 | 0.32 |
| Wdbc | 0.15 | 0.78 | 0.21 | 0.09 | 0.19 | 0.25 |
| Yeast1vs7 | 0.15 | 0.76 | 0.35 | 0.25 | 0.26 | 0.67 |
| Yeast1 | 0.15 | 0.81 | 0.49 | 0.57 | 2.49 | 1.67 |
| Vehicle2 | 0.16 | 0.78 | 0.40 | 0.30 | 0.61 | 0.86 |

## 5 Conclusion

This paper proposes a multi-oversampling with evidence fusion (MOEF) method for imbalanced data classification based on Dempster-Shafer theory. MOEF implements multi-oversampling to characterize the uncertainty of generated samples in the oversampling process. Moreover, MOEF quantifies the degree of inconsistency in data distribution post-oversampling, weakening the negative impact of data distribution alterations on classification. The experiments on synthetic and several real imbalanced datasets have verified the effectiveness of MOEF compared to typical methods. Moreover, we also investigate the influence of the parameter on the classification performance of MOEF and provide guidance on parameter settings. In the future, we will extend the application scope of MOEF to other real-world tasks.

**Acknowledgment.** This work was supported by the Fundamental Research Funds for the Central Universities under Grant G2023KY05102.

## References

1. Xu, Y., Yu, Z., Chen, C.L.P.: Classifier ensemble based on multiview optimization for high-dimensional imbalanced data classification. IEEE Trans. Neural Netw. Learn. Syst. **35**(1), 870–883 (2024)
2. Bai, L., Ju, T., Wang, H., Lei, M., Pan, X.: Two-step ensemble under-sampling algorithm for massive imbalanced data classification. Inf. Sci. 120351 (2024)
3. Tian, H., Zhang, Z., Martin, A., Liu, Z.: Reliability-based imbalanced data classification with Dempster-Shafer theory. In: Le Hegarat-Mascle, S., Bloch, I., Aldea, E. (eds.) BELIEF 2022. LNCS, vol. 13506, pp. 77–86. Springer, Cham (2022). https://doi.org/10.1007/978-3-031-17801-6_8
4. Zhang, Y., Liu, G., Luan, W.: An approach to class imbalance problem based on stacking and inverse random under sampling methods. In: 2018 IEEE 15th International Conference on Networking, Sensing and Control (ICNSC), pp. 1–6. IEEE (2018)
5. Lin, W.C., Tsai, C.F., Hu, Y.H., et al.: Clustering-based undersampling in class-imbalanced data. Inf. Sci. **409**, 17–26 (2017)
6. He, H., Garcia, E.A.: Learning from imbalanced data. IEEE Trans. Knowl. Data Eng. **21**(9), 1263–1284 (2009)
7. Chawla, N.V., Bowyer, K.W., Hall, L.O., et al.: SMOTE: synthetic minority over-sampling technique. J. Artif. Intell. Res. **16**, 321–357 (2002)
8. Han, H., Wang, W.Y., Mao, B.H.: Borderline-SMOTE: a new over-sampling method in imbalanced data sets learning. In: Huang, D.S., Zhang, X.P., Huang, G.B. (eds.) ICIC 2005. LNCS, vol. 3644, pp. 878–887. Springer, Heidelberg (2005). https://doi.org/10.1007/11538059_91
9. Chao, X., Zhang, L.: Few-shot imbalanced classification based on data augmentation. Multimed. Syst. **29**(5), 2843–2851 (2023)
10. Dempster, A.P.: Upper and Lower probabilities induced by a multivalued mapping. Ann. Statist. **83**, 325–339 (1967)
11. Shafer, G.: A Mathematical Theory of Evidence. Princeton University Press, Princeton (1976)

12. Zhang, Z., Tian, H., Yan, L.Z., Martin, A., Zhou, K.: Learning a credal classifier with optimized and adaptive multiestimation for missing data imputation. IEEE Trans. Syst. Man Cybern. Syst. **52**(7), 1–13 (2021)
13. Zhang, Z., Liu, Z., Ning, L., Martin, A., Xiong, J.: Representation of imprecision in deep neural networks for image classification. IEEE Trans. Neural Netw. Learn. Syst. (2023). https://doi.org/10.1109/TNNLS.2023.3329712
14. Niu, J., Liu, Z.G., Lu, Y., Wen, Z.: Evidential combination of classifiers for imbalanced data. IEEE Trans. Syst. Man Cybern. **52**(12), 7642–7653 (2022)
15. Zhang, Z., Ye, S., Zhang, Y., Ding, W., Wang, H.: Belief combination of classifiers for incomplete data. IEEE-CAA J. Automatica Sin. **9**(4), 652–667 (2022)
16. Jia, X., Zhao, M., Di, Y., Yang, Q., Lee, J.: Assessment of data suitability for machine prognosis using maximum mean discrepancy. IEEE Trans. Ind. Electron. **65**(7), 5872–5881 (2017)
17. Wong, T.T.: Linear approximation of f-measure for the performance evaluation of classification algorithms on imbalanced data sets. IEEE Trans. Knowl. Data Eng. **34**(2), 753–763 (2022)

# An Evidence-Based Framework For Heterogeneous Electronic Health Records: A Case Study In Mortality Prediction

Yucheng Ruan[1,2], Ling Huang[1], Qianyi Xu[1], and Mengling Feng[1,2(✉)]

[1] Saw Swee Hock School of Public Health, National University of Singapore, Singapore, Singapore
ephfm@nus.edu.sg
[2] Institute of Data Science, National University of Singapore, Singapore, Singapore

**Abstract.** Electronic Health Records (EHRs), characterized by their centralization of patient comprehensive disease and history information, hold significant promise to improve healthcare quality and efficiency. However, the heterogeneous nature of EHRs potentially affects the accuracy and reliability of predictive models. Many conventional methods analyze these data without explicitly considering their heterogeneity, potentially diminishing performance. Leveraging on the concepts of multimodal information analysis and the Dempster-Shafer theory, we propose an evidence-based learning framework that utilizes multi-sourced encoders to address the heterogeneity in EHRs and combines the multi-sourced evidence using Dempster's combination rule. Our framework significantly outperforms conventional EHR analysis methods, demonstrating higher effectiveness on two tabular encoders in mortality prediction.

**Keywords:** Dempster-Shafer theory · Information fusion · Mortality prediction · Electronic health records

## 1 Introduction

Electronic health record (EHR) systems function as comprehensive repositories for patient health data and play a crucial role in facilitating improved healthcare delivery. Recent advances in deep learning (DL) have promoted the study of identifying predictive patterns within EHR databases under the DL framework such as MLP [1]. However, EHR data are inherently heterogeneous due to the variability of patient data sources across different healthcare systems. For instance, demographic particulars are derived from patient medical questionnaires, while medication and diagnosis information is documented in medical records, and laboratory test results are produced by medical equipment. This heterogeneity presents significant challenges for data analysis and can potentially diminish the performance of DL models [4]. Motivated by the recent development of multimodal learning [7], we believe that even within a single modality, there is significant potential for modeling and analyzing heterogeneous tabular

EHRs from different sources independently and combining them effectively to yield promising performance. Furthermore, the data collection process inherently introduces data biases and noise from factors such as different measuring devices and diverse responses to questionnaires among individuals. Consequently, the uncertainty of source data is introduced due to the imperfect data collection process.

The Dempster-Shafer (DS) theory, also referred to as belief function theory or evidence theory, offers a robust framework for modeling, reasoning with, and combining imprecise information [2,11]. An illustrative application of this theory is observed in evidence fusion, where the outputs of classifiers are transformed into mass functions and aggregated using Dempster's combination rule [9,13,15]. Its recent success in medical data analysis, particularly in the field of medical image analysis, underscores its significant potential for effectively modeling and combining imperfect medical data as well [8]. Given its proven capability, we believe that DS theory can also be extended into the analysis of heterogeneous EHRs and offer a robust framework for improving analytical efficacy in healthcare informatics.

In this paper, we introduce an evidence-based learning framework with multi-sourced encoders for analyzing heterogeneous EHR data. We first strategically categorize EHR features according to their information sources, such as demographics, medications, and laboratory tests. Independent encoders are then developed for each source data to extract representative deep features. These deep features are subsequently mapped into uncombined mass functions through a prototype-based evidence-mapping module under the DS framework. A heterogeneous evidence fusion module is applied later to integrate these mass functions for final decision-making using Dempster's combination rule. Furthermore, our framework augments the primary learning objective that relies on combined mass functions, by incorporating auxiliary losses to the encoder outputs. This operation is crucial for ensuring that representations from each data source are learned sufficiently. Numerical experiments conducted on the MIMIC-III dataset [10] demonstrate the superiority of the proposed framework compared to other conventional and SOTA EHR analysis methods [5].

The rest of the paper is structured as follows: Sect. 2 presents the methodology, with an exposition of DS theory and the proposed framework. Section 3 demonstrates the experimental setup and the results. Section 4 concludes the paper.

## 2 Methodology

We first introduce the Dempster-Shafer theory in Sect. 2.1 and then detail our proposed framework in Sect. 2.2.

### 2.1 Dempster-Shafer (DS) Theory

Let $\Omega = \{\omega_1, \omega_2, \cdots, \omega_M\}$ be a finite set of hypotheses about some question, called the *frame of discernment*. Evidence about a variable taking values in $\Omega$ can be represented by a *mass function*: $2^\Omega$ to [0,1] such that $m(\emptyset) = 0$ and

$$\sum_{A \subseteq \Omega} m(A) = 1, \tag{1}$$

which is called a *mass function*. For any hypothesis $A \subseteq \Omega$, the quantity $m(A)$ is interpreted as a share of a unit mass of belief allocated to the hypothesis that the truth is in $A$, and which cannot be allocated to any strict subset of $A$ based on the available evidence. Set $A$ is called a *focal set* of $m$ if $m(A) > 0$. A mass function is said to be Bayesian if its focal sets are singletons, and logical if it has only one focal set. Two mass functions $m_1$ and $m_2$ representing independent items of evidence can be combined conjunctively by *Dempster's combination rule* [11] $\oplus$ as

$$(m_1 \oplus m_2)(A) = \frac{\sum_{B \cap C = A} m_1(B) m_2(C)}{1 - \sum_{B \cap C = \emptyset} m_1(B) m_2(C)}, \tag{2}$$

for all $A \neq \emptyset$.

After aggregating all available evidence, we adopt the pignistic transformation, as proposed by Smets in the Transferable Belief Model [12], to make a final decision based on the combined mass function. This approach bases decisions on the pignistic probability distribution with the following expression:

$$p(\omega) = \sum_{A \subseteq \Omega : \omega \in A} \frac{m(A)}{|A|}, \forall \omega \in \Omega. \tag{3}$$

### 2.2 Overview of Proposed Framework

In this section, we provide the details of the proposed framework for mortality prediction, which includes multi-sourced encoder modules, a prototype-based evidence mapping (PEM) module, a heterogeneous evidence fusion (HEF) module, and an augmented learning algorithm. The overview of our proposed framework is illustrated in Fig. 1.

**Multi-sourced Encoders.** For data from a single source, we design an independent encoder to generate deep feature representations. This encoder can be any deep learning framework, ranging from classical models such as MLP to SOTA architectures such as ResNet [5], which have been successfully utilized in tabular data modeling. Let $x = (x_1, x_2, \cdots, x_K)$ be the training sample in the EHR dataset, where $K$ is the total number of data sources. Here, $x_j (j = 1, 2, \cdots, K)$ is the training sample from data source $j$. Assuming that $Encoder_j$ is the encoder for data source $j$, we can obtain the encoded deep feature $r_j$ by

$$r_j = Encoder_j(x_j). \tag{4}$$

Despite potential variations in input dimensions across encoders, we ensure consistency in their output dimensions by setting them to a constant value denoted by $P$ for simplicity, thus $r_j \in \mathbb{R}^P$.

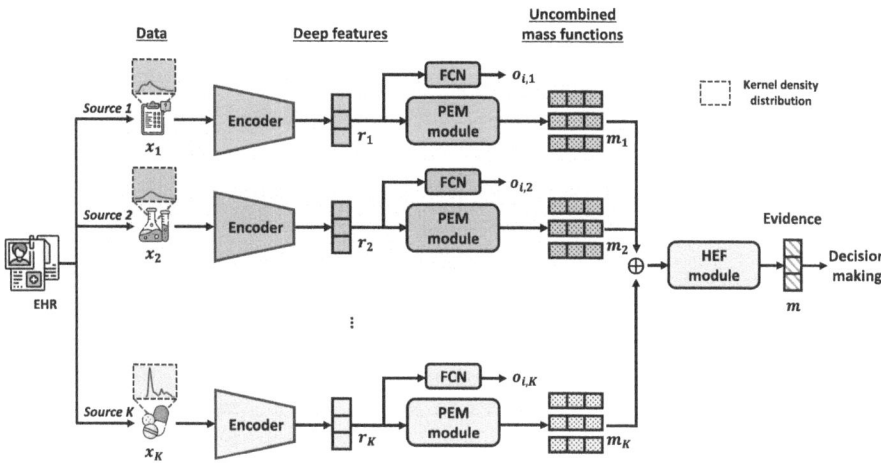

**Fig. 1.** The overview of the proposed evidence-based learning framework for heterogeneous EHRs. The data from the same source (e.g., questionnaires, lab tests, diagnoses, etc.) are first modeled with independent encoders; the output deep features of encoders are then fed into the prototype-based evidence mapping (PEM) module to calculate uncombined mass functions. Finally, the uncombined mass functions from all data sources are combined in the heterogeneous evidence fusion (HEF) module to generate predictive evidence for decision-making.

**Prototype-Based Evidence Mapping (PEM).** Denœux [3] proposed an evidential neural network (ENN), in which mass functions are computed under the DS framework of evidence. The first application of ENN in medical data analysis can be found in [6]. The essential idea of ENN is to consider each prototype as a piece of evidence, which is discounted based on its distance to the input vector. The evidence from different prototypes is then aggerated by Dempster's combination rule (2). Inspired by ENN and the medical application in [7], we propose a prototype-based evidence mapping (PEM) module to calculate uncombined mass functions from independent encoders for further exploration.

The PEM module consists of one input layer and one hidden layer. The input layer is composed of $H$ units ($H$ is the number of prototypes), whose weights vectors are prototypes $\pi_{j,1}, \pi_{j,2}, \cdots, \pi_{j,H}$ in input space. For source $j$, the activation of unit $h$ in the input layer is

$$s_{j,h} = \beta_{j,h}\exp(-\gamma_{j,h}d_{j,h}^2), \tag{5}$$

where $d_{j,h} = \|r_j - \pi_{j,h}\|$ denotes the Euclidean distance between input vector $r_j$ and prototype $\pi_{j,h}$, $\gamma_{j,h} > 0$ is a scale parameter, and $\beta_{j,h} \in [0,1]$ is an extra parameter. The hidden layer computes mass functions $m_{j,h}$ (evidence) of each prototype $\pi_{j,h}$, as follows:

$$m_{j,h}(\{\omega_c\}) = u_{j,h}^{(c)} s_{j,h}, \quad c = 1, 2, \cdots, M, \tag{6a}$$

$$m_{j,h}(\Omega) = 1 - s_{j,h}, \tag{6b}$$

where $u_{j,h}^{(c)}$ is the membership degree of prototype $h$ to class $\omega_c$, and $\sum_{c=1}^{M} u_{j,h}^{(c)} = 1$. Therefore, the PEM module outputs the following uncombined mass functions of source $j$:

$$m_j = (m_{j,1}, m_{j,2}, \cdots, m_{j,H})^T \in \mathbb{R}^{H \times (M+1)}, \tag{7}$$

with

$$m_{j,h} = (m_{j,h}(\{\omega_1\}), m_{j,h}(\{\omega_2\}), \cdots, m_{j,h}(\{\omega_M\}), m_{j,h}(\Omega)) \in \mathbb{R}^{M+1}. \tag{8}$$

**Heterogeneous Evidence Fusion (HEF).** In the heterogeneous evidence fusion module, the $K \times H$ uncombined mass functions from all data sources are treated collectively, which are then aggregated by Dempster's combination rule using Eq. 2. A combined mass function $m$ is computed as the orthogonal sum of the $K \times H$ mass functions, expressed as $m = m_{1,1} \oplus m_{1,2} \oplus \cdots \oplus m_{1,H} \oplus m_{2,1} \oplus \cdots \oplus m_{K,H} \in \mathbb{R}^{M+1}$. The combined mass functions represent the degrees of belief about the given class with $m(\{\omega_c\})$, as well as its prediction uncertainty with $m(\Omega)$.

**Augmented Learning Algorithm.** We optimize the proposed framework based on an augmented learning algorithm, with which the predictive performance based on transformed evidence is optimized as the main loss function, and several auxiliary cross-entropy losses are combined to augment the feature representation ability of the independent encoders.

Let $p_i = (p_i(\omega_1), \cdots, p_i(\omega_c), \cdots, p_i(\omega_M))$ be the final probability after the pignistic transformation (3) for training sample $i$, and $y_i = (y_{i,1}, y_{i,2}, \cdots, y_{i,M})$ denotes the one-hot encoding for corresponding ground-truth labels. The main loss function $\mathcal{L}_p$ is computed as:

$$\mathcal{L}_p = -\frac{1}{N} \sum_{i=1}^{N} \sum_{c=1}^{M} w_c y_{i,c} \log(p_i(\omega_c)), \tag{9}$$

where $N$ is the number of training samples, and $w_c$ is the weight assigned to each class to address the class imbalance issue.

The auxiliary cross-entropy losses are proposed to optimize the feature representation performance of the multi-sourced encoders. Giving $r_{i,j}$ as the outputs from encoder $j$ for the $i$-th training example, an additional fully connected network (FCN) is utilized to generate the logits $o_{i,j} = (o_{i,j}^{(1)}, o_{i,j}^{(2)}, \cdots, o_{i,j}^{(M)})$ for calculating the cross-entropy loss $\mathcal{L}_{ce}^j$ with ground-truth $y_i$ for source $j$. Here, the cross-entropy loss $\mathcal{L}_{ce}^j$ is calculated as:

$$\mathcal{L}_{ce}^j = -\frac{1}{N} \sum_{i=1}^{N} \sum_{c=1}^{M} w_c y_{i,c} \log \frac{\exp(o_{i,j}^{(c)})}{\sum_{b=1}^{M} \exp(o_{i,j}^{(b)})}. \tag{10}$$

Ultimately, the overall loss function $\mathcal{L}_{overall}$ is defined as follows:

$$\mathcal{L}_{overall} = \mathcal{L}_p + \alpha \sum_{j=1}^{K} \mathcal{L}_{ce}^j, \tag{11}$$

where $\alpha$ is a hyperparameter that controls the balance between the main loss and the auxiliary cross-entropy losses.

## 3 Experiment

In this section, we outline the experimental settings as follows: Sect. 2.1 discusses the dataset, Sect. 2.2 describes the evaluation metrics, Sect. 3.1 covers baselines, and Sect. 3.2 elaborates on the implementation details. Section 3.3 presents the numerical results and analyses of the proposed framework and the established baselines.

### 3.1 Dataset

We used the MIMIC-III dataset, a publicly available collection of de-identified health records from patients in Beth Israel Deaconess Medical Center [10]. This dataset has 38,648 patient samples with 38 variables, including 4 demographic variables (e.g., race, gender, age), 21 vital sign variables (e.g., blood pressure, heart rate), and 13 questionnaire variables (e.g., diabetes, hypertension). Therefore, we have patient data from three different sources. The prediction label is patients' mortality in ICU. We preprocessed the dataset by imputing missing values using the mean for continuous features and the mode for categorical features. Subsequently, the dataset was divided into training, validation, and testing sets at a 3:1:1 ratio.

### 3.2 Evaluation Metrics

The evaluation criteria most commonly used to assess mortality predictive performance are area under ROC (AUROC) and accuracy (ACC). We also add balanced accuracy (BACC) to fairly evaluate the models on the imbalanced dataset. Furthermore, we also calculate the Brier score (Brier), expected calibration error (ECE), and negative log-likelihood (NLL) [6] to evaluate the reliability of prediction results.

### 3.3 Baselines

Since our proposed framework is compatible with any SOTA encoders, we test the performance of our proposal on one classical baseline encoder MLP and one SOTA baseline encoder ResNet [5]. We also compare some conventional heterogeneous data analysis methods using the same baseline encoders:

- Single encoder (*Single*): All variables in the tabular EHRs were considered as one single input, and a single encoder was applied for mortality prediction.
- Multi-sourced Concatenation (*M-Concat*): The outputs from multi-sourced encoders were concatenated directly, followed by an additional fully-connected layer for mortality prediction.

- Multi-sourced MeanPooling (*M-MeanPooling*): Instead of concatenating features obtained by multi-sourced encoders, we employed a mean pooling strategy over the independent representations for mortality prediction.
- Multi-sourced Attention (*M-Attention*): We also extended the attention mechanism from Transformer architecture [14] to combine the independent deep features for mortality prediction.

### 3.4 Implementation Details

Both the MLP and ResNet encoders in this study have three blocks, with the number of hidden and output neuron nodes set to 16. For the other hyperparameters, we followed the default settings in [5]. For gradient updating, we adopt the Adam optimizer with a learning rate set at 1e-3. We set the batch size to 128 and the maximum epochs to 100 with early stopping to prevent overfitting. Furthermore, the balancing parameter $\alpha$ was set to 0.45 and 1, respectively, in MLP and ResNet by the *grid search* hyper-parameters estimator. For simplicity, the number of prototypes was set identically for each source, with 5 prototypes in models using MLP and 10 prototypes in models using ResNet. In addition, the parameters were initialized randomly. To address the class imbalance, $w_c$ for the death and aliveness class was set at 7.5 and 1, respectively, according to the positive-to-negative case imbalance ratio. To achieve fair comparisons, each model was trained five times using different random seeds, and both the average values and standard error of results were reported.

### 3.5 Results and Analyses

Tables 1 and 2 present the comparison results of model performance on mortality prediction when MLP and ResNet were used as the encoders, respectively.

When MLP was used as the encoder, we observed significant performance improvements in BACC, AUROC, and ACC by 0.20%, 0.44%, and 0.80%, respectively, compared to the second-best methods. Additionally, our model reduced the Brier score, ECE, and NLL by about 2.22%, 1.16%, and 2.31%, respectively. Similar improvements were noted when using ResNet as the encoder.

We can further conclude that compared to the method using a single encoder, the multi-sourced methods yielded promising performance improvements both in prediction accuracy and reliability. This observation aligns with our expectation in that the inherent heterogeneity of EHRs may lead to a degradation in performance when all features are modeled collectively.

Additionally, our experimental results show that methods using ResNet as the encoder exhibit superior overall performance compared to those using MLP. This highlights the importance of robust feature representation encoder on predictive performance.

**Table 1.** Means and standard errors of the proposed and compared methods using the MLP encoder. The best and second-best results are in bold and underlined.

| Model | BACC↑ | AUROC↑ | ACC↑ | Brier↓ | ECE↓ | NLL↓ |
|---|---|---|---|---|---|---|
| Single | 0.7458±0.0014 | 0.8266±0.0007 | 0.7243±0.0017 | 0.1693±0.0009 | 0.2492±0.0020 | 0.5028±0.0026 |
| M-Concat | 0.7468±0.0023 | 0.8337±0.0014 | 0.7588±0.0039 | 0.1623±0.0022 | 0.2413±0.0034 | 0.4945±0.0060 |
| M-MeanPooling | 0.7478±0.0016 | 0.8343±0.0007 | 0.7578±0.0030 | 0.1638±0.0019 | 0.2421±0.0032 | 0.4988±0.0053 |
| M-Attention | 0.7481±0.0017 | 0.8328±0.0014 | 0.7360±0.0063 | 0.1715±0.0033 | 0.2544±0.0066 | 0.5119±0.0082 |
| Ours | **0.7496±0.0016** | **0.8380±0.0003** | **0.7649±0.0015** | **0.1587±0.0008** | **0.2385±0.0015** | **0.4831±0.0022** |

**Table 2.** Means and standard errors of the proposed and compared methods using the ResNet encoder. The best and second-best results are in bold and underlined.

| Model | BACC↑ | AUROC↑ | ACC↑ | Brier↓ | ECE↓ | NLL↓ |
|---|---|---|---|---|---|---|
| Single | 0.7511±0.0020 | 0.8333±0.0013 | 0.7240±0.0060 | 0.1694±0.0012 | 0.2470±0.0030 | 0.5011±0.0031 |
| M-Concat | **0.7553±0.0024** | 0.8388±0.0012 | 0.7535±0.0042 | 0.1653±0.0022 | 0.2466±0.0051 | 0.5019±0.0059 |
| M-MeanPooling | 0.7515±0.0011 | 0.8379±0.0008 | 0.7536±0.0026 | 0.1648±0.0012 | 0.2498±0.0020 | 0.5015±0.0034 |
| M-Attention | 0.7529±0.0020 | 0.8383±0.0010 | 0.7565±0.0061 | 0.1596±0.0028 | **0.2310±0.0063** | 0.4795±0.0069 |
| Ours | 0.7552±0.0029 | **0.8417±0.0008** | **0.7736±0.0037** | **0.1548±0.0027** | 0.2284±0.0053 | **0.4726±0.0072** |

## 4 Conclusion

In this study, we introduce an evidence-based learning framework with multi-sourced encoders to manage the heterogeneity of EHRs. The proposed framework leverages the DS theory for effective evidence mapping and information fusion, demonstrating its superiority in handling heterogeneous data sources and enhancing decision-making performance. Numerical experiments verified the effectiveness of the proposed multi-sourced framework in mortality prediction with both improved prediction accuracy and reliability. Future work will consider the analysis of wider heterogeneous medical data (e.g., time series, texts, and image data) and more medical tasks (e.g., ICU readmission prediction). Additionally, we will further evaluate our approach using more powerful feature extraction architectures, such as Transformers [14].

**Acknowledgments.** This research is supported by A*STAR, CISCO Systems (USA) Pte. Ltd, and National University of Singapore under its Cisco-NUS Accelerated Digital Economy Corporate Laboratory (Award I21001E0002) and the National Research Foundation Singapore under AI Singapore Programme (Award Number: AISG-GC-2019-001-2B).

## References

1. Che, Z., Purushotham, S., Khemani, R., Liu, Y.: Interpretable deep models for ICU outcome prediction. In: AMIA Annual Symposium Proceedings, vol. 2016, p. 371. American Medical Informatics Association (2016)
2. Dempster, A.P.: Upper and lower probability inferences based on a sample from a finite univariate population. Biometrika **54**(3–4), 515–528 (1967)

3. Denoeux, T.: A neural network classifier based on Dempster-Shafer theory. IEEE Trans. Syst. Man Cybern.-Part A Syst. Hum. **30**(2), 131–150 (2000)
4. Fu, S., et al.: Assessment of the impact of EHR heterogeneity for clinical research through a case study of silent brain infarction. BMC Med. Inform. Decis. Mak. **20**, 1–12 (2020)
5. Gorishniy, Y., Rubachev, I., Khrulkov, V., Babenko, A.: Revisiting deep learning models for tabular data. In: Advances in Neural Information Processing Systems, vol. 34, pp. 18932–18943 (2021)
6. Huang, L., Ruan, S., Decazes, P., Denœux, T.: Lymphoma segmentation from 3D PET-CT images using a deep evidential network. Int. J. Approximate Reasoning **149**, 39–60 (2022)
7. Huang, L., Ruan, S., Decazes, P., Denoeux, T.: Deep evidential fusion with uncertainty quantification and contextual discounting for multimodal medical image segmentation. arXiv preprint arXiv:2309.05919 (2023)
8. Huang, L., Ruan, S., Denœux, T.: Application of belief functions to medical image segmentation: a review. Inf. fusion **91**, 737–756 (2023)
9. Huang, L., Ruan, S., Denœux, T.: Semi-supervised multiple evidence fusion for brain tumor segmentation. Neurocomputing **535**, 40–52 (2023)
10. Johnson, A.E., et al.: MIMIC-III, a freely accessible critical care database. Sci. Data **3**(1), 1–9 (2016)
11. Shafer, G.: A Mathematical Theory of Evidence, vol. 42. Princeton University Press, Princeton (1976)
12. Smets, P., Kennes, R.: The transferable belief model. Artif. Intell. **66**, 191–243 (1994)
13. Tong, Z., Xu, P., Denœux, T.: Fusion of evidential CNN classifiers for image classification. In: Denœux, T., Lefèvre, E., Liu, Z., Pichon, F. (eds.) BELIEF 2021. LNCS, vol. 12915, pp. 168–176. Springer, Cham (2021). https://doi.org/10.1007/978-3-030-88601-1_17
14. Vaswani, A., et al.: Attention is all you need. In: Advances in Neural Information Processing Systems, vol. 30 (2017)
15. Xu, P., Davoine, F., Bordes, J.B., Zhao, H., Denœux, T.: Multimodal information fusion for urban scene understanding. Mach. Vis. Appl. **27**, 331–349 (2016)

# Conflict Management in a Distance to Prototype-Based Evidential Deep Learning

Mihreteab Negash Geletu[1,2](✉)[iD], Dănuţ-Vasile Giurgi[1](✉)[iD],
Thomas Josso-Laurain[1][iD], Maxime Devanne[1][iD],
Jean-Philippe Lauffenburger[1][iD], and Jean Dezert[3][iD]

[1] IRIMAS-UR7499, Université de Haute-Alsace, Mulhouse, France
{mihreteab-negash.geletu,vasile.giurgi,danut-vasile.giurgi,
thomas.josso-laurain,maxime.devanne,jean-philippe.lauffenburger}@uha.fr
[2] AAiT-SECE, Addis Ababa University, Addis Ababa, Ethiopia
mihreteab.negash@aau.edu.et
[3] The French Aerospace Lab, ONERA, Palaiseau, France
jean.dezert@onera.fr

**Abstract.** In the domain of autonomous vehicles, perception tasks are very complex, and deep learning can be coupled with evidence theory for uncertainty management of perception models. If the pieces of evidence involved in the merging process of the deep learning-based model are discordant, the results can be degraded. Therefore, verifying the conflicting level of sources and alleviating it when possible, gives the capability to increase efficiency of following steps of the workflow: fusion rules and decision making. This paper highlights scenarios where high conflict values occur within an evidential neural network architecture. The cause of conflict is analyzed and a conflict management method is proposed, allowing the appropriate use of fusion rules and decision-making based on belief functions. Thus, particularly in, the distance to prototypes approach, a parameter rectification is proposed. The experimental results are obtained using a lidar-camera cross-fusion architecture with evidential formulation based on Dempster-Shafer's theory. The model is investigated on a road detection task and it uses the KITTI dataset.

**Keywords:** Conflict management · Evidential deep learning · Sensor fusion · Belief functions

## 1 Introduction

In data fusion, the sensing modalities provide sometimes imperfect and inherent deficiencies in the data. In addition, the models add epistemic uncertainties to the existence of possible imprecise data coming from these sources of information.

Thus, the effectiveness of the model can be determined by its design or by the fusion of the sources. To ensure the quality of the combination features involved in the data fusion, apart from data consistency, the level of conflict amidst the merging elements should be also minimized.

Evidential deep learning (EDL) architectures are inspired by classical neural network architectures combined with the theory of evidence [1]. This involves integrating belief functions at a certain level of the architecture to improve the reasoning and obtain better representations of uncertain information.

For autonomous driving, several works are pursued combining sources of information [2], and theory of evidence [3,4]. Moreover, in other common classification tasks, neural networks and belief functions are integrated by replacing the probabilistic inference with an evidential formulation [5]. The cross-fusion architecture based on belief functions reasoning [8] leverages the concept of distance to prototypes (D2P), inspired by Denœux's evidential classifier [11]. Instead of a probabilistic approach, an evidential formulation based on belief functions is employed, and decisions rely on Dempster-Shafer's (DS) rule of combination. Approaches based on prototypes are present in [6], but conflict is not examined.

This work addresses conflict investigation in the distance to prototypes approach from the evidential formulation of a deep learning model in the task of road detection (segmentation). The high degree of conflict among Basic Belief Functions (BBAs) associated with prototypes is identified and a solution to minimize the conflict is proposed. Within the computation of mass functions, a minimum threshold on a parametric value is set to reduce conflict. The remaining part of the paper focuses on evidential formulation (theory of evidence and conflict measure for evidential deep learning architectures) in Sect. 2, conflict analysis (architecture and conflict investigation, as well as their impact on the fusion rules) in Sect. 3, results (experimental results with conflicting situations and a rectification) in Sect. 4, and conclusions in Sect. 5.

## 2 Evidential Formulation

### 2.1 Basics in Belief Functions Theory

Belief functions serve for decision-making when considering uncertainties, reasoning for imprecise data, and conflict management in the theory of evidence. Dempster-Shafer's (DS) theory stands out as a recognized technique to address these challenges. The fundamentals of the approach begin by denoting: $\Omega = \{\omega_1, \omega_2, \omega_3, ..., \omega_n\}$ to represent the *frame of discernment* (FoD). This universe comprises singular entities, each representing distinctive elements named singletons [1]. Continuously, *basic belief assignments* (BBAs) or mass functions $m(\cdot)$ are attributable to the group of singletons. An element $A$ is called a *focal element* of $m(\cdot)$, if $m(A) > 0$. Assuredly, if the focal element is unique and represents total ignorance, it is known as *vacuous* ($\Omega$). Generally, this measurement of evidence is characterized by a distribution $m : 2^\Theta \to [0,1]$, conforming:

$$m(\emptyset) = 0, \quad \text{and} \quad \sum_{A \subseteq \Omega} m(A) = 1, \tag{1}$$

$$Bel(A) = \sum_{B \subseteq A} m(B), \ Pl(A) = \sum_{B \cap A \neq \emptyset} m(B) = 1 - Bel(\bar{A}) \quad (2)$$

where the empty set (1) stands for the conflict between elements of evidence. The last two statements (2) define *a belief function* or the full support of evidence ($Bel$), respectively a *plausibility function* ($Pl$) or the extent of a certain belief, without doubting it. One way to combine independent sources of evidence e.g. $m_1(\cdot)$ and $m_2(\cdot)$ is to use Dempster's rule [1]:

$$(m_1 \oplus m_2)(A) = \frac{1}{1-K} \sum_{B \cap C = A} m_1(B) m_2(C), \text{ where } K = \sum_{B \cap C = \emptyset} m_1(B) m_2(C) \quad (3)$$

For all $A \subseteq \Omega$, $A \neq \emptyset$, and $(m_1 \oplus m_2)(\emptyset) = 0$. This allows finding $K$ as the degree (weight) of conflict between two mass functions by using the conjunctive rule [1], regardless of its non-idempotency [9]. In definition, a conflict in evidence theory is represented by the contradiction between two or more BBAs, such that, they support opposite pieces of evidence [9]. Another way to express conflict is to consider plausibility functions [10] as expressed below. More details about conflict measures can be found in [9].

$$\text{Conf}(m_1, m_2) = 1 - \frac{Pl_1^T \cdot Pl_2}{\|Pl_1\| \|Pl_2\|}, \text{ where } Pl_1^T \cdot Pl_2 \in 2^n \text{ space of } Pl \text{ functions} \quad (4)$$

## 2.2 Evidential Deep Learning

Introduced by [11], the D2P is a method consisting of three phases: 1) calculating the distances ($d^i$) (5) to a $n$ number of prototypes ($p^i$ that has a specific membership to different classes) with respect to the features (patterns formed by the architecture before the final decision), 2) assigning mass functions ($m^i(\omega_j)$) which are computed from the membership degree to a class $\omega_j$ and the previously calculated distance (6) and the fusion of combined masses using Dempster's rule as in (3).

$$d^i = \|x - p^i\|, \quad i = 1, \cdots, n \in \mathbb{N} \quad (5)$$

$$m^i(\omega_j) = \alpha^i u_j^i \phi^i(d^i), \quad j = 1, 2, \quad m^i(\Omega) = 1 - \alpha^i \phi^i(d^i), \quad \phi^i(d^i) = \exp(-\gamma^i (d^i)^2), \gamma^i > 0 \quad (6)$$

The elements of $p^i$: $\alpha^i, u_j^i$, and $\gamma^i$ represent constrained [11] weights of the evidential architecture. These parameters are mapped to real number valued variables $\eta^i, \xi^i$, and $\beta_j^i$:

$$\gamma^i = (\eta^i)^2, \ \alpha^i = \frac{1}{1 + \exp(-\xi^i)}, \ u_j^i = \frac{(\beta_j^i)^2 + \epsilon}{\sum_{k=1}^{2}((\beta_k^i)^2 + \epsilon)} \quad (7)$$

where $\epsilon$, a small positive number, is added to keep the membership values $u_j^i$ positive. Within this approach, evidential neural networks can be constructed by plugging the specific D2P as the DS layer to replace the Bayesian inference. Finally, the evidential neural network model outputs mass functions to be exploited.

## 3 Conflict Analysis

### 3.1 Conflict-Aware Evidential Deep Learning for Sensor Fusion

The conflict analysis is done for an evidential deep learning model called Lite-CF-Evi architecture [8]. It is considered with 4 prototypes. The model is a convolutional neural network that encapsulates two cross-fused pipelines, one for LiDAR and one for the camera, followed by an evidential formulation, as illustrated in Fig. 1.

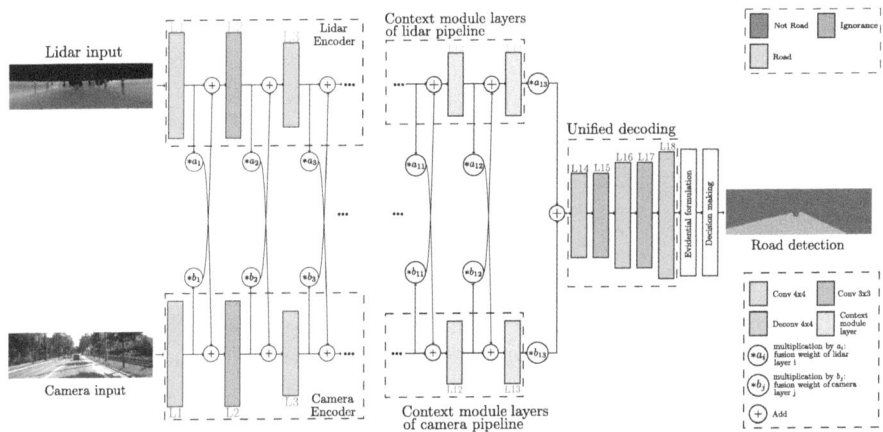

**Fig. 1.** Evidential Lite-CF Architecture: Road Detection [8].

Within the Lite-CF-Evi architecture, the D2P approach is used for transferring the features to mass functions to be fused. Once the BBAs are combined, they are involved in the decision-making based on belief interval distance (dBI) decision rule [13].

For the training, KITTI benchmark [12] is considered, specifically the road detection dataset composed of 289 frames. Thus, data is split into 260 images for training and 29 images for validation, for each pipeline: camera (RGB images), and LiDAR (pre-processed frames: lidar projection and up-sampled 2D dense depth map images Fig. 1, [14]). Since the dataset does not provide the ground truth for the test set, the model is trained on the training split and evaluated on the validation split. While pursuing experiments according to the previous diagram, the results obtained with the evidential formulation outperform the same architecture, based on probabilistic inference [8]. It is known that DS theory is powerful in the work with belief functions and it can generate beneficial reasoning. However, in specific cases, the DS approach can also lead to counter-intuitive performances. Nonetheless, this technique can come with challenges when implemented. Highly contradictory elements (i.e. BBAs) end up concealed, despite the apparent smoothness of fusion, leaving unnoticed vacuousness and conflict (see Sect. 3.2). The high degree of conflict is also observed in a semantic segmentation application of a pets dataset [15] using an evidential implementation of a segmentation network called U-net [7]. It is observed in a region where there is no visual ambiguity, and this conflict is similarly made latent by the DS rule.

In this way, a deeper investigation before the fusion level is conducted to ensure the consistency of combination elements. This work focuses on investigating the D2P scheme, by analyzing the conflict involved in BBAs associated with prototypes, as well as the evolution of these BBAs within the training process.

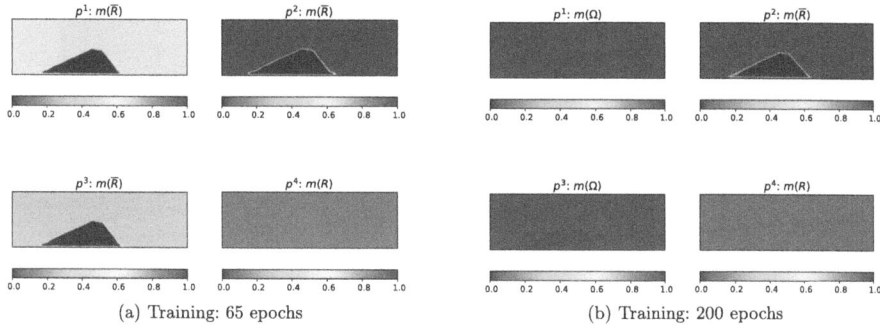

**Fig. 2.** BBA Visualisations: $R$ - road, $\overline{R}$ - not-road, and $\Omega$ - ignorance.

### 3.2 Preliminary Visuals Into Conflicting BBAs

Considering the aforementioned model with 4 prototypes in the D2P approach, the mass functions associated with prototypes are visualized and analyzed for several model setups. Different numbers of epochs are considered. In Fig. 2, the left side (Fig. 2a), shows BBAs associated with prototypes after 65 epochs of training, in an early stage, respectively after 200 epochs (Fig. 2b). Each subplot represents a BBA support for one class of the FoD, coming from a specific prototype $(p^i)$. Prototypes are representative feature vectors, and in line with the membership degrees, they are supposed to settle for an individual class within training. It can be seen from the figure that BBAs evolve as the training epoch increases. Inspecting the left side, Fig. 2a, illustrates how three prototypes $(p^1, p^2, p^3)$ have BBAs that support the not-road class $(m(\overline{R}))$, while the last prototype $p^4$ substantiate only values for the road class ($m(R)$), showing that mass functions are conflicting. For the right side, Fig. 2b, with a more mature level of training, has two prototypes ($p^1, p^3$) that show vacuousness values $m(\Omega)$, while the other two prototypes ($p^2, p^4$) illustrate highly conflicting BBAs: $m(\overline{R})$ against $m(R)$. Following the training, the prototypes show support to their classes based on the membership degree, however, some of the mass functions associated with these prototypes indicate conflict (left side image) or both conflict and vacuousness (right side image). This shows how they converge to non-representative values for their class labels. Thus there are BBA values of a prototype that strongly support one class (i.e. road), while another prototype's BBAs firmly support exactly the opposite class for the same pixel. Consequently, in terms of multi-source fusion and the theory of evidence, merging strongly conflicting elements or vacuous sources can degrade performance, meaning that the fusion part has no positive impact on the final decision, regardless of the applied technique.

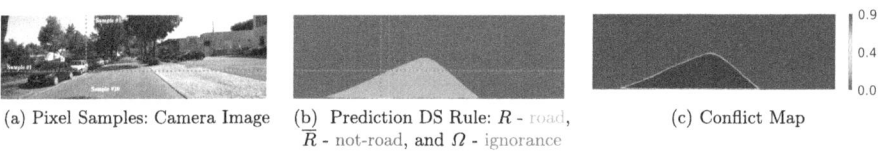

(a) Pixel Samples: Camera Image    (b) Prediction DS Rule: $R$ - road, $\overline{R}$ - not-road, and $\Omega$ - ignorance    (c) Conflict Map

Fig. 3. Sample extractions for conflict investigation.

### 3.3 D2P Parameters Analysis and Conflict Rectification

In the development of the evidential cross-fusion, the predictions are inspected just before Dempster's combination rule. The workflow before this point is conserved (i.e., feature extraction, and construction of the uncombined mass functions). The analysis shows that some areas are easily predicted as specific classes even though the BBAs are highly conflicting. Because, it is known that in some situations Dempster's rule has some counter-intuitive properties, and it is not effective in handling conflicting pieces of evidence (see discussions and counter-example given in [17]). In the analysis, the parameters associated with prototypes (i.e., $\alpha^i$, $u_j^i$, and $\gamma^i$) in the BBA construction step (6) are also exposed.

For analysis, samples of pixel points along a vertical and horizontal lines of a frame are considered (see Fig. 3a). Here, uncombined constructed BBAs (from the 4 prototypes), the degree of conflict $K$ among the BBAs, and their fused results with DS are presented in Table 1 (86 samples that are shown in Fig. 3a are analyzed, however, the table presents only for 3 samples because of space limit). In addition, the table also gives $\eta^i, \xi^i$ and $\beta_j^i$ parameters used when constructing masses. As can be seen from the table, the combined BBAs of the vertical sample 1 have strong support for the not-road class, when Dempster's rule is utilized, where is finally assigned within decision-making. This is a correct classification given the not-road ground truth. The computed degree of conflict $K$ value, however, raises questions. Since $K$ has a high value, Dempster's rule becomes ineffective in combining the constructed BBAs. This allows the not-road class to have strong support from the combined BBA, while the BBA associated with the prototype $p^3$ is in conflict. Hence, this result is counter-intuitive. However, the deep learning-based evidential model training takes advantage of the weakness of Dempster's rule in combining conflicting BBAs. Since the training tries to make the prediction output value (i.e., predicted BBAs) close to the ground truth without considering the degree of conflict, the strong support for the not-road class from Dempster's rule is taken as a good stance even though it is obviously counter-intuitive for this scenario. This indicates that the cause of the high degree of conflict is training the model without constraining the model parameters from producing highly conflicting BBAs. The conflicting BBA, in this case, is the one that comes from prototype $p^3$. This conflict occurs because of the value of $\gamma^3 = (\eta^3)^2 = 0$. This makes any feature vector to support a specific class, in this case, road, irrespective of its distance to the prototype. Hence, it is called a "dummy" BBA and it conflicts with any BBA that supports the not-road class. However, the location of the sample on the camera image does not seem to have ambiguity, at least to a human viewer. The same analysis follows for the horizontal sample number 1. Hence, conflict can result. However, in this pixel, the conflict is not due to ambiguity. It is again because of the $\gamma^3$ value being 0. Thus, regardless of the associated feature vector distance to the prototype $p^3$, conflict occurs. When the degree of conflict is 0, like in vertical sample number 20, $\gamma^3 = 0$ is still problematic. Because the associated BBA is thus, not based on D2P. The BBA represents evidence that has not been measured. Thus, this reflection needs to be considered and rectified.

To the best of the authors' knowledge, this is the first time the problem of a high degree of conflict among BBAs associated with prototypes has been investigated. In [11,16] prototypes and associated parameters are initialized using clustering and heuristic. This may have prevented the issue of a high degree of conflict from occurring. Because, starting from a pre-worked initialization, the training may have converged without falling into situations that create conflicting prototypes. However, this can not be stated for sure. Since, in the automatic optimization, the training may converge at any time into a local minima with a latent high degree of conflict. Because there is nothing in the training to constrain the parameter search space from having values that induce highly conflicting BBAs. Therefore, methods need to be proposed to resolve the high degree of conflict among BBAs, that has occurred.

Since the cause of the observed conflict is the value of $\gamma^i = 0$, training trials are proposed by making $\gamma^i$ strictly positive. Hence, in the set of equations from (7), $\gamma^i$ is modified by adding a positive value:

$$\gamma^i = (\eta^i)^2 + \delta, \text{ where, } \delta > 0. \tag{8}$$

BBAs associated with a pixel are also not equally relevant, because some are less informative (or have a small value of $\xi$, implying a small $\alpha^i$ value). The associated values of these parameters with each prototype ($p^i$) are the following: $\xi^i : (p^1 = 0.39, p^2 = -3.27, p^3 = 2.14, p^4 = 5.33)$, and $\alpha^i : (p^1 = 0.596, p^2 = 0.037, p^3 = 0.895, p^4 = 0.995)$. The value of $\alpha^2 = 0.037$, which is associated with prototype $\boldsymbol{p^2}$ is too small. Therefore, referring Eq. 6, as a minimum limit (i.e., $\phi^2(d^2) = 1$), the associated BBA will have $m^2(\Omega) = 0.963$. This is manifested in the extracted BBAs associated with the prototype $\boldsymbol{p^2}$ as given in Table 1, which shows the minimum value of $m^2(\Omega) = 0.963$ and the maximum value $m^1(\Omega) = 1$. Following up, the effect of $\alpha^i$ in penalizing less relevant BBAs as per the formulation (Eq. 6) is revealed. This can be used to ablate prototypes related to less relevant BBAs.

Table 1. Model extractions on illustrative samples.

| Sample # groundtruth | BBAs before combination $p^i$ : $m^i(\omega_1)$, $m^i(\omega_2)$, $m^i(\Omega)$ | $K$ | Combined BBA $m(\omega_1)$, $m(\omega_2)$, $m(\Omega)$ | Decision | Parameters (prototype $p^i$) $p^i$ : $\eta^i$, $\xi^i$, $\lfloor \beta_1^i, \beta_2^i \rfloor$ |
|---|---|---|---|---|---|
| 1 (vertical) Not-road ($\omega_1$) | $p^1$ : 5.97e−01, 1.50e−09, 4.03e−01 | 0.89 | DS: 0.98, 0.02, 0.00 | Not-road | $p^1$ : 0.07, 0.39, [1.99e−01, −9.10e−36] |
| | $p^2$ : 3.67e−02, 3.79e−10, 9.63e−01 | | | | |
| | $p^3$ : 8.91e−09, 8.94e−01, 1.06e−01 | | | | |
| | $p^4$ : 9.95e−01, 1.18e−08, 4.85e−03 | | | | |
| 20 (vertical): Road ($\omega_2$) | $p^1$ : 0, 0, 1 | 0. | DS: 0.00, 0.89, 0.11 | Road | $p^2$ : −0.08, −3.27, [9.84e−02, 2.99e−37] |
| | $p^2$ : 0, 0, 1 | | | | |
| | $p^3$ : 8.91e−09, 8.94e−01, 1.06e−01 | | | | |
| | $p^4$ : 0, 0, 1 | | | | $p^3$ : 0., 2.14, [−1.97e−35, −1.00e−01] |
| 1 (horizontal): Not road ($\omega_1$) | $p^1$ : 5.97e−01, 1.50e−09, 4.03e−01 | 0.89 | DS: 0.98, 0.02, 0.00 | Not-road | $p^4$ : 0.23, 5.33, [−9.19e−02, −1.30e−36] |
| | $p^2$ : 3.67e−02, 3.79e−10, 9.63e−01 | | | | |
| | $p^3$ : 8.91e−09, 8.94e−01, 1.06e−01 | | | | |
| | $p^4$ : 9.95e−01, 1.18e−08, 4.85e−03 | | | | |

## 4 Experimental Results

To avoid the observed high degree of conflict among BBAS, hyper-parameter tuning of the lower threshold of $\gamma^i$ (i.e., $\delta$ in Eq. (8)) is made. First, $\delta = 0.01$ is tried. The high degree of conflict is slightly reduced, while in one of the prototypes, the value of $\gamma^i$ is strictly positive, but still close to zero. This is again the reason for the observed high degree of conflict (0.8) in the not-road area. Figure 4a shows the heat map of the degree of conflict on the frame of the samples.

When the value of $\delta$ is increased to 0.1, a further decrease in the value of the degree of conflict is observed on the sample points. The corresponding heat map of the same frame is also given in Fig. 4b. The figure shows a different distribution of degree of conflict from the two previous cases (i.e., Fig. 3c and Fig. 4a). In this case, with $\delta = 0.1$, the high degree of conflict belongs to the road edge area. This is more sound as the area is ambiguous. It decreases slightly into the not-road area, and significantly into the road area. It is significant to observe that implementing a minimum threshold on the value of $\gamma^i$ can avoid the occurrence of a "dummy" BBA and diminish the degree of conflict.

The value of $\delta$ is increased to 1, and Fig. 4c gives the heat map of the corresponding degree of conflict. In this case, the degree of conflict is further decreased

**Table 2.** Model extractions on illustrative samples when $\gamma^i = (\eta^i)^2 + 1$.

| Sample # groundtruth | BBAs before combination $p^i$ : $m^i(\omega_1)$, $m^i(\omega_2)$, $m^i(\Omega)$ | K | Combined BBA $m(\omega_1)$, $m(\omega_2)$, $m(\Omega)$ | Decision | Parameters (prototype $p^i$) $p^i$ : $\eta^i$, $\xi^i$, $[\beta_1^i, \beta_2^i]$ |
|---|---|---|---|---|---|
| 1 (vertical) Not-road ($\omega_1$) | $p^1$ : 5.81e−09, 1.97e−01, 8.03e−01 $p^2$ : 9.92e−01, 4.90e−08, 7.53e−03 $p^3$ : 2.71e−03, 1.08e−10, 9.97e−01 $p^4$ : 1.15e−03, 6.40e−11, 9.99e−01 | 0.20 | DS: 0.99, 0.00, 0.01 | Not-road | $p^1$ : −8.96e−34, 3.04, [1.11e−36, −5.81e−02] |
| 20 (vertical) Road ($\omega_2$) | $p^1$ : 2.82e−08, 9.54e−01, 4.59−02 $p^2$ : 1.75e−01, 8.61e−09, 8.25e−01 $p^3$ : 2.22e−04, 8.87e−12, 1.00 $p^4$ : 9.24e−05, 5.16e−12, 1.00 | 0.17 | DS:0.01, 0.94, 0.05 | Road | $p^2$ : 2.91e−01, 4.89, [4.50e−02, −1.07e−36] $p^3$ : −6.16e−01, −5.90, [−5.00e−02, 8.64e−38] |
| 1 (horizontal) Not road ($\omega_1$) | $p^1$ : 5.97e−09, 2.02e−01, 7.98e−01 $p^2$ : 9.92e−01, 4.89e−08, 7.84e−03 $p^3$ : 2.70e−03, 1.08e−10, 9.97e−01 $p^4$ : 1.14e−03, 6.38e−11, 9.99e−01 | 0.20 | DS: 0.99, 0.00, 0.01 | Not-road | $p^4$ : −6.41e−01, −6.76, [4.23e−02, −5.62e−37] |

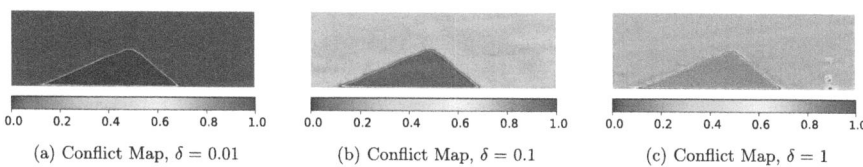

(a) Conflict Map, $\delta = 0.01$  (b) Conflict Map, $\delta = 0.1$  (c) Conflict Map, $\delta = 1$

**Fig. 4.** Conflict maps for different threshold values.

and it becomes more uniform. The maximum degree of conflict is around 0.4, and it is in the area of the road edge. It is expected to get a comparatively high degree of conflict in this area because it is ambiguous. However, in the not-road and road area the value is small, it is about 0.2. This is a sound result as one expects a set of experts to have a low degree of conflict (uniform kind) in plain scenarios, and comparatively higher in ambiguous ones. A comparatively higher degree of conflict is also observed in a scattered zone (see Fig. 4c) around the camera image's lower right edge. Visual comparison with the corresponding camera image (refer to Fig. 3a) does not provide many clues. However, this could be because of the model's imperfection due to various factors. To consolidate the result, similar BBAs as in Table 1 are extracted. Thus, Table 2 gives a short and precise detail of how a minimum threshold on $\gamma^i$ impacts the result. As can be observed in this table, Dempster's rule behavior appears more satisfactory because the degree of conflict is significantly reduced.

Ultimately, once the $\gamma^i$ values are adjusted (i.e., $\delta = 1$, the final road detection evaluation is computed in terms of MaxF, Precision, and Recall metrics [18] as in Table 3. These results illustrate that the rectification does not negatively impacted the model performance, rather some little improvement is observed.

**Table 3.** Comparison of Evaluation Metrics for category urban road (cumulative).

| Metric | Before Rectification | After Rectification |
|---|---|---|
| MaxF | 96.99 | **97.48** |
| Precision | 96.79 | **96.80** |
| Recall | 97.19 | **98.17** |

## 5 Conclusion

This work highlights some investigations into the distance-to-prototypes approach of a cross-fused evidential neural network model. The task of the architecture is to provide road detection features while taking advantage of the use of reasoning based on belief functions. Within the architecture, 4 prototypes are considered and therefore analyzed from the degree of conflict perspective. It is observed that after the training, some of the BBAs associated with prototypes become highly conflicting which does not help later within the fusion part. Consequently, a method to rectify is proposed highlighting a minimum threshold on $\gamma^i$ value, to avoid highly conflicting prototypes. Given experimental results, the thresholding of the $\gamma^i$ value reduces the conflict between the prototypes, providing more representative mass functions for the fusion level, and evaluation metrics remain unaffected. Rather, though slight, improvement is observed. In future works, multiple models and alternative rectification methods will be investigated, including loss regularization. Finally, within this approach, the uncombined BBAs become more trustworthy for merging, reducing the counter-intuitive behavior of Dempster's rule.

**Acknowledgments.** The authors extend their gratitude to the French National Research Agency ANR JCJC EviDeep project, the Ethiopian Ministry of Education and the French Embassy in Ethiopia, the University of Haute-Alsace, and the Pierre-et-Jeanne Spiegel Foundation for their support in the execution of this project.

**Disclosure of Interests.** The authors have no competing interests to declare that are relevant to the content of this article.

## References

1. Shafer, G.: A Mathematical Theory of Evidence. Princeton University Press, Princeton (1976)
2. Caltagirone, L., Bellone, M., Svensson, L., Wahde, M.: LIDAR-Camera Fusion for Road Detection Using Fully Convolutional Neural Networks (2018)
3. Xu, P., Davoine, F., Denœux, T.: Evidential combination of pedestrian detectors. In: British Machine Vision Conference, pp. 1–14 (2014)
4. Capellier, E., et al.: A deep evidential learning for arbitrary lidar object classification in the context of autonomous driving. In: Proceedings of 2019 IEEE Intelligent Vehicles Symposium (IV), pp. 1304–1311 (2019)

5. Tong, Z., Xu, P., Denœux, T.: Evidential fully convolutional network for semantic segmentation. Appl. Intell. **51**(9), 6376–6399 (2021)
6. Li, O., Liu, H., Chen, C., Rudin, C.: Deep learning for case-based reasoning through prototypes: a neural network that explains its predictions. In: Proceedings of the AAAI Conference on Artificial Intelligence (2018)
7. Ronneberger, O., Fischer, P., Brox, T.: U-net: convolutional networks for biomedical image segmentation. In: Navab, N., Hornegger, J., Wells, W., Frangi, A. (eds.) MICCAI 2015. LNCS, vol. 9351, pp. 234–241. Springer, Cham (2015). https://doi.org/10.1007/978-3-319-24574-4_28
8. Geletu, M.N., Giurgi, D-V., Josso-Laurain, T., Devanne, M., Wogari, M.M., Lauffenburger, J-P.: Evidential deep learning-based multi-modal environment perception for intelligent vehicles. In: Intelligent Vehicles Symposium (IV), Anchorage, USA (2023)
9. Martin, A., Jousselme, A.-L., Osswald, C.: Conflict measure for the discounting operation on belief functions. In: International Conference on Information Fusion, Cologne, Germany (2008)
10. Jousselme, A.-L., Maupin, P.: Distances in evidence theory: comprehensive survey and generalizations. Int. J. Approximate Reasoning **53**, 118–145 (2011)
11. Denœux, T.: A neural network classifier based on Dempster-Shafer theory. IEEE Trans. Syst. Man Cybern.-Part A: Syst. Hum. **30**(2), 131–150 (2000)
12. Geiger, A., et al.: Vision meets robotics: the KITTI dataset. Int. J. Robot. Res. (IJRR) **32**, 1231–1237 (2013)
13. Dezert, J., Han, D., Tacnet, J.-M., Carladous, S., Yang, Y.: Decision-making with belief interval distance. In: Proceedings of Belief 2016 International Conference, Prague, CZ, 21–23 September (2016)
14. Premebida, C., et al.: Pedestrian detection combining RGB and dense lidar data. In: Proceedings of 2014 IEEE/RSJ International Conference on Intelligent Robots and Systems, pp. 4112–4117 (2014)
15. Parkhi, O.M., et al.: Cats and dogs. In: IEEE Conference on Computer Vision and Pattern Recognition (2012)
16. Huang, L., Ruan, S., Decazes, P., Denœux, T.: Lymphoma segmentation from 3D PET-CT images using a deep evidential network. Int. J. Approximate Reasoning **149**, 39–60 (2022)
17. Tchamova, A., Dezert, J.: On the behavior of dempster's rule of combination and the foundations of Dempster-Shafer theory. In: IEEE IS 2012, Sofia, Bulgaria, 6–8 September (2012)
18. Fritsch, J., Kuehnl, T., Geiger, A.: A new performance measure and evaluation benchmark for road detection algorithms. In: International Conference on Intelligent Transportation Systems (ITSC) (2013)

# A Novel Privacy Preserving Framework for Training Dempster-Shafer Theory-Based Evidential Deep Neural Network

Anh-Tu Tran[1(✉)], Van-Nam Huynh[2], and Viet-Hung Dang[3]

[1] Academy of Cryptography Techniques, Hanoi, Vietnam
tutran@actvn.edu.vn
[2] School of Knowledge Science, Japan Advanced Institute of Science and Technology, Nomi, Ishikawa, Japan
[3] Vietnamese Security Network Joint Stock Company, Hanoi, Vietnam

**Abstract.** Evidential Deep Neural Networks (EDNNs), based on Dempster-Shafer (DS) theory, offer an advanced method for managing uncertainty in deep learning. Like other deep learning models, EDNNs need large datasets, often containing sensitive information, highlighting the need for strong data privacy measures. Our research introduces Pri-DSENN, an innovative framework designed to enhance privacy in the EDNN training process according to DS principles. Pri-DSENN uses a secure multiparty computation (SMC) protocol combined with federated learning (FL) to significantly improve data security. Our experiments with the CIFAR10 and CIFAR100 datasets confirm that integrating SMC with FL in DS-based EDNNs preserves high classification accuracy while effectively handling unclear patterns, ensuring advanced decision-making and robust data privacy and security.

**Keywords:** Privacy-preserving deep learning · federated learning · Dempster Shafer Theory · Evidential Deep Neural Network

## 1 Introduction

Deep learning's widespread application has significantly advanced technology, but it struggles with accurate classification under high uncertainty, such as ambiguous features and outlier detection [4]. Recent efforts have begun to address this by incorporating deep learning outputs into the DS framework [2]. A key development is the evidential deep-learning classifier, which integrates DS theory with deep CNNs for set-valued classification [10]. This classifier uses a deep CNN to extract data features, which are then processed through a DS-based layer to generate mass functions. These functions enable set-valued classification by evaluating the utility of different class sets, with the entire network trained end-to-end.

However, integrating deep learning across sectors raises significant privacy concerns due to the reliance on large amounts of personal and sensitive data.

This data collection and processing risk unauthorized access and misuse, raising ethical issues about user consent and data protection [3]. This has prompted a shift towards privacy-centric approaches in using EDNNs, aiming to leverage big data while adhering to strict privacy standards [6].

Privacy-preserving deep learning has emerged to address these concerns in collaborative environments [1]. Techniques like federated learning (FL) [11], differential privacy (DP) [12], and homomorphic encryption (HE) [7] are gaining traction. These methods protect user privacy by minimizing the exposure of sensitive information during training, aligning deep learning advancements with the need for strong privacy protection. However, current research often overlooks the challenges of privacy preservation for uncertain data and the complexities of EDNNs. Integrating privacy-preserving techniques into EDNNs is essential for enhancing data security and privacy, extending their utility in sectors plagued by data ambiguity. This calls for researchers to develop specialized privacy-preserving mechanisms for EDNNs, pushing the boundaries beyond existing methods to ensure that advancements in handling uncertain data also uphold privacy standards.

**Our Contributions.** In our study, we introduce *Pri-DSENN*, a new approach designed to enhance user privacy while also improving the accuracy and robustness of models when dealing with uncertain data. By combining FL with a novel SMC protocol called *MaskElGamal*, we help protect sensitive information and reduce the chances of indirect data leaks. This protocol allows participants to easily add up private input vectors as floating-point numbers, ensuring privacy against honest-but-curious adversaries, even with up to $n - 2$ colluding parties.

The structure of this paper is as follows: After the introduction, we cover preliminary concepts including EDNNs, FL, and the ElGamal encryption scheme. Section 3 details our Pri-DSENN framework and the MaskElGamal protocol. Section 4 describes the implementation and empirical results on the CIFAR10 and CIFAR100 datasets. We conclude with reflections and implications in Sect. 5.

## 2 Preliminary

### 2.1 Dempster Shafer Theory-Based Evidential Deep Neural Network

EDNNs create a novel classification framework by combining the ENN classifier with a DNN structure, adding a DS and utility layer at the end [9,10]. This evidential deep-learning classifier handles multi-valued classifications and evaluates class uncertainties through belief functions in three stages: First, a sample is processed through a DNN to extract key features, producing a $P$-dimensional vector for the DS layer, ensuring reliable classification accuracy. Second, the DS layer converts the feature vector into mass functions using Dempster's rule, resulting in an $M + 1$ mass vector representing confidence in class probabilities and uncertainty. This layer adjusts mass distribution based on feature clarity, assessing evidence adequacy and aiding model training. Finally, the mass vector guides the expected utility layer, generating an expected-utility vector that supports multi-valued classification by considering all possible actions, demonstrating the classifier's ability to handle various classification scenarios.

### 2.2 Federated Learning

A Federated Learning (FL) system includes a central server $S$ and multiple nodes $\mathcal{U} = U_1, U_2, \ldots, U_N$, each with its own dataset $D_i$. The server initializes a global model $W^0$ and shares it with the nodes. In each iteration, nodes locally train the model to produce $W_i^t$, which they send back to $S$ for aggregation into an updated global model $W^{t+1}$. This process involves three phases: initiation (sharing the global model with nodes), local training (nodes train with their private data), and integration (aggregating local models into a refined global model). This cycle repeats, continually improving the global model.

In federated learning, the aggregation algorithm is crucial. FedAvg [5], along with variants like FedProx and FedMA, shows strong performance [8]. FedAvg, often excelling in practical applications, averages local model updates to calculate the global model:

$$W^{t+1} \leftarrow \sum_{i=1}^{N} \frac{m_i}{M} W_i^t \qquad (1)$$

Here, $m_i$ is the data count in node $U_i$, $M$ is the total data count, and $W_i^{(t)}$ is the local model parameters at epoch $t$.

### 2.3 Elgamal Cryptosystem

In a cyclic group $G$ of order $p$ with a generator $g$, solving discrete logarithms is considered difficult. A private key $x$ is randomly chosen from $\{1, 2, \ldots, p-1\}$, and the corresponding public key is $h = g^x$. To encrypt a message $m$, a sender selects a random $k$ from $\{1, 2, \ldots, p-1\}$ and computes the ciphertext $C = (C_1 = m \cdot h^k, C_2 = g^k)$. To decrypt $C$, the recipient uses the private key $x$ to find $m$ by calculating $m = C_1 \cdot (C_2^x)^{-1}$.

ElGamal encryption's semantic security is based on the decisional Diffie-Hellman assumption, which relies on the difficulty of the discrete logarithm problem in the cyclic group $G$.

## 3 Privacy Preserving Framework for Training Dempster Shafer Theory-Based Evidential Deep Neural Network

### 3.1 Threat Model

In this study, we assume that participants $U_1, \ldots, U_i, \ldots, U_n$ hold private datasets and use EDNNs within the FedAVG framework to address data uncertainty. Each participant has a fixed number of samples, keeping the total data count constant. The primary challenge is securely aggregating parameter vectors:

$$W^{t+1} \leftarrow \sum_{i=1}^{n} W_i^t \qquad (2)$$

The goal is to compute the sum vector $V$ without revealing the individual vectors $W_i$ at each global epoch $t$:

$$V = \sum_{i=1}^{n} W_i \qquad (3)$$

We propose a novel approach to ensure privacy during EDNN training. Our method uses a secure protocol for vector sum aggregation in a semi-honest environment, where participants follow protocols but may try to infer shared values. Our framework also addresses potential collusion among participants and between participants and the central server, enhancing model resilience against such threats.

### 3.2 MaskElGamal: Secure Multipary Vector Sum Protocol with Masking Matrix and ElGamal-Based Cryptography System

Before beginning the protocol, several assumptions and requirements regarding the public parameters, and the client's public and private keys are shared between the server and clients.

- Each client holds a secret vector containing a list of floating-point numbers, which is converted into a square matrix $v_i$ of size $d \times d$ by arranging the numbers into rows. If the matrix is incomplete, 0s are added to fill it out.
- A cyclic group $G = \mathbb{Z}_p$ is chosen where the Discrete Logarithm Problem is NP-hard, with $p$ being a large prime number and $g$ a generator of $G$. An invertible matrix $H$ of size $d \times d$ is also selected. The parameters $H$, $p$, and $g$ are known to all clients. Operations within the $\mathbb{Z}_p$ field are assumed to be modulo operations.
- Each client $U_i$ has two secret key matrices $x_i$ and $y_i$ of size $d \times d$, with elements $x_i^{(jk)}$ and $y_i^{(jk)}$ in $\{1, 2, \ldots, p-1\}$. Additionally, each client creates a random mask matrix $r_i$ of size $d \times d$, where each element $r_i^{(jk)}$ is an integer in $[0, \alpha]$. This mask matrix is removed during the aggregation process at the server.

After generating and agreeing upon the parameters, the participants and the server will proceed with computations and communications as described in Fig. 1.

**Proof of Correctness.** We can easily demonstrate the correctness of the protocol through standard arithmetic transformations. $Q$ represents the sum of the random noise masks $r_i$, and thus, after performing the computations, the final result is the sum of the secret vectors from the participants.

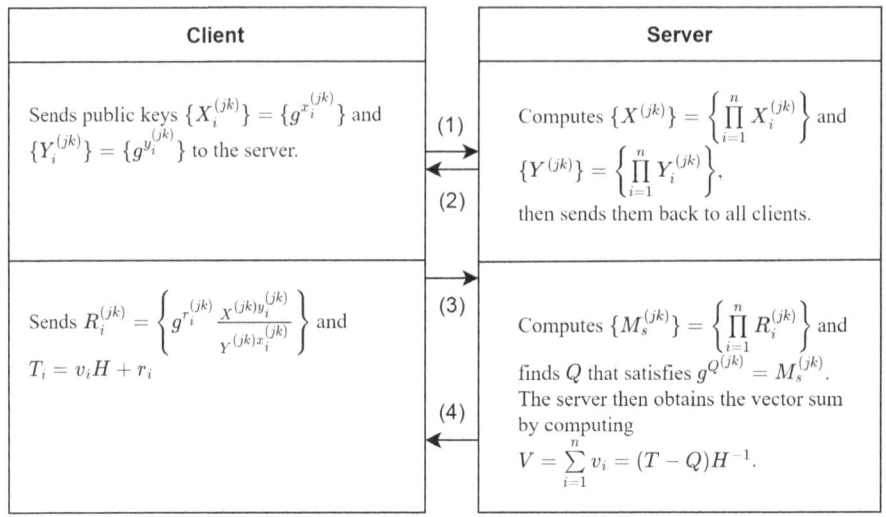

**Fig. 1.** MaskElGamal protocol for secure vector sum

$$V = (T - Q)H^{-1} = \left(\sum_{i=1}^{N} v_i\right) HH^{-1} = \sum_{i=1}^{N} v_i.$$

**Privacy Analysis.** In this section, we analyze the security of the proposed protocol. We assume that all participating clients are semi-honest, meaning they comply fully with the protocol and the previously mentioned assumptions.

**Theorem 1.** *The multi-party sum protocol using El-Gamal is secure against any semi-honest participating clients.*

*Proof.* In the initialization phase, each client $U_i$ sends their public key matrices $X_i$ and $Y_i$, with each element derived from secret keys chosen randomly from $\{1, 2, \ldots, p-1\}$. Due to the discrete logarithm problem, it is difficult to recover the secret keys from these public keys.

In the computation phase, each party sends two messages, $R_i$ and $T_i$. Each entry of $R_i$ is given by:

$$R_i = g^{r_i} \frac{X^{y_i}}{Y^{x_i}}.$$

Determining the secret elements in this formula is equivalent to solving discrete logarithmic problems. Thus, the entries in the mask matrix remain secure and cannot be deduced from the public value.

Considering $T_i = v_i H + r_i$, with $r_i$ being a secret matrix, recovering $v_i$ is equivalent to solving a system of linear equations with $2 \times d \times$ variables and only $d \times d$ equations. The probability of predicting all integer values of $r_i$ is $\frac{1}{k^{d \times d}}$, where $k$ is the integer space. With large $d$ and $k$, accurately reconstructing $v_i$ from the public value is nearly impossible.

Therefore, a passive attacker or semi-honest client cannot recover the secret key matrices $x_i, y_i$, mask matrix $r_i$, or secret message matrix $v_i$ of any client $U_i$ from the public values.

**Theorem 2.** *The multi-party sum protocol using El-Gamal protects the confidential data of any participating client even if there are n-2 colluding members (and colluding with the server).*

*Proof.* Without loss of generality, we assume that two client $P_1$ and $P_2$ have zero collusion, the remaining members $U_i | i \in I = \{3, 4, \ldots, N\}$ collude with each other and with the server $S$.

In the protocol under consideration, the clients transmit only the matrices $T_i, R_i, X_i, Y_i$ to the aggregation server, where the matrices $R_i, X_i, Y_i$ are generated at random.

From the matrices $R_i | i \in I$, the adversary can not compute $R_1$ or $R_2$. From the matrices $R_i | i \in I$, the adversarial can not compute $R_1$ or $R_2$. Similarly, using matrices $T_i | i \in I$ also can not compute $T_1$ and $T_2$ since we do not have efficient information. Therefore, $T_i, R_i, X_i, Y_i$ is independent.

### 3.3 Pri-DSENN: Privacy Preserving Framework for Training Dempster Shafer Theory-Based Evidential Deep Neural Network

Based on the secure aggregation protocol for real number vectors mentioned above, we can integrate it with Federated Learning (FL) to develop a protocol that ensures privacy during the training of deep learning networks. This protocol is described in Fig. 2.

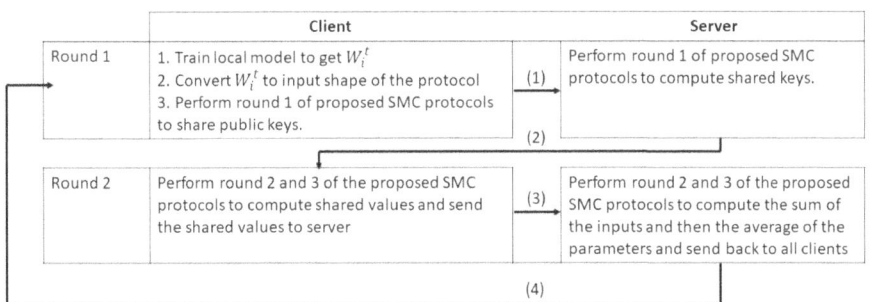

**Fig. 2.** Privacy Preserving Framework with SMC

## 4 Experiments and Results

### 4.1 Dataset and Experiments Setup

To evaluate the accuracy of the Pri-DSENN training protocol, we used two commonly used datasets for image data evaluation: CIFAR-10 and CIFAR-100. The execution environment included a Core i7 processor, 32 GB RAM, and an NVIDIA GeForce RTX 3060, utilizing libraries such as TensorFlow 2.15.0, Python 3.11, Clang 16.0.0, Bazel 6.1.0, cuDNN 8.9, and CUDA 12.2.

**Deep Neural Network Architectures.** We utilized the DSEDNN architectural framework, an enhanced version of the CNN FitNet 4 model. The final feature maps are flattened and passed through a two-layer DS ensemble with 200 neurons each. The first DS layer uses ReLU activation, and the second uses softmax for classification into 11 classes for CIFAR10 dataset. Figure 3 illustrates the network architecture.

Introduction to this protocol can be applied to the DSENN network. The training process for these DSENN networks is referred to as the PriDSENN framework. It is a framework that allows for privacy-preserving training of EDNN networks.

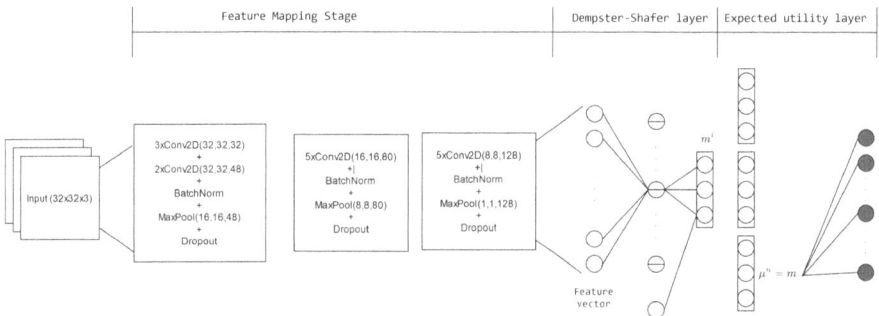

**Fig. 3.** Deep neural network architectures

This configuration includes 1,031,056 parameters (1,030,544 trainable and 512 non-trainable). For the CIFAR-100 dataset, a similar architecture was used, differing only in the output layer for specific classification tasks.

**Data Partition.** Two data partitioning paradigms are implemented: Independent and Identically Distributed (IID) and non-IID. To analyze the impact of distribution levels, data is segmented for varying client numbers: 5, 10, 15, and 20. This allows a systematic investigation of how data heterogeneity affects the client-based training process.

## 4.2 Results

**Overall Performance.** To evaluate the accuracy of training the EDNN model with the proposed protocol, we set the baseline accuracy for CIFAR-10 trained on the entire dataset at 0.9320, while the baseline accuracy for CIFAR-100 was 0.7231.

In the IID setting, CIFAR-10 exhibited convergence behavior after 200 epochs, achieving a maximum accuracy of 0.9189. Similarly, in the Non-IID setting, CIFAR-10 also showed convergence tendencies after 200 epochs, attaining a peak accuracy of 0.9000.

Similar results were observed for CIFAR-100. In the IID setting, CIFAR-100 achieved a maximum accuracy of 0.7189, while in the Non-IID setting, the best result was an accuracy of 0.6900. The experimental results demonstrate the effectiveness of the proposed PriDSENN in achieving high accuracy on both CIFAR-10 and CIFAR-100 datasets, even under Non-IID data distribution conditions. The algorithm's ability to handle data heterogeneity without sacrificing performance highlights its potential for practical applications in federated learning scenarios (Fig. 4).

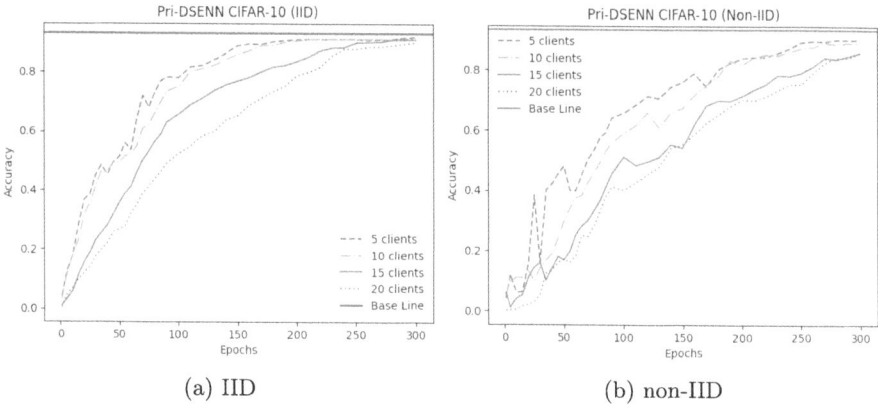

(a) IID  (b) non-IID

**Fig. 4.** CIFAR10

**Deep Model with and Without DS Layer Under Privacy Enhancement.** Pri-DSENN was evaluated using two datasets, CIFAR10 and CIFAR100, with and without the DS layer. The experiments were conducted with the best-performing Pri-DSENN model with five participating clients due to hardware infrastructure limitations.

On the CIFAR10 dataset, the accuracy converged after 100 epochs, reaching around 0.91. Without the DS layer, the method achieved an accuracy of 0.9120 after 300 epochs.

Likewise, on the CIFAR100 dataset, the results converged after 200 epochs. The best accuracy with the DS layer was 0.7289, and without the DS layer, the accuracy was 0.7050 (Figs. 5 and 6).

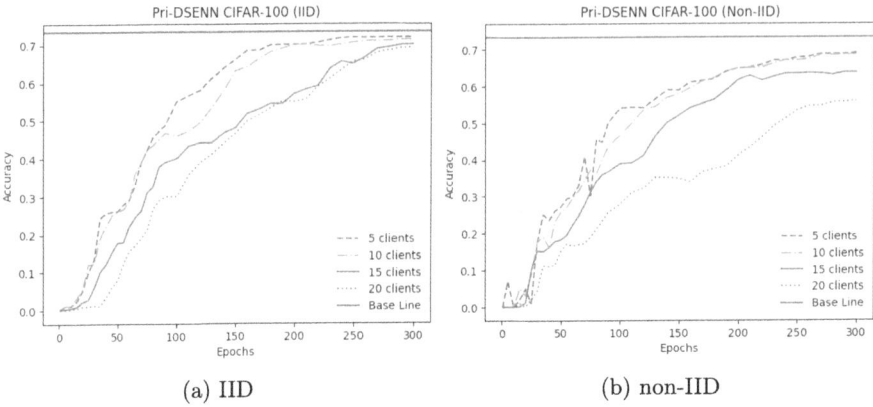

**Fig. 5.** CIFAR100

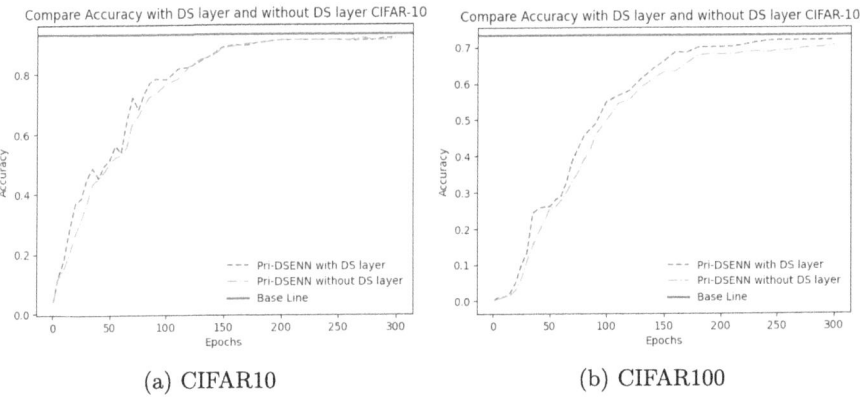

**Fig. 6.** Compare Accuracy with DS layer and without DS layer

These results indicate that training the model with Pri-DSENN does not rely heavily on the DS layer to achieve the best training performance.

## 5 Conclusion

This paper introduced a novel distributed training framework that maintains privacy standards and matches centralized training performance. This framework can applied to ensure the privacy when training EDNNs. Our evaluations on CIFAR 10 and CIFAR 100 datasets confirm that our model's privacy features do not compromise accuracy. By integrating SMC and FL, we demonstrated that EDNN can effectively manage uncertainty while protecting data privacy. The promising results provide a strong foundation for future applications of EDNNs. Looking ahead, we plan to expand our research to include various network models and explore different mass functions within DS layers. This work advances privacy-preserving machine learning and paves the way for secure, efficient computational technologies.

**Acknowledgments.** Anh Tu Tran is partly supported by a PhD Fellowship from Japan Advanced Institute of Science and Technology (JAIST). This work was sponsored by the Office of Naval Research (ONR), and the Office of Naval Research Global (ONRG), United States under grant number N62909-23-1-2058. The views and conclusions contained herein are those of the authors only and should not be interpreted as representing those of the U.S. Government.

# References

1. Boulemtafes, A., Derhab, A., Challal, Y.: A review of privacy-preserving techniques for deep learning. Neurocomputing **384**, 21–45 (2020)
2. Du, S., Du, S., Liu, B., Zhang, X.: Incorporating DeepLabv3+ and object-based image analysis for semantic segmentation of very high resolution remote sensing images. Int. J. Digit. Earth **14**(3), 357–378 (2021)
3. European Commission: Regulation (EU) 2016/679 of the European Parliament and of the Council of 27 April 2016 on the protection of natural persons with regard to the processing of personal data and on the free movement of such data, and repealing Directive 95/46/EC (General Data Protection Regulation) (Text with EEA relevance) (2016). https://eur-lex.europa.eu/eli/reg/2016/679/oj
4. Gawlikowski, J., et al.: A survey of uncertainty in deep neural networks. Artif. Intell. Rev. **56**(Suppl 1), 1513–1589 (2023)
5. McMahan, B., Moore, E., Ramage, D., Hampson, S., y Arcas, B.A.: Communication-efficient learning of deep networks from decentralized data. In: Artificial Intelligence and Statistics, pp. 1273–1282. PMLR (2017)
6. Mireshghallah, F., Taram, M., Vepakomma, P., Singh, A., Raskar, R., Esmaeilzadeh, H.: Privacy in deep learning: a survey. arXiv preprint arXiv:2004.12254 (2020)
7. Pulido-Gaytan, B., et al.: Privacy-preserving neural networks with homomorphic encryption: Challenges and opportunities. Peer-to-Peer Netw. Appl. **14**(3), 1666–1691 (2021)
8. Qi, P., Chiaro, D., Guzzo, A., Ianni, M., Fortino, G., Piccialli, F.: Model aggregation techniques in federated learning: a comprehensive survey. Future Gener. Comput. Syst. (2023)
9. Tong, Z., Xu, P., Denœux, T.: ConvNet and Dempster-Shafer theory for object recognition. In: Ben Amor, N., Quost, B., Theobald, M. (eds.) SUM 2019. LNCS (LNAI), vol. 11940, pp. 368–381. Springer, Cham (2019). https://doi.org/10.1007/978-3-030-35514-2_27
10. Tong, Z., Xu, P., Denoeux, T.: An evidential classifier based on Dempster-Shafer theory and deep learning. Neurocomputing **450**, 275–293 (2021)
11. Zhang, C., Xie, Y., Bai, H., Yu, B., Li, W., Gao, Y.: A survey on federated learning. Knowl.-Based Syst. **216**, 106775 (2021)
12. Zhao, J., Chen, Y., Zhang, W.: Differential privacy preservation in deep learning: challenges, opportunities and solutions. IEEE Access **7**, 48901–48911 (2019)

# Statistical Inference

# Large-Sample Theory for Inferential Models: A Possibilistic Bernstein–von Mises Theorem

Ryan Martin[✉] and Jonathan P. Williams[✉]

Department of Statistics, North Carolina State University, Raleigh, NC 27695, USA
{rgmarti3,jwilli27}@ncsu.edu

**Abstract.** The inferential model (IM) framework offers alternatives to the familiar probabilistic (e.g., Bayesian and fiducial) uncertainty quantification in statistical inference. Allowing uncertainty quantification to be imprecise makes exact validity/reliability possible. But is imprecision and exact validity compatible with attainment of statistical efficiency? This paper gives an affirmative answer to this question via a new possibilistic Bernstein–von Mises theorem that parallels a fundamental result in Bayesian inference. Among other things, our result demonstrates that the IM solution is asymptotically efficient in the sense that, asymptotically, its credal set is the smallest that contains the Gaussian distribution with variance equal to the Cramér–Rao lower bound.

**Keywords:** Asymptotics · Bayesian · belief · fiducial · relative likelihood

## 1 Introduction

Efron (2013) writes that "the most important unresolved problem in statistical inference is the use of Bayes theorem in the absence of prior information." Numerous attempts have been made, including Bayesian inference with default priors (Berger et al. 2009; Jeffreys 1946), fiducial inference in Fisher's original sense (Fisher 1935; Zabell 1992) and its generalizations (Dempster 1967, 2008; Fraser 1968; Hannig et al. 2016), and the various imprecise-probabilistic proposals (e.g., Augustin et al. 2014; Berger 1984; Dubois 2006; Walley 1991). Many of these solutions offer a large-sample result as follows: roughly, the associated credible sets achieve the nominal frequentist coverage probability when the sample size, $n$, is large. Statisticians insist on methods that are reliable in the sense of tending to report "correct inferences" across repeated uses, so results like this are high-value. The *résultat extraordinaire* is the *Bernstein–von Mises theorem* which says that, under certain conditions, the Bayesian or fiducial posterior distribution is approximately Gaussian and efficient in the sense that the variance equals the Cramér–Rao lower bound; here "approximately" is in the sense that the total variation distance between the two distributions is vanishing in probability as $n \to \infty$. For details, see van der Vaart (1998, Ch. 10), Ghosh et al. (2006, Ch. 4), or Hannig et al. (2016).

Still, the above have not proved to be fully satisfactory, perhaps because large-sample confidence sets do not tell the whole story. Indeed, the false confidence theorem (Balch et al. 2019; Martin 2019) implies that there always exists false hypotheses to which the (Bayesian or fiducial) posterior will tend to assign relatively high probability. To avoid this inherent unreliability, precise probabilities must be replaced by data-dependent imprecise probabilities. The inferential model (IM) framework, including the original developments in Martin and Liu (2013, 2015) and the more recent generalizations in Martin (2021, 2022a, 2022b), does just this. The IM is provably valid in a sense that implies safety from false confidence and, e.g., it provides exact finite-sample confidence sets. But one might guess that in order to be provably valid, the IM must sacrifice on efficiency—that is, the aforementioned IM's confidence sets must be larger than the efficient Bayesian or fiducial credible sets. It has been empirically demonstrated, however, that, despite the imprecision and exact validity, the IM's solutions are in fact no less (and sometimes more) efficient than these others. But no general theoretical statement concerning the IM's efficiency is currently available.

Our paper presents a novel possibilistic version of the Bernstein–von Mises theorem that applies to the class of IMs in Martin (2015, 2018, 2022b). Specifically, we show here that a version of the IM contour scaled by the observed Fisher information is accurately approximated by a Gaussian possibility measure. In other words, when $n$ is large, the IM's credal set is the smallest that contains the Gaussian distribution with variance equal to the Cramér–Rao lower bound. This result confirms our conjecture that, despite the IM's imprecision and exact validity, there is no loss of efficiency asymptotically.

## 2 Background

### 2.1 Possibility Theory

Possibility theory is among the simplest imprecise probability theories, corresponding to *consonant* belief structures (e.g., Shafer 1976, Ch. 10). Other key references include Zadeh (1978), Dubois and Prade (1988), and Dubois (2006). The simplicity of possibilistic uncertainty quantification comes from its parallels to precise probability theory. A necessity–possibility measure pair $(\underline{\Pi}, \overline{\Pi})$ that's intended to quantify uncertainty about an uncertain $Z$ in $\mathbb{Z}$ is determined by a possibility contour $\pi: \mathbb{Z} \to [0, 1]$, with $\sup_{z \in \mathbb{Z}} \pi(z) = 1$, via the rules

$$\overline{\Pi}(B) = \sup_{z \in B} \pi(z) \quad \text{and} \quad \underline{\Pi}(B) = 1 - \sup_{z \notin B} \pi(z), \quad B \subseteq \mathbb{Z}.$$

So, where ordinary probability is determined by integrating a density function, possibility is determined by optimizing a contour. The above values are often interpreted subjectively as (coherent) upper and lower probabilities associated with the proposition "$Z \in B$" or as Shaferian degrees of belief.

One way to elicit a possibility measure, specifically relevant to us here, is via the *probability-to-possibility transform* (e.g., Dubois et al. 2004; Hose 2022). If

$p$ is a probability density function, which determines a random variable $Z \sim \mathsf{P}$, then the probability-to-possibility transform defines $\pi$ as

$$\pi(z) = \mathsf{P}\{p(Z) \leq p(z)\}, \quad z \in \mathbb{Z}.$$

This defines the "best approximation" of $\mathsf{P}$ by a possibility measure in the sense that its corresponding credal set is the smallest one that contains $\mathsf{P}$. If $p$ is the (multivariate) standard normal density on $\mathbb{Z}$, the probability-to-possibility transform has contour defined as $\gamma$ as shown in Fig. 1(a); the corresponding Gaussian possibility measure defined via optimization is denoted by $\overline{\varGamma}$.

## 2.2 Inferential Models

The original IM constructions (e.g., Martin and Liu 2013, 2015) used (nested) random sets and, hence, the connection to possibility theory was indirect. A more streamlined version in Martin (2022b) directly defines the IM's possibility contour using the probability-to-possibility transform; see, also, Martin (2015, 2018). An advantage of this new and direct construction is that it avoids the ambiguity in specifying both a data-generating equation and a so-called "predictive random set"—one only needs the model/likelihood function.

The statistical model assumes that $X^n = (X_1, \ldots, X_n)$ consists of iid samples from a distribution $\mathsf{P}_\Theta$ depending on an unknown/uncertain $\Theta \in \mathbb{T}$ to be inferred. The model and observed data $X^n = x^n$ together determine a likelihood function $\theta \mapsto L_{x^n}(\theta)$ and a corresponding relative likelihood function

$$R(x^n, \theta) = \frac{L_{x^n}(\theta)}{\sup_\vartheta L_{x^n}(\vartheta)}.$$

The relative likelihood itself defines a data-dependent possibility contour that has been widely studied (e.g., Denœux 2006, 2014; Shafer 1982; Wasserman 1990). But the relative likelihood lacks a standard scale, i.e., what constitutes a "small" relative likelihood depends on aspects of the individual application. A (literally) uniform scale of interpretation across applications can easily be obtained via what Martin (2022a) calls "validification," which is a sort of possibilistic transform. In particular, for observed data $X^n = x^n$, the possibilistic IM's contour is

$$\pi_{x^n}(\theta) = \mathsf{P}_\theta\{R(X^n, \theta) \leq R(x^n, \theta)\}, \quad \theta \in \mathbb{T}.$$

This is the same contour obtained in Martin (2015, 2018) using a so-called "generalized association" based on the relative likelihood and a simple interval predictive random set. The corresponding possibility measure is

$$\overline{\varPi}_{x^n}(H) = \sup_{\theta \in H} \pi_{x^n}(\theta), \quad H \subseteq \mathbb{T}.$$

Critical to the IM developments is the so-called *validity property* which can be succinctly described by the following expression:

$$\sup_{\Theta \in \mathbb{T}} \mathsf{P}_\Theta\{\pi_{X^n}(\Theta) \leq \alpha\} \leq \alpha, \quad \text{for all } \alpha \in [0,1].$$

This is precisely the universal scaling that the relative likelihood itself is missing, i.e., "$\pi_{x^n}(\theta) \leq \alpha$" has the same inferential meaning/force in every application. It also implies that the IM is safe from false confidence—that is, the IM will *not* tend to assign high confidence, $\overline{\varPi}_{X^n}(H)$, to false hypotheses $H$ about $\Theta$. Finally, it has some important and familiar statistical consequences:

– a test that rejects the hypothesis "$\Theta \in H$" when $\overline{\varPi}_{x^n}(H) \leq \alpha$ will control the frequentist Type I error rate at level $\alpha$, and
– the set $C_\alpha(x^n) = \{\theta \in \mathbb{T} : \pi_{x^n}(\theta) > \alpha\}$ is a $100(1-\alpha)\%$ frequentist confidence set in the sense that its coverage probability is at least $1 - \alpha$.

Given that the IM is inherently imprecise and offers finite-sample guarantees, the reader might think that this comes with an associated loss of efficiency compared to those methods whose justification relies on asymptotic considerations. The result in the next section aims to debunk this myth.

## 3 A Possibilistic Bernstein–von Mises Theorem

### 3.1 Preview

To build some intuition, consider a simple exactly Gaussian model where $X^n = (X_1, \ldots, X_n)$ is an iid sample from $\mathsf{N}(\Theta, \sigma^2)$, where the mean $\Theta$ is unknown but $\sigma^2$ is known. The maximum likelihood estimator is $\hat{\theta}_{x^n} = \bar{x}$, the sample mean, and the relative likelihood is $R(x^n, \theta) = \exp\{-\frac{n}{\sigma^2}(\hat{\theta}_{x^n} - \theta)^2\}$. From here it follows by the definition of the Gaussian possibility contour $\gamma$ in Sect. 2.1 that the exactly IM's possibility contour is Gaussian:

$$\pi_{x^n}(\theta) = \mathsf{P}_\theta\{R(X^n, \theta) \leq R(x^n, \theta)\} = \gamma\{n^{1/2}(\theta - \hat{\theta}_{x^n})/\sigma\}, \quad \theta \in \mathbb{R}.$$

Switching to a local parametrization, namely, $\theta = \hat{\theta}_{x^n} + \sigma n^{-1/2} z$, gives

$$\check{\pi}_{x^n}(z) := \pi_{x^n}(\hat{\theta}_{x^n} + \sigma n^{-1/2} z) = \gamma(z), \quad z \in \mathbb{R},$$

i.e., the localized/standardized IM contour is exactly standard Gaussian. Then, for a generic $H \subseteq \mathbb{T}$, the corresponding possibility measure is

$$\overline{\varPi}_{x^n}(H) = \overline{\varGamma}\{n^{1/2}(H - \hat{\theta}_{x^n})/\sigma\},$$

where $\overline{\varGamma}$ is the Gaussian possibility measure from Sect. 2.1 and, for constants $a$ and $b$, the set $aH + b$ is defined as $aH + b = \{a\theta + b : \theta \in H\}$.

Once we step away from an exact Gaussian model, there is no longer an exact correspondence between the possibilistic IM's solution and the Gaussian possibility measure. It's a similar story in the Bayesian and (generalized) fiducial case. But the probabilistic Bernstein–von Mises theorem implies that, under certain conditions, as $n \to \infty$, the suitably centered and scaled Bayesian posterior distribution will be approximately Gaussian. Our main result below demonstrates that the same holds true for the possibilistic IM solution.

## 3.2 Main Result

The regularity conditions stated below are exactly those assumed in textbook treatments of the asymptotic normality of maximum likelihood estimators—in particular, Schervish (1995, Theorem 7.63).

REGULARITY CONDITIONS. The probability measures $\mathsf{P}_\theta$, indexed by $\theta \in \mathbb{T} \subseteq \mathbb{R}^d$, have density functions $p_\theta$ with respect to a fixed dominating measure. Recall that $\Theta$ denotes the special "true" parameter value.

1. The support of $\mathsf{P}_\theta$ doesn't depend on $\theta$.
2. $\theta \mapsto \log p_\theta(x)$ is twice continuously differentiable for almost all $x$.
3. Differentiation can be passed under the integral sign.
4. The second derivative of $\theta \mapsto \log p_\theta(X_1)$ is Lipschitz continuous in a neighborhood of $\Theta$ with Lipschitz constant that has finite $\mathsf{P}_\Theta$-expectation.

Under the above regularity conditions, we establish a possibilistic Bernstein–von Mises theorem for the IM solution presented in Sect. 2.2. That is, a centered and scaled version of the IM's possibility contour, namely,

$$\check{\pi}_{X^n}(z) = \pi_{X^n}\bigl(\hat{\theta}_{X^n} + J_{X^n}^{-1/2} z\bigr), \quad z \in \mathbb{R}^d, \tag{1}$$

where $J_{X^n}$ denotes the observed Fisher information matrix, can be well approximated by the standard Gaussian possibility contour.

**Theorem 1.** *Let $X^n = (X_1, \ldots, X_n)$ consist of iid observations from $\mathsf{P}_\Theta$, where $\{\mathsf{P}_\theta : \theta \in \mathbb{T}\}$ is the posited model and $\Theta$ is in the interior of $\mathbb{T}$. If the above regularity conditions hold at $\Theta$, and the maximum likelihood estimator $\hat{\theta}_{X^n}$ is consistent, then the possibilistic IM's solution satisfies*

$$\sup_{z \in K} \bigl|\check{\pi}_{X^n}(z) - \gamma(z)\bigr| \to 0, \quad \text{in } \mathsf{P}_\Theta\text{-probability as } n \to \infty,$$

*where $K \subset \mathbb{R}^d$ is an arbitrary compact set, $\check{\pi}_{X^n}$ is as in (1), and $\gamma$ is the Gaussian contour. This, in turn, implies that*

$$\bigl|\overline{\Pi}_{X^n}(H) - \overline{\Gamma}\{J_{X^n}^{1/2}(H - \hat{\theta}_{X^n})\}\bigr| \to 0, \quad \text{in } \mathsf{P}_\Theta\text{-probability as } n \to \infty,$$

*for all compact $H \subset \mathbb{T}$.*

The proof of Theorem 1 is lengthy and will be presented elsewhere. While it follows immediately from Wilks (1938) that $\pi_{X^n}(\Theta) \to \gamma(Z)$ in distribution under $\mathsf{P}_\Theta$, for $Z \sim \mathsf{N}_d(0, I)$, getting the required in-probability convergence uniformly over parameter values requires significant care.

The take-away message is that there is now a mathematically rigorous sense in which this possibilistic IM—despite its imprecision and exact validity—is also statistically efficient. This also shows that the connection between the possibilistic IMs and Bayes/fiducial solutions highlighted in Martin (2023) for a special class of models holds much more broadly, asymptotically. Finally, other asymptotically equivalent scaling schemes can be employed in (1), e.g., replacing $J_{X^n}^{-1/2}$ with $n^{-1/2}$, and this leads to a limiting Gaussian possibility with underlying covariance matrix matching the Cramér–Rao lower bound.

## 4 Numerical Illustrations

Here we show two examples where the exact IM contour can be found and visually compared with the Gaussian approximation suggested in Theorem 1.

*Example 1.* Let $X^n = (X_1, \ldots, X_n)$ consists of iid $\mathsf{Ber}(\Theta)$ random variables, with $\Theta \in \mathbb{T} = [0,1]$ uncertain. The relative likelihood, which depends on data only through the sum $S = \sum_{i=1}^n X_i$, is

$$R(X^n, \theta) \equiv R(S, \theta) = \left(\frac{n\theta}{S}\right)^S \left(\frac{n-n\theta}{n-S}\right)^{n-S}, \quad \theta \in [0,1],$$

with the two boundary cases settled as $R(0,\theta) = (1-\theta)^n$ and $R(n,\theta) = \theta^n$. There is no closed-form expression for the relative likelihood-based IM possibility contour, but it is easy to evaluate numerically. Indeed, if $\bar{x}$ is the sample proportion and $g_\theta(s)$ is the $\mathsf{Bin}(n, \theta)$ probability mass function, then

$$\pi_{x^n}(\theta) = \sum_{s=0}^n \mathbf{1}\{R(s,\theta) \leq R(n\bar{x}, \theta)\} g_\theta(s), \quad \theta \in [0,1],$$

with $\mathbf{1}(\cdot)$ the indicator. The Gaussian approximation, $\theta \mapsto \gamma\{J_{x^n}^{1/2}(\theta - \hat{\theta}_{x^n})\}$, suggested in Sect. 3.2 specializes in this Bernoulli case to

$$\theta \mapsto \gamma\left[\frac{n^{1/2}(\theta - \bar{x})}{\{\bar{x}(1-\bar{x})\}^{1/2}}\right], \quad \theta \in [0,1].$$

Figure 1(a) shows the exact (standardized) possibilistic IM contour for $\Theta$, at four different sample sizes, along with the limiting standard Gaussian contour. As Theorem 1 suggests, the approximation improves as $n$ increases.

*Example 2.* Let $X^n = (X_1, \ldots, X_n)$ consist of iid $\mathsf{Beta}(\Theta, 1)$ random variables, where $\Theta > 0$ is unknown. Recall that the $\mathsf{Beta}(\theta, 1)$ has PDF and CDF $x \mapsto \theta x^{\theta-1}$ and $x \mapsto x^\theta$, respectively, for $x \in (0,1)$. It can be shown that the maximum likelihood estimator is $\hat{\theta}_{x^n} = (-\frac{1}{n}\sum_{i=1}^n \log x_i)^{-1}$ and that the relative likelihood function is $R(x^n, \theta) = g(x^n, \theta)^n \exp\{-g(x^n, \theta)\}$, with

$$g(x^n, \theta) = -\sum_{i=1}^n \log x_i^\theta.$$

The above CDF formula implies that $X_i^\Theta \sim \mathsf{Unif}(0,1)$, which, in turn, implies that $g(X^n, \Theta) \sim \mathsf{Gamma}(n, 1)$. Consequently, the exact IM contour is

$$\pi_{x^n}(\theta) = \mathsf{P}\{G^n e^{-G} \leq R(x^n, \theta)\}, \quad \text{with} \quad G \sim \mathsf{Gamma}(n, 1).$$

There is no closed-form expression for this, but it can easily be evaluated via Monte Carlo simulation. Since the observed Fisher information is $J_{x^n} = n/\hat{\theta}_{x^n}^2$, we can now readily evaluate the standardized IM contour $z \mapsto \check{\pi}_{x^n}(z)$ and compare to the Gaussian approximation. Figure 1(b) shows the contours for four different values of $n$, and the Gaussian approximation improves as $n$ increases.

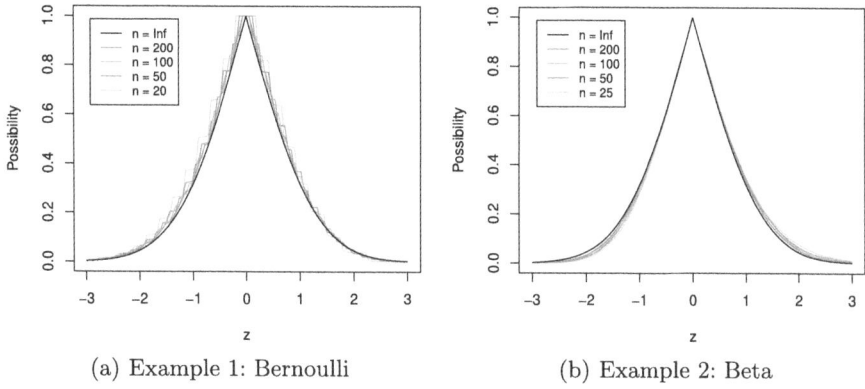

(a) Example 1: Bernoulli   (b) Example 2: Beta

**Fig. 1.** Possibility contours for the two illustrative examples.

## 5 Application

Data presented in Table 8.4 of Ghosh et al. (2006, p. 252) concern the relationship between exposure to chloracetic acid and the death of mice. A total of $n = 120$ mice were exposed, ten at each of the twelve dose levels (denoted by $x$), and a binary death indicator is measured (denoted by $y$). Figure 2(a) shows a plot of the data, with jitter in $x$ to show the replications, along with a simple logistic regression model fit. Let $\Theta = (\Theta_0, \Theta_1)$ denote the uncertain value of the logistic regression model parameter, the "intercept" and "slope" pair, respectively.

Concerning inference on $\Theta$ itself, computation of the "exact" IM contour requires lots of Monte Carlo evaluations and, hence, is very expensive. The Gaussian approximation is incredibly easy to evaluate, and Fig. 2(b) shows, as Theorem 1 predicts, that the approximation is quite accurate.

The IM also allows for marginal inference on any relevant feature $\Phi = f(\Theta)$ of $\Theta$. Of particular interest here is the median lethal dose, $\Phi = -\Theta_0/\Theta_1$, i.e., the exposure level at which the probability of death is 0.5. The dotted lines in Fig. 2(a) show that the maximum likelihood estimator is $\hat{\phi}_n = 0.244$. For inference on $\Phi$, a marginal possibility contour is obtained via optimization, i.e., $\pi_n^{\mathrm{MDL}}(\phi) = \sup_{\theta: -\theta_0/\theta_1 = \phi} \pi_n(\theta)$. This would be virtually impossible to obtain from the exact IM contour, but it is easy to get from the Gaussian approximation.

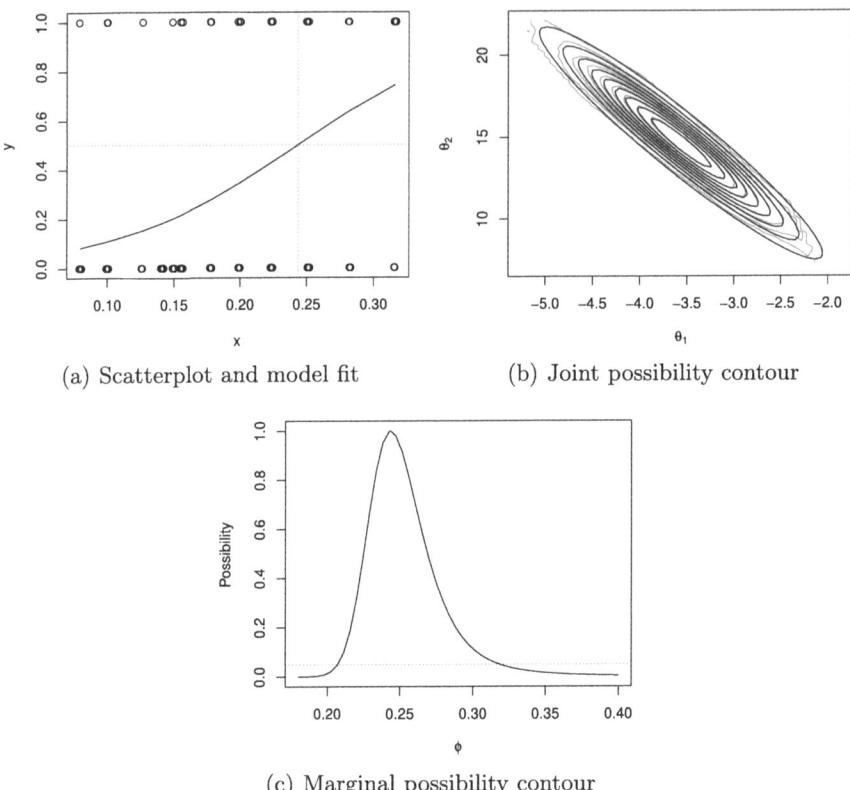

(a) Scatterplot and model fit  (b) Joint possibility contour

(c) Marginal possibility contour

**Fig. 2.** Results for the logistic regression example in Sect. 5. In Panel (b), the (Monte Carlo approximation of the) exact IM contour is red and the Gaussian approximation is black. In Panel (c), the line at 0.05 determines a 95% confidence interval for $\Phi$. (Color figure online)

## 6 Conclusion

This paper establishes a possibilistic Bernstein–von Mises theorem which, in addition to providing new insights and connections to the more familiar Bayesian and fiducial solutions, answers theoretically and practically important questions concerning the IM's efficiency. In particular, it debunks the myth that the IM's inherent imprecision and exact validity comes at the cost of efficiency.

Follow-up work will consider extensions of our present result to cases that involve nuisance parameters and/or partial prior information. For marginal inference, there are two ways to construct a possibilistic IM: one is as in Sect. 5, via direct optimization; the other is to profile out the nuisance parameters in the relative likelihood, then apply the probability-to-possibility transform. Our conjecture is that the latter is more efficient than the former, and we hope that an asymptotic analysis along the lines here will shed light on this matter.

**Acknowledgments.** The authors thank the reviewers for their helpful comments. RM's research is supported by the U.S. National Science Foundation, SES-2051225.

**Disclosure of Interests.** The authors have no competing interests to declare that are relevant to the content of this article.

# References

Augustin, T., Walter, G., Coolen, F.P.A.: Statistical inference. In: Introduction to Imprecise Probabilities. Wiley Series in Probability and Statistics, pp. 135–189. Wiley, Chichester (2014)

Balch, M.S., Martin, R., Ferson, S.: Satellite conjunction analysis and the false confidence theorem. Proc. Roy. Soc. A **475**(2227), 20180565 (2019)

Berger, J.O.: The robust Bayesian viewpoint. In: Robustness of Bayesian Analyses, Volume 4 of Stud. Bayesian Econometrics, pp. 63–144. North-Holland, Amsterdam (1984). With comments and with a reply by the author

Berger, J.O., Bernardo, J.M., Sun, D.: The formal definition of reference priors. Ann. Stat. **37**(2), 905–938 (2009)

Dempster, A.P.: Upper and lower probabilities induced by a multivalued mapping. Ann. Math. Stat. **38**, 325–339 (1967)

Dempster, A.P.: The Dempster-Shafer calculus for statisticians. Int. J. Approximate Reasoning **48**(2), 365–377 (2008)

Denœux, T.: Constructing belief functions from sample data using multinomial confidence regions. Int. J. Approximate Reasoning **42**(3), 228–252 (2006)

Denœux, T.: Likelihood-based belief function: justification and some extensions to low-quality data. Int. J. Approximate Reasoning **55**(7), 1535–1547 (2014)

Dubois, D.: Possibility theory and statistical reasoning. Comput. Stat. Data Anal. **51**(1), 47–69 (2006)

Dubois, D., Foulloy, L., Mauris, G., Prade, H.: Probability-possibility transformations, triangular fuzzy sets, and probabilistic inequalities. Reliab. Comput. **10**(4), 273–297 (2004)

Dubois, D., Prade, H.: Possibility Theory. Plenum Press, New York (1988)

Efron, B.: Discussion: "confidence distribution, the frequentist distribution estimator of a parameter: a review" [mr3047496]. Int. Stat. Rev. **81**(1), 41–42 (2013)

Fisher, R.A.: The fiducial argument in statistical inference. Ann. Eugenics **6**, 391–398 (1935)

Fraser, D.A.S.: The Structure of Inference. Wiley, New York (1968)

Ghosh, J.K., Delampady, M., Samanta, T.: An Introduction to Bayesian Analysis. Springer, New York (2006)

Hannig, J., Iyer, H., Lai, R.C.S., Lee, T.C.M.: Generalized fiducial inference: a review and new results. J. Am. Stat. Assoc. **111**(515), 1346–1361 (2016)

Hose, D.: Possibilistic reasoning with imprecise probabilities: statistical inference and dynamic filtering. Ph.D. thesis, University of Stuttgart (2022)

Jeffreys, H.: An invariant form for the prior probability in estimation problems. Proc. Roy. Soc. Lond. Ser. A **186**, 453–461 (1946)

Martin, R.: Plausibility functions and exact frequentist inference. J. Am. Stat. Assoc. **110**(512), 1552–1561 (2015)

Martin, R.: On an inferential model construction using generalized associations. J. Stat. Plann. Inference **195**, 105–115 (2018)

Martin, R.: False confidence, non-additive beliefs, and valid statistical inference. Int. J. Approximate Reasoning **113**, 39–73 (2019)

Martin, R.: An imprecise-probabilistic characterization of frequentist statistical inference. arXiv:2112.10904 (2021)

Martin, R.: Valid and efficient imprecise-probabilistic inference with partial priors, I. First results. arXiv:2203.06703 (2022a)

Martin, R.: Valid and efficient imprecise-probabilistic inference with partial priors, II. General framework. arXiv:2211.14567 (2022b)

Martin, R.: Fiducial inference viewed through a possibility-theoretic inferential model lens. In: Miranda, E., Montes, I., Quaeghebeur, E., Vantaggi, B. (eds.) Proceedings of the Thirteenth International Symposium on Imprecise Probability: Theories and Applications, Volume 215 of Proceedings of Machine Learning Research, pp. 299–310. PMLR (2023)

Martin, R., Liu, C.: Inferential models: a framework for prior-free posterior probabilistic inference. J. Am. Stat. Assoc. **108**(501), 301–313 (2013)

Martin, R., Liu, C.: Inferential Models, Volume 147 of Monographs on Statistics and Applied Probability. CRC Press, Boca Raton (2015)

Schervish, M.J.: Theory of Statistics. Springer, New York (1995)

Shafer, G.: A Mathematical Theory of Evidence. Princeton University Press, Princeton (1976)

Shafer, G.: Belief functions and parametric models. J. Roy. Stat. Soc. Ser. B **44**(3), 322–352 (1982). With discussion

van der Vaart, A.W.: Asymptotic Statistics. Cambridge University Press, Cambridge (1998)

Walley, P.: Statistical Reasoning with Imprecise Probabilities, Volume 42 of Monographs on Statistics and Applied Probability. Chapman & Hall Ltd., London (1991)

Wasserman, L.A.: Belief functions and statistical inference. Can. J. Stat. **18**(3), 183–196 (1990)

Wilks, S.S.: The large-sample distribution of the likelihood ratio for testing composite hypotheses. Ann. Math. Stat. **9**, 60–62 (1938)

Zabell, S.L.: R.A. Fisher and the fiducial argument. Stat. Sci. **7**(3), 369–387 (1992)

Zadeh, L.A.: Fuzzy sets as a basis for a theory of possibility. Fuzzy Sets Syst. **1**(1), 3–28 (1978)

# Variational Approximations of Possibilistic Inferential Models

Leonardo Cella[1]($^{(\boxtimes)}$) and Ryan Martin[2]

[1] Department of Statistical Sciences, Wake Forest University,
Winston-Salem, NC 27109, USA
cellal@wfu.edu
[2] Department of Statistics, North Carolina State University,
Raleigh, NC 27695, USA
rgmarti3@ncsu.edu

**Abstract.** Inferential models (IMs) offer reliable, data-driven, possibilistic statistical inference. But despite IMs' theoretical/foundational advantages, efficient computation in applications is a major challenge. This paper presents a simple and apparently powerful Monte Carlo-driven strategy for approximating the IM's possibility contour, or at least its $\alpha$-level set for a specified $\alpha$. Our proposal utilizes a parametric family that, in a certain sense, approximately covers the credal set associated with the IM's possibility measure, which is reminiscent of variational approximations now widely used in Bayesian statistics.

**Keywords:** Confidence regions · credal set · Monte Carlo · statistical inference · stochastic approximation

## 1 Introduction

For a long time, despite Bayesians' foundational advantages, few statisticians were actually using Bayesian methods. The computational burden for any serious Bayesian analysis was simply too high. Things changed significantly when Monte Carlo methods brought Bayesian solutions within reach. Things changed again more recently with the advances in various approximate Bayesian computational methods, in particular, the variational approximations in Blei et al. (2017) and the references therein. The once clear lines between what was computationally feasible for Bayesians and for others have now been blurred, reinvigorating Bayesians' efforts in modern applications. Dennis Lindley predicted that "[statisticians] will all be Bayesians in 2020" (Smith 1995)—his prediction did not come true, but the Bayesian community is stronger now than ever.

While Bayesian and frequentist are currently the two mainstream schools of thought in statistical inference, these are not the only perspectives. For example, the Dempster–Shafer theory of belief functions originated as an improvement to and generalization of both Bayesian inference and Fisher's fiducial argument.

Of particular interest to us here are the recent advances in *inferential models* (IMs, Martin 2021, 2022b; Martin and Liu 2013, 2015), a framework that offers Bayesian-like, data-dependent, possibilistic quantification of uncertainty about unknowns but with built-in, frequentist-like reliability guarantees. IMs and other new/non-traditional frameworks are currently facing the same computational challenges that Bayesians faced years ago. That is, we know what we want to compute and why, but we are lacking the tools to do so efficiently. Monte Carlo methods are still useful, but the imprecision that is central to the IM's reliability guarantees implies that Monte Carlo methods alone are not enough. Similar to Lindley's prediction, for Efron's speculation about fiducial and related methods—"Maybe Fisher's biggest blunder will become a big hit in the 21st century!" (Efron 1998)—to come true, imprecision-accommodating advances in Monte Carlo computations are imperative. This paper offers a simple idea that we hope can be further developed into a general tool for computationally efficient and provably, statistically reliable possibilistic inference.

Our idea leverages a well-known characterization of a possibility measure's credal set in terms of the probabilities assigned to the associated contour's upper level sets. If our goal is simply to identify those upper level sets—which, in the IM context, are confidence regions—then it can be done using Monte Carlo sampling from the "most diffuse" member of that credal set. Akin to variational Bayes, we propose to cover that credal set with a parametric family and then numerically solve for the parameter corresponding to our best approximation of that "most diffuse" member. This is inspired by the recent work in Jiang et al. (2023) and the seemingly unrelated developments in Syring and Martin (2019).

## 2 Background: Possibilistic IMs

The first IMs (e.g., Martin and Liu 2013, 2015) were formulated in terms of nested random sets and their corresponding belief functions, with certain connections to Dempster–Shafer theory. A more recent IM construction presented in Martin (2022b) defines the IM's possibility contour using a probability-to-possibility transform applied to the relative likelihood. This section briefly reviews this possibilistic IM construction, its properties, and its shortcomings.

Suppose that $X^n = (X_1, \ldots, X_n)$ consists of iid samples from a distribution $\mathsf{P}_\Theta$ depending on an unknown/uncertain $\Theta \in \mathbb{T}$. The model and observed data $X^n = x^n$ together determine a likelihood function $\theta \mapsto L_{x^n}(\theta)$ and a corresponding relative likelihood function

$$R(x^n, \theta) = \frac{L_{x^n}(\theta)}{\sup_\vartheta L_{x^n}(\vartheta)}.$$

The relative likelihood itself is a data-dependent possibility contour (e.g., Denoeux 2006, 2014; Shafer 1982; Wasserman 1990), which has a number of nice properties. What it lacks, however, is a calibration that gives (a) meaning to "possibility values" it assigns to different parameter values and (b) sufficient structure to establish frequentist-style error rate control guarantees. Fortunately, it is at least conceptually straightforward to achieve this calibration by applying what Martin (2022a) calls "validification," which is just a version of the probability-to-possibility transform (e.g., Dubois et al. 2004; Hose 2022). In particular, for observed data $X^n = x^n$, the possibilistic IM's contour is

$$\pi_{x^n}(\theta) = \mathsf{P}_\theta\{R(X^n, \theta) \leq R(x^n, \theta)\}, \quad \theta \in \mathbb{T}, \tag{1}$$

and the corresponding possibility measure is $\overline{\Pi}_{x^n}(H) = \sup_{\theta \in H} \pi_{x^n}(\theta)$, for $H \subseteq \mathbb{T}$. Critical to the IM developments is the so-called *validity property*:

$$\sup_{\Theta \in \mathbb{T}} \mathsf{P}_\Theta\{\pi_{X^n}(\Theta) \leq \alpha\} \leq \alpha, \quad \text{for all } \alpha \in [0, 1]. \tag{2}$$

Aside from providing meaning or inferential force to the numerical values returned by the possibilistic IM, the validity property (2) also ensures its safety from false confidence (Balch et al. 2019; Martin 2019) and has some more familiar statistical consequences. Of particular relevance here is that the set

$$C_\alpha(x^n) = \{\theta \in \mathbb{T} : \pi_{x^n}(\theta) > \alpha\}, \tag{3}$$

indexed by a confidence level $\alpha \in [0, 1]$, is a $100(1 - \alpha)\%$ frequentist confidence set in the sense that its coverage probability is at least $1 - \alpha$, i.e.,

$$\sup_{\Theta \in \mathbb{T}} \mathsf{P}_\Theta\{C_\alpha(X^n) \not\ni \Theta\} \leq \alpha, \quad \alpha \in [0, 1].$$

For further details about the IM's properties, its connection to Bayesian inference, and its extension to prediction, see Martin (2022b, 2023b).

The IM construction and corresponding theoretical properties are quite clean. Where things start to get messy, however, is when it comes to computation of the IM's possibility contour, the corresponding confidence set (3), etc. The point is that rarely do we have the sampling distribution of the relative likelihood $R(X^n, \theta)$, under $\mathsf{P}_\theta$, available in closed form to facilitate exact computation of $\pi_{x^n}(\theta)$ for any $\theta$. So, instead, the go-to strategy is to approximate that sampling distribution using Monte Carlo at each value of $\theta$ on a sufficiently fine grid. That is, the possibility contour is approximated as

$$\pi_{x^n}(\theta) \approx \frac{1}{M} \sum_{m=1}^{M} 1\{R(X^n_{m,\theta}, \theta) \leq R(x^n, \theta)\}, \quad \theta \in \mathbb{T},$$

where $1(\cdot)$ is the indicator function and $X^n_{m,\theta}$ consists of $n$ iid samples from $\mathsf{P}_\theta$ for $m = 1, \ldots, M$. The above computation is feasible at one or a few different $\theta$ values, but frequently this needs to be carried out on a sufficiently fine grid covering the (relevant area) of the possibly multi-dimensional parameter space $\mathbb{T}$. For example, the confidence region in (3) requires that we can solve the equation $\pi_{x^n}(\theta) = \alpha$ and the naive approach is to compute the contour over a huge grid and then keep those that (approximately) solve the aforementioned equation.

This amounts to lots of wasted computations. Simple tweaks to this most-naive approach can be employed in certain cases, e.g., importance sampling, but these adjustments require problem-specific considerations, so this does not offer any substantial improvements in computational efficiency. The next section proposes a new and general strategy that is much more efficient.

## 3 Variational Approximations

There are a number of different strategies one can employ to approximate the possibility contour. In addition to the Monte Carlo-based strategy described above, there are analytical approximations available based on large-sample theory (Martin and Williams 2024). The goal here is to strike a balance between the "exact" but expensive Monte Carlo-based approximation and the rough but cheap large-sample approximation. To strike this balance, we must focus on a specific feature of the IM solution, in particular, the confidence sets $C_\alpha(x^n)$ in (3). Our specific proposal resembles the variational approximations that are now widely used in Bayesian analysis, where the approximation is obtained by minimizing (an upper bound on) the distance/divergence of the exact posterior distribution from a relatively simple posited family of distributions.

Following Destercke and Dubois (2014), Couso et al. (2001), and others, the possibilistic IM's credal set, $\mathscr{C}(\overline{\Pi}_{x^n})$, has a remarkable characterization:

$$\mathsf{Q}_{x^n} \in \mathscr{C}(\overline{\Pi}_{x^n}) \iff \mathsf{Q}_{x^n}\{C_\alpha(x^n)\} \geq 1-\alpha, \quad \text{for all } \alpha \in [0,1].$$

That is, a data-dependent probability measure $\mathsf{Q}_{x^n}$ is consistent with $\overline{\Pi}_{x^n}$ if and only if, for each $\alpha \in [0,1]$, it assigns at least $1-\alpha$ probability to the IM's confidence set $C_\alpha(x^n)$ in (3). The best inner probabilistic approximation of the possibilistic IM, if it exists, corresponds to a $\mathsf{Q}^\star_{x^n}$ such that $\mathsf{Q}^\star_{x^n}\{C_\alpha(x^n)\} = 1-\alpha$ for all $\alpha \in [0,1]$. For a certain class of statistical models, Martin (2023a) identified this best inner approximation with Fisher's fiducial distribution. Beyond this special class of models, however, it is not clear if a best inner approximation exists and, if so, how to find it. A less ambitious goal is to find, for a fixed choice of $\alpha$, a probability distribution $\mathsf{Q}^\star_{x^n,\alpha}$ such that

$$\mathsf{Q}^\star_{x^n,\alpha}\{C_\alpha(x^n)\} = 1-\alpha. \qquad (4)$$

Our goal here is to develop a general strategy for finding, for a given $\alpha \in (0,1)$, a probability distribution $\mathsf{Q}^\star_{x^n,\alpha}$ that (at least approximately) solves the equation in (4). Once identified, we can reconstruct relevant features of the possibilistic IM via (Bayesian-like) Monte Carlo sampling from this $\mathsf{Q}^\star_{x^n,\alpha}$.

We propose to start with a parametric class of (data-dependent) probability distributions $\mathscr{Q} = \{\mathsf{Q}^\xi_{x^n} : \xi \in \varXi\}$, e.g., a Gaussian distribution with mean and covariance matrix depending in a particular way on the data and on $\xi$. More specifically, since the possibility contour's mode is at the maximum likelihood estimator $\hat{\theta}_{x^n}$, it makes sense to fix the Gaussian family $\mathscr{Q}$'s mean at $\hat{\theta}_{x^n}$ but allow the covariance matrix to depend on both data and a hyperparameter $\xi > 0$

via, say, $\mathrm{cov}(Q_{x^n}^\xi) = \xi^2 J_{x^n}^{-1}$, where $J_{x^n}$ is the observed Fisher information matrix determined by the model. A "right" choice of $\mathscr{Q}$ is context-dependent.

Given a suitable choice of $\mathscr{Q}$, our proposed procedure is as follows. Define an objective function

$$f_\alpha(\xi) = Q_{x^n,\alpha}^\star(\{\theta : \pi_{x^n}(\theta) > \alpha\}) - (1-\alpha), \qquad (5)$$

so that solving (4) boils down to finding a root of $f$. If the probability on the right-hand side could be evaluated in closed-form, then one could apply any of the standard root-finding algorithms to solve this, e.g., Newton–Raphson. However, this probability typically cannot be evaluated in closed-form; instead, $f$ can be approximated via Monte Carlo with $\hat{f}$ defined as

$$\hat{f}_\alpha(\xi) = \frac{1}{M} \sum_{m=1}^{M} 1\{\pi_{x^n}(\Theta_m^\xi) > \alpha\} - (1-\alpha), \qquad (6)$$

where $\Theta_1^\xi, \ldots, \Theta_M^\xi \stackrel{\text{iid}}{\sim} Q_{x^n}^\xi$. Presumably, the aforementioned samples are cheap for every $\xi$ because the family $\mathscr{Q}$ has been specified by the user. But only having an unbiased estimator of the objective function requires some adjustments to the numerical routine. In particular, rather than Newton–Raphson we must use a *stochastic approximation* algorithm (e.g., Kushner and Yin 2003; Martin and Ghosh 2008; Robbins and Monro 1951; Syring and Martin 2019, 2021) that is adapted to noisy function evaluations, such as $\hat{f}_\alpha$. The basic Robbins–Monro algorithm, for instance, seeks the root of (5) through the updates

$$\xi_{t+1} = \xi_t \pm w_{t+1} \hat{f}_\alpha(\xi_{t+1}), \quad t \geq 0,$$

where "$\pm$" depends on whether $\xi \mapsto f_\alpha(\xi)$ is decreasing or increasing, and $(w_t)$ is a deterministic step size sequence that satisfies

$$\sum_{t=1}^{\infty} w_t = \infty \quad \text{and} \quad \sum_{t=1}^{\infty} w_t^2 < \infty.$$

A summary of our proposed approximation is given in Algorithm 1.

Robbins and Monro (1951) showed that, under certain conditions, the sequence $(\xi_t)$ converges in probability to the root $\xi^\star = \xi^\star(x^n, \alpha)$, with $f_\alpha(\xi^\star) = 0$. If $\hat{\xi}$ is the value returned when the algorithm reaches convergence, then we set $\widehat{Q}_{x^n,\alpha} = Q_{x^n,\alpha}^{\hat{\xi}}$. This distribution should be at least a decent approximation of the IM possibility measure's inner approximation, i.e., the "most diffuse" member of $\mathscr{C}(\overline{\Pi}_{x^n})$. Consequently, the probability-to-possibility transform applied to $\widehat{Q}_{x^n,\alpha}$ should be a decent approximation to the exact IM contour $\pi_{x^n}$, at least in terms of their respective upper $\alpha$-cuts. The illustrations presented in Sect. 4 below show that the proposed approximation is quite accurate, not just "decent," in cases where we can evaluate the exact IM and do a direct visual comparison. Extensions of the proposed algorithm are discussed in Sect. 5.

**Algorithm 1:** Variational approximation of the IM's $\alpha$-cut

requires: data $x^n$ and ability to evaluate $\pi_{x^n}(\cdot)$;
initialize: $\alpha$-level, class $\mathscr{Q} = \{\mathsf{Q}_{x^n}^\xi : \xi \in \Xi\}$, guess $\xi_0$, step size sequence $(w_t)$, Monte Carlo sample size $M$, and convergence threshold $\varepsilon > 0$;
set: stop = FALSE, $t = 0$;
while !stop do
$\quad$ set $\xi = \xi_t$;
$\quad$ sample $\Theta_1^\xi, \ldots, \Theta_M^\xi \overset{\text{iid}}{\sim} \mathsf{Q}_{x^n}^\xi$;
$\quad$ evaluate $\hat{f}_\alpha(\xi)$ as in (6);
$\quad$ update $\xi_{t+1} = \xi_t \pm w_{t+1}\,\hat{f}_\alpha(\xi_t)$;
$\quad$ if $|\xi_{t+1} - \xi_t| < \varepsilon$ then
$\quad\quad$ $\hat{\xi} = \xi_{t+1}$;
$\quad\quad$ stop = TRUE;
$\quad$ else
$\quad\quad$ $t \leftarrow t+1$;
$\quad$ end
end
return $\hat{\xi}$;

## 4 Illustrations

Our modest goal is to provide proof-of-concept for the proposed approximation. We do so here with a few low-dimensional illustrations where we can visualize both the exact and approximation IM contours and directly assess the quality of the approximation. Towards this, for all but Example 4 below, we use the normal variational family $\mathscr{Q}$ with mean $\hat{\theta}_{x^n}$ and covariance $\xi^2 J_{x^n}^{-1}$ as described above, with $\xi$ to be determined. All of the examples display the variational approximation $\widehat{\mathsf{Q}}_{x^n,\alpha}$ based on $\alpha = 0.1$, $M = 200$ Monte Carlo samples, step size sequence $w_t = 2(1+t)^{-1}$, and convergence threshold $\varepsilon = 0.005$.

*Example 1.* Suppose the data is $X \sim \text{Bin}(n, \Theta)$ with PMF $p_\theta$. The exact IM possibility contour based on $X = x$ is

$$\pi_x(\theta) = \sum_{s=0}^n \mathbf{1}\{R(s,\theta) \leq R(x,\theta)\}\,p_\theta(s),$$

where $R(s, \theta)$ is the binomial relative likelihood of $\theta$ for $s$ successes; the right-hand side of the above display can be evaluated numerically without Monte Carlo. This and the contour corresponding to the variational approximation are shown in Fig. 1(a) for $n = 15$ and $x = 6$ successes. Note that the two contours closely agree, especially at level $\alpha = 0.1$.

*Example 2.* Suppose $X^n$ consists of iid bivariate normal pairs with zero means and unit variances. Inference on the unknown correlation $\Theta$ is a surprisingly challenging goal (e.g., Basu 1964). Figure 1(b) shows the exact IM contour, based

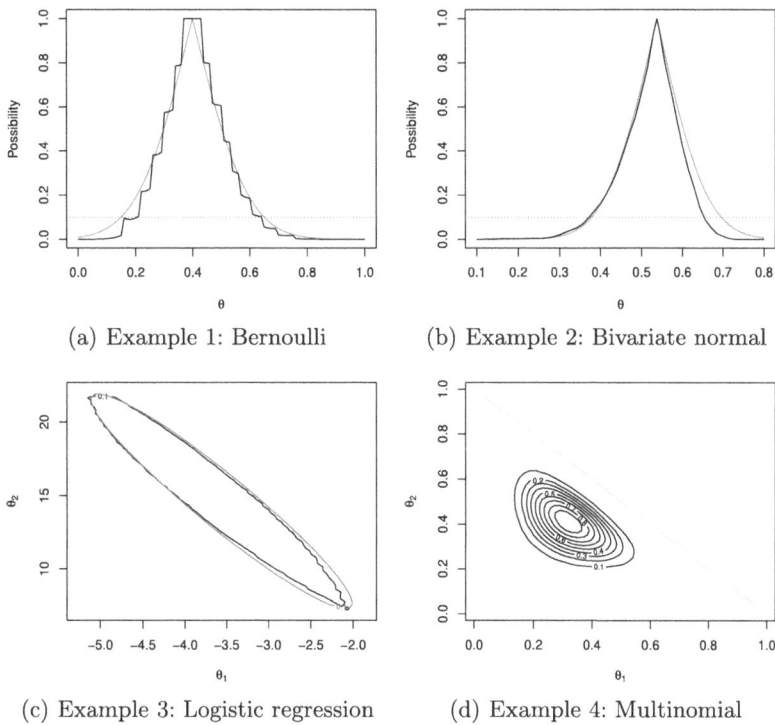

**Fig. 1.** Exact (black) and approximate (red) IM contours, the latter based on $\widehat{\mathsf{Q}}_{x^n,\alpha}$ with $\alpha = 0.1$; Panel (d) shows only the approximate contour (black), which is fully supported on the lower triangle corresponding to the probability simplex. (Color figure online)

on a naive Monte Carlo implementation of (1), and the approximation for simulated data of size $n = 50$ with true correlation 0.5. The exact contour has some asymmetry that the normal approximation cannot perfectly accommodate, but it makes up for this imperfection with a slightly wider upper 0.1-level set.

*Example 3.* The data presented in Table 8.4 of Ghosh et al. (2006) concerns the relationship between exposure to chloroacetic acid and mouse mortality. A simple logistic regression model is fitted to relate the binary death indicator with the levels of exposure to chloroacetic acid for the dataset's $n = 120$ mice. Figure 1(c) presents the 0.1-level set of the exact IM possibility contour for the regression coefficients, based on a naive Monte Carlo implementation of (1), alongside the variational approximation. The two contours line up almost perfectly.

*Example 4.* Consider $n$ iid data points sampled from $\{1, 2, \ldots, K\}$ with unknown probabilities $\Theta = (\Theta_1, \ldots, \Theta_K)$. The frequency table is a sufficient statistic, having a multinomial distribution with parameters $(n, K, \Theta)$. Here we use a Dirichlet variational family $\mathsf{Q}_{x^n}^\xi$, centered at the maximum likelihood estimator with precision $n\xi$; a Gaussian variational approximation could also be used here, but the Dirichlet approximation highlights our proposal's flexibility. Figure 1(d) shows the approximate IM contour based on $K = 3$ and counts $X = (8, 10, 7)$. The exact IM contour is virtually impossible to evaluate, because naive Monte Carlo is slow, the contours our noisy when based on Monte Carlo sample sizes that are too small, and the discrete nature of the data gives it the unusual shape. Here, however, we get a smooth approximation in a matter of seconds.

## 5 Conclusion

In a similar spirit to the variational approximations that are now widely used in Bayesian statistics, and building on recent ideas presented in Jiang et al. (2023), we develop here a strategy to approximate the possibilistic IM's contour function—or at least its $\alpha$-cuts/level sets for specified $\alpha$—using ordinary Monte Carlo sampling and stochastic approximation. A few examples are presented to highlight the accuracy and overall potential of our proposal.

A number of important and practically relevant extensions to the proposed method can and will be explored. First, it is important to be able to handle cases where $\mathcal{Q}$ is indexed by a multivariate $\xi \in \Xi$. No conceptual change is necessary to accommodate this, but the root-finding representation of the solution needs to be re-expressed as an optimization. Second, statistical inference problems often involve nuisance parameters and eliminating these efficiently requires care. The IM framework facilitates this marginalization, and our conjecture is that the proposed variational approximation strategy can be directly applied to approximate the corresponding "marginal IM contour." Third, recent efforts (e.g., Martin 2022b) have focused on incorporating incomplete or partial prior information into the possibilistic IM, the only downside being that the evaluation of the contour $\pi_{x^n}$ is often even more complicated than in the vacuous-prior case. Efficient, numerical approximations are even more important in such cases and, fortunately, we expect that the proposal will carry over directly into this more general setting. Finally, scaling up our proposed approximation in order to handle high-dimensional problems—where partial priors are likely needed for regularization—is a natural next step in this line of research.

**Acknowledgments.** Thanks to the reviewers for their helpful comments. RM is partially supported by the U.S. National Science Foundation, SES–2051225.

**Disclosure of Interests.** The authors have no competing interests to declare that are relevant to the content of this article.

## References

Balch, M.S., Martin, R., Ferson, S.: Satellite conjunction analysis and the false confidence theorem. Proc. Roy. Soc. A **475**(2227), 20180565 (2019)

Basu, D.: Recovery of ancillary information. Sankhyā Ser. A **26**, 3–16 (1964)

Blei, D.M., Kucukelbir, A., McAuliffe, J.D.: Variational inference: a review for statisticians. J. Am. Stat. Assoc. **112**(518), 859–877 (2017)

Couso, I., Montes, S., Gil, P.: The necessity of the strong α-cuts of a fuzzy set. Int. J. Uncertainty Fuzziness Knowl.-Based Syst. **9**(2), 249–262 (2001)

Denœux, T.: Constructing belief functions from sample data using multinomial confidence regions. Int. J. Approximate Reasoning **42**(3), 228–252 (2006)

Denœux, T.: Likelihood-based belief function: justification and some extensions to low-quality data. Int. J. Approximate Reasoning **55**(7), 1535–1547 (2014)

Destercke, S., Dubois, D.: Special cases. In: Introduction to Imprecise Probabilities. Wiley Series in Probability and Statistics, pp. 79–92. Wiley, Chichester (2014)

Dubois, D., Foulloy, L., Mauris, G., Prade, H.: Probability-possibility transformations, triangular fuzzy sets, and probabilistic inequalities. Reliab. Comput. **10**(4), 273–297 (2004)

Efron, B.: R. A. Fisher in the 21st century. Stat. Sci. **13**(2), 95–122 (1998)

Ghosh, J.K., Delampady, M., Samanta, T.: An Introduction to Bayesian Analysis. Springer, New York (2006)

Hose, D.: Possibilistic reasoning with imprecise probabilities: statistical inference and dynamic filtering. Ph.D. thesis, University of Stuttgart (2022)

Jiang, Y., Liu, C., Zhang, H.: Finite sample valid inference via calibrated bootstrap. Under review (2023)

Kushner, H.J., Yin, G.G.: Stochastic Approximation and Recursive Algorithms and Applications, 2nd edn. Springer, New York (2003)

Martin, R.: False confidence, non-additive beliefs, and valid statistical inference. Int. J. Approximate Reasoning **113**, 39–73 (2019)

Martin, R.: An imprecise-probabilistic characterization of frequentist statistical inference. arXiv:2112.10904 (2021)

Martin, R.: Valid and efficient imprecise-probabilistic inference with partial priors, I. First results. arXiv:2203.06703 (2022a)

Martin, R.: Valid and efficient imprecise-probabilistic inference with partial priors, II. General framework. arXiv:2211.14567 (2022b)

Martin, R.: Fiducial inference viewed through a possibility-theoretic inferential model lens. In: Miranda, E., Montes, I., Quaeghebeur, E., Vantaggi, B., (eds.) Proceedings of the Thirteenth International Symposium on Imprecise Probability: Theories and Applications, Volume 215 of *Proceedings of Machine Learning Research*, pp. 299–310. PMLR (2023a)

Martin, R.: Valid and efficient imprecise-probabilistic inference with partial priors, III. Marginalization. arXiv:2309.13454 (2023b)

Martin, R., Ghosh, J.K.: Stochastic approximation and Newton's estimate of a mixing distribution. Stat. Sci. **23**(3), 365–382 (2008)

Martin, R., Liu, C.: Inferential models: a framework for prior-free posterior probabilistic inference. J. Am. Stat. Assoc. **108**(501), 301–313 (2013)

Martin, R., Liu, C.: Inferential Models, Volume 147 of Monographs on Statistics and Applied Probability. CRC Press, Boca Raton (2015)

Martin, R., Williams, J.P.: Large-sample theory for inferential models: a possibilistic Bernstein–von Mises theorem. arXiv:2404.15843 (2024)

Robbins, H., Monro, S.: A stochastic approximation method. Ann. Math. Stat. **22**, 400–407 (1951)

Shafer, G.: Belief functions and parametric models. J. Roy. Stat. Soc. Ser. B **44**(3), 322–352 (1982). With discussion

Smith, A.: A conversation with Dennis Lindley. Stat. Sci. **10**(3), 305–319 (1995)

Syring, N., Martin, R.: Calibrating general posterior credible regions. Biometrika **106**(2), 479–486 (2019)

Syring, N., Martin, R.: Stochastic optimization for numerical evaluation of imprecise probabilities. In: Cano, A., De Bock, J., Miranda, E., Moral, S. (eds.) Proceedings of the Twelveth International Symposium on Imprecise Probability: Theories and Applications, Volume 147 of Proceedings of Machine Learning Research, pp. 289–298. PMLR (2021)

Wasserman, L.A.: Belief functions and statistical inference. Can. J. Stat. **18**(3), 183–196 (1990)

# Decision Theory via Model-Free Generalized Fiducial Inference

Jonathan P. Williams[1(✉)] and Yang Liu[2]

[1] Department of Statistics, North Carolina State University, Raleigh, NC 27695, USA
`jwilli27@ncsu.edu`
[2] Department of Human Development and Quantitative Methodology, University of Maryland, College Park, MD 20742, USA
`yliu87@umd.edu`

**Abstract.** Building on the recent development of the model-free generalized fiducial (MFGF) paradigm (Williams 2023) for predictive inference with finite-sample frequentist validity guarantees, in this paper, we develop an MFGF-based approach to decision theory. The MFGF paradigm establishes a formal connection between fiducial inference, conformal prediction, and imprecise probability theory. Beyond the utility of the new tools we contribute to the field of decision theory, our work builds on these connections. In our paper, we establish pointwise and uniform consistency of an *MFGF upper risk function* as an approximation to the true risk function via the derivation of nonasymptotic concentration bounds, and our work serves as the foundation for future investigations of the properties of the MFGF upper risk.

**Keywords:** Choquet integral · empirical risk · possibility theory · upper prevision · upper probability

## 1 Introduction

Decision theory is an important topic in statistical inference, where the goal is to determine an optimal decision rule based on observed data. Such problems are typically mathematically specified via minimizing an expected value of a loss function. The choice of loss function is application and context specific, and is considered to be given by the practitioner. Our goal in this article is to develop a decision theory based on imprecise-probabilistic inference inherited from the model-free generalized fiducial (MFGF) inference paradigm.

The MFGF paradigm was introduced in Williams (2023) as a mechanism for building model-free predictive distributions based on imprecise probability theory. It offers general purpose inferential tools and yields prediction sets with finite-sample, frequentist coverage guarantees. The MFGF paradigm establishes a formal connection between three important disciplines focused on the development of uncertainty quantification; namely, fiducial inference (Fisher 1935; Hannig et al. 2016), conformal prediction (Vovk et al. 2022), and imprecise probability theory (Augustin et al. 2014; Walley 1991). Thus, the contribution of our paper is the development of a new approach to decision theory, and one that builds on the connections established by the MFGF paradigm.

Fiducial inference for decision theory has previously been considered in Taraldsen and Lindqvist (2024), and references therein, based on [precise] probability calculus. In the context of conformal prediction, a few papers now exist under the umbrella of "conformal risk control" (Angelopoulos et al. 2023); and imprecise-probabilistic developments in decision theory are surveyed in Huntley et al. (2014) and Denœux (2019). Additionally relevant are the decision-theoretic developments in the inferential models context in Martin (2021).

## 2 Background on MFGF

Assume the variables $Y_1, \ldots, Y_{n+1} \stackrel{iid}{\sim} \mathsf{P}$ are continuous and $\mathcal{Y}$-valued, and let $\Psi : \mathcal{Y}^n \times \mathcal{Y} \to \mathcal{R}$ be a nonconformity measure (i.e., a measurable function that is invariant to the order of its first $n$ components) of interest. In this case, the random variables $t_i(Y_i) := \Psi(Y_{-i}^{n+1}, Y_i)$, for $i \in \{1, \ldots, n+1\}$, are exchangeable, in which $Y_{-i}^{n+1} := (Y_1, \ldots, Y_{i-1}, Y_{i+1}, \ldots, Y_{n+1})$; and so the position or *rank* of $t_i(Y_i)$ in the order statistics (in ascending order) of the sample $t_1(Y_1), \ldots, t_{n+1}(Y_{n+1})$ follows a discrete uniform distribution over $\{1, \ldots, n+1\}$.

From a generalized fiducial (GF) inference perspective this rank statistic represents a pivot that can be used to quantify uncertainty in $Y_{n+1}$ via a data generating *association*:

$$\text{rank}\{t_{n+1}(Y_{n+1})\} = V \sim \text{uniform}\{1, \ldots, n+1\}. \tag{1}$$

This data generating *association* is agnostic to the choice of model $Y \sim \mathsf{P}$, and is investigated in Williams (2023) in the context of MFGF predictive inference for $Y_{n+1}$ given a sample $Y_1, \ldots, Y_n$. The GF inference algorithm as in Hannig et al. (2016) is then applied by first replacing the unobserved true auxiliary variable in (1) with an independent copy, $V^\star \sim \text{uniform}\{1, \ldots, n+1\}$.

Next, the GF *switching principle* is applied to obtain an imprecise GF distribution of $Y_{n+1}$ as a distribution over the random *focal sets*

$$A_n(V^\star) := \arg\min_{y \in \mathcal{Y}}\{|\text{rank}[t_{n+1}(y)] - V^\star|\} = \{y : \text{rank}[t_{n+1}(y)] = V^\star\},$$

with respect to the imprecise GF probability measure denoted by $\mu : 2^\mathcal{Y} \to [0,1]$. In particular, for each $k \in \{1, \ldots, n+1\}$,

$$\mu\{A_n(V^\star) = A_n(k)\} = \frac{1}{n+1},$$

derived from the discrete uniform auxiliary random variable $V^\star$.

As developed in Williams (2023), predictive inference for $Y_{n+1}$ can be carried out in an imprecise fashion via the construction of belief/plausibility functions, or in a more typical [precise] fashion via a choice of some *compatible* probability measure, i.e., contained in the *credal set* of $\mu$,

$$\mathscr{C}(\mu) := \{\text{probability measures } \Delta : \Delta(B) \leq \overline{\mu}(B), \text{ for any measurable set } B\}$$

associated with the plausibility function:

$$\overline{\mu}(B) := \sum_{j=1}^{n+1} \mu\{A_n(j)\} \cdot 1\{A_n(j) \cap B \neq \emptyset\}.$$

As shown in Williams (2023), prediction sets that achieve at least $1 - \alpha$ frequentist coverage, for any finite sample size $n$, can be constructed on $\mathcal{Y}$, for any level $1 - \alpha \in [0, 1]$, as the union $\Omega_n(k) := \cup_{1 \leq j \leq k} A_n(j)$, where $k$ is the smallest integer such that $k \geq (1 - \alpha)(n + 1)$:

$$\mathsf{P}\{Y_{n+1} \in \Omega_n(k)\} = \mathsf{P}[\operatorname{rank}\{t_{n+1}(Y_{n+1})\} \leq k] = \frac{k}{n+1} \geq 1 - \alpha,$$

where the second equality follows by the exchangeability of $t_1(Y_1), \ldots, t_{n+1}(Y_{n+1})$. Alternatively, these prediction sets can be defined as the upper-level sets of the contour function given in Definition 1:

**Definition 1.** *A GF transducer function is a contour function* $f_n : \mathcal{Y} \to \mathcal{R}^+$ *defined by* $f_n(y) := \mu\{\Omega_n(V^\star) \ni y\}$. *See Williams (2023) for more details.*

See Fig. 1 for graphical illustrations of two GF transducers based on a set of $n = 4$ hypothetical observed data points.

Lastly, an important property of this GF distribution is that it coincides with frequentist probability on the focal sets, i.e., for every $k \in \{1, \ldots, n+1\}$,

$$\mathsf{P}\{Y_{n+1} \in A_n(k)\} = \frac{1}{n+1} = \mu\{A_n(V^\star) = A_n(k)\},$$

and by Lemma 8 in Williams (2023) $\Delta \in \mathscr{C}(\mu)$ if and only if $\forall k \in \{1, \ldots, n+1\}$,

$$\Delta\{A_n(k)\} = \frac{1}{n+1} = \mu\{A_n(k)\} = \mathsf{P}\{A_n(k)\}. \qquad (2)$$

## 3 Decision Theory with Imprecision

Given a loss function $\ell : \Theta \times \mathcal{Y} \to \mathcal{R}^+$ and a data generating model $Y \sim \mathsf{P}$, traditional decision theory (e.g., Dey and Williams 2023; Shalev-Shwartz and Ben-David 2014) begins by defining the true risk function

$$\mathsf{R}(\vartheta) := \int \ell(\vartheta, y) \, d\mathsf{P}(y), \qquad (3)$$

which is approximated by the empirical risk, $\mathsf{R}_n(\vartheta) := n^{-1} \sum_{i=1}^{n} \ell(\vartheta, y_i)$. Alternatively, from the above MFGF developments, we could approximate the true risk by $\mathsf{R}_\Delta(\vartheta) := \int \ell(\vartheta, y) \, d\Delta(y)$, for any $\Delta \in \mathscr{C}(\mu)$. Any probability measure $\Delta \in \mathscr{C}(\mu)$, i.e., satisfying (2), can be regarded as a precise-probabilistic approximation to the imprecise MFGF distribution, and is guaranteed to produce prediction sets that achieve at least $1 - \alpha$ frequentist coverage, for any finite sample size $n$, at any level $1 - \alpha \in [0, 1]$.

Every probability measure $\Delta \in \mathscr{C}(\mu)$ is, however, similarly compatible with the observed data, and so relying on the risk associated with any single $\Delta \in \mathscr{C}(\mu)$ is not justified without further assumptions on the data generating mechanism. Moreover, the false confidence theorem (Balch et al. 2019; Carmichael and Williams 2018; Martin 2019) makes it clear that statistical inference based on any probability measure is provably unreliable: while prediction sets with finite-sample validity can be derived from precise-probabilistic inference, there always exists *some* false assertion having arbitrarily large probability, arbitrarily often (i.e., with arbitrarily large frequentist probability). Alternatively, as advocated in Martin (2021), we can define an imprecise analogue of the risk via an upper prevision/expectation as

$$\overline{\mathsf{R}}_\mu(\vartheta) := \sup\{\mathsf{R}_\Delta(\vartheta) \,:\, \Delta \in \mathscr{C}(\mu)\}. \tag{4}$$

Since the credal set $\mathscr{C}(\mu)$ is defined by a mass function over focal sets, this is also the form of the upper expectation given in Denœux (2019). In any case, this construction makes it possible to consider a measure of risk that reflects all probability measures in the credal set $\mathscr{C}(\mu)$, and as discussed in Martin (2021), this imprecision likely plays a fundamental role in the construction of decision-theoretic, finite-sample validity criteria. In what follows here, however, we investigate consistency properties of this MFGF upper risk in simple settings to facilitate our intuitions and to motivate a future, extended version of this manuscript where we endeavor to establish decision-theoretic, finite-sample validity properties.

### 3.1 MFGF Upper Risk Function

Most importantly, for the MFGF upper risk to be reliable, it must approximate in some sense the true risk function as defined in Eq. (3), i.e., an expectation with respect to the true but unknown distribution of the data, P. A standard notion for such an approximation is via the construction of a nonasymptotic concentration bound to precisely determine a probabilistic upper bound on the rate at which the MFGF upper risk will concentrate in a neighborhood of the true risk, if at all. As we will argue in establishing Theorems 1 and 2, the MFGF upper risk is, up to an error on the order of $1/n$, equivalent to the empirical risk for bounded data. Based on this mathematical reasoning, it can be established that the MFGF upper risk will approximate the true risk in settings where the empirical risk approximates the true risk, e.g., when the uniform convergence property holds, as in Definition 2.

As it turns out, the MFGF upper risk in (4) reduces to

$$\overline{\mathsf{R}}_\mu(\vartheta) = \frac{1}{n+1} \sum_{j=1}^{n+1} \sup_{y \in A_n(j)} \{\ell(\vartheta, y)\}. \tag{5}$$

To facilitate an explicit expression for (5) we take the nonconformity score as the identity function, i.e., $t_i(Y_i) = Y_i$, so that $A_n(j) = (Y_{(j-1)}, Y_{(j)})$ for $j \in$

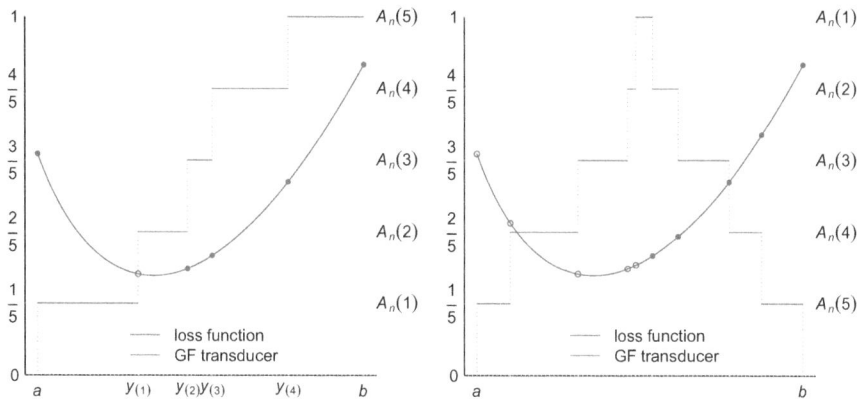

**Fig. 1.** The left panel corresponds to $\Psi_i(Y_{-i}^{n+1}, Y_i) = Y_i$ (i.e., a Dempster-Hill procedure), and the right panel corresponds to $\Psi_i(Y_{-i}^{n+1}, Y_i) = |Y_i - \text{mean}(Y_{-i}^{n+1})|$. The suprema of the loss function in Eq. (5) must occur at boundaries of the focal sets, leading to the expression given in (6); we highlight those suprema as filled dots, whose average amounts to the MFGF upper risk.

$\{1, \ldots, n+1\}$, where $Y_{(0)} := a$ and $Y_{(n+1)} := b$. Recalling that $\mu\{A_n(j)\} = \frac{1}{n+1}$, this special setup of the MFGF distribution is equivalent to the *Dempster-Hill procedure* (Coolen 1998; Hill 1968; Vovk 2024) for bounded data. Finally, we assume that $\ell(\vartheta, \cdot) : [a, b] \to \mathcal{R}^+$ is a convex loss function for every $\vartheta \in \Theta$, meaning that it is also convex on $A_n(j)$ for every $j \in \{1, \ldots, n+1\}$. In this case,

$$\sup_{y \in A_n(j)} \{\ell(\vartheta, y)\} = \max\{\ell(\vartheta, Y_{(j-1)}), \ell(\vartheta, Y_{(j)})\}, \tag{6}$$

in such a manner that

$$\overline{\mathsf{R}}_\mu(\vartheta) = \frac{1}{n+1}\left[\left\{\sum_{j=0}^{n+1} \ell(\vartheta, Y_j)\right\} - \min_{0 \le i \le n+1}\{\ell(\vartheta, Y_i)\}\right]$$

$$= \frac{1}{n+1}\left[n\mathsf{R}_n(\vartheta) + \underbrace{\ell(\vartheta, a) + \ell(\vartheta, b) - \min_{0 \le i \le n+1}\{\ell(\vartheta, Y_i)\}}_{=: M(\vartheta)}\right]. \tag{7}$$

See the left panel of Fig. 1 for a simple graphical illustration of this setup based on $n = 4$ hypothetical observed data points. The GF transducer is a non-decreasing step function having jumps at the sample order statistics $y_{(1)}, \ldots, y_{(4)}$, each corresponding to the boundary of a focal set.

**Theorem 1 (pointwise consistency of the MFGF upper risk).** *Assuming $\ell(\vartheta, \cdot) : [a, b] \to \mathcal{R}^+$ is a convex loss function for every $\vartheta \in \Theta$, for every $\epsilon > 0$ and for all $n \ge 3M/\epsilon - 1$,*

$$\mathsf{P}\{|\overline{\mathsf{R}}_\mu(\vartheta) - \mathsf{R}(\vartheta)| > \epsilon\} \le 2e^{-\frac{2}{9}n\epsilon^2/L^2(\vartheta)},$$

where $M := \sup_{\vartheta \in \Theta} \ell(\vartheta, a) + \sup_{\vartheta \in \Theta} \ell(\vartheta, b)$ and $L(\vartheta) := \sup_{y \in [a,b]} \ell(\vartheta, y) - \inf_{y \in [a,b]} \ell(\vartheta, y)$.

*Proof.* By the triangle inequality,

$$|\overline{\mathsf{R}}_\mu(\vartheta) - \mathsf{R}(\vartheta)| \leq |\overline{\mathsf{R}}_\mu(\vartheta) - n\mathsf{R}_n(\vartheta)/(n+1)| + \mathsf{R}_n(\vartheta)/(n+1) + |\mathsf{R}_n(\vartheta) - \mathsf{R}(\vartheta)|,$$

and we construct probabilistic bounds separately for each of the three terms on the right side of the inequality.

First observe that $M(\vartheta)$, as defined in (7), is nonnegative:

$$\min_{0 \leq i \leq n+1} \{\ell(\vartheta, Y_i)\} = \min\{\ell(\vartheta, a), \ell(\vartheta, Y_1), \ldots, \ell(\vartheta, Y_n), \ell(\vartheta, b)\}$$
$$\leq \min\{\ell(\vartheta, a), \ell(\vartheta, b)\},$$

and so

$$M(\vartheta) = \max\{\ell(\vartheta, a), \ell(\vartheta, b)\} + \min\{\ell(\vartheta, a), \ell(\vartheta, b)\} - \min_{0 \leq i \leq n+1}\{\ell(\vartheta, Y_i)\} \quad (8)$$
$$\geq \max\{\ell(\vartheta, a), \ell(\vartheta, b)\}.$$

Then

$$0 \leq \overline{\mathsf{R}}_\mu(\vartheta) - \frac{n\mathsf{R}_n(\vartheta)}{n+1} = \frac{M(\vartheta)}{n+1} \leq \frac{\ell(\vartheta, a) + \ell(\vartheta, b)}{n+1} \leq \frac{M}{n+1}.$$

Thus, by Eq. (7),

$$\mathsf{P}\{|\overline{\mathsf{R}}_\mu(\vartheta) - n\mathsf{R}_n(\vartheta)/(n+1)| > \epsilon/3\} \leq 1\{M/(n+1) > \epsilon/3\}. \quad (9)$$

Next, by the convexity of $\ell(\vartheta, \cdot)$ over the closed and bounded interval $[a, b]$,

$$\mathsf{P}\{\mathsf{R}_n(\vartheta)/(n+1) > \epsilon/3\} \leq 1\big[\max\{\ell(\vartheta, a), \ell(\vartheta, b)\}/(n+1) > \epsilon/3\big] \quad (10)$$
$$\leq 1\{M/(n+1) > \epsilon/3\}.$$

where the second inequality is due to Eq. (8). Finally, considering the bounds $\inf_{y \in [a,b]} \ell(\vartheta, y) \leq \ell(\vartheta, Y_i) \leq \sup_{y \in [a,b]} \ell(\vartheta, y)$, from Hoeffding's inequality it follows that

$$\mathsf{P}\{|\mathsf{R}_n(\vartheta) - \mathsf{R}(\vartheta)| > \epsilon/3\} \leq 2e^{-\frac{2}{9}n\epsilon^2/L^2(\vartheta)}. \quad (11)$$

The result of the theorem is established by combining bounds (9), (10), and (11) via the triangle inequality given at the beginning of the proof. □

**Definition 2 (uniform convergence property).** *A data generating distribution* $\mathsf{P}$ *along with a loss function* $\ell : \Theta \times \mathcal{Y} \to \mathcal{R}^+$ *is said to have the uniform convergence property if there exists a function* $g : \mathcal{R}^+ \times \mathcal{R}^+ \to \mathcal{R}$ *such that for any* $\epsilon > 0$ *and* $\alpha > 0$, *if* $n \geq g(\epsilon, \alpha)$, *then*

$$\mathsf{P}\Big\{\sup_\vartheta |\mathsf{R}_n(\vartheta) - \mathsf{R}(\vartheta)| > \epsilon\Big\} < \alpha.$$

*The function $g$ is referred to as the witness function. See, e.g., Dey and Williams (2023) for more details.*

**Theorem 2 (uniform consistency of the MFGF upper risk).** *Assume that $\ell(\vartheta, \cdot) : [a, b] \to \mathcal{R}^+$ is a convex loss function for every $\vartheta \in \Theta$, and that the uniform convergence property holds, as in Definition 2. Then for every $\epsilon > 0$, for every $\alpha > 0$, and for all $n \geq \max\{g(\epsilon, \alpha), 3M/\epsilon - 1\}$,*

$$\mathsf{P}\left\{\sup_{\vartheta} |\overline{\mathsf{R}}_\mu(\vartheta) - \mathsf{R}(\vartheta)| > \epsilon\right\} < \alpha.$$

*Proof.* Following the proof of Theorem 1, observe that the probabilistic bounds in (9) and (10) actually hold uniformly over $\vartheta \in \Theta$:

$$\mathsf{P}\left\{\sup_{\vartheta \in \Theta} |\overline{\mathsf{R}}_\mu(\vartheta) - n\mathsf{R}_n(\vartheta)/(n+1)| > \epsilon/3\right\} \leq 1\{M/(n+1) > \epsilon/3\}$$

and

$$\mathsf{P}\left\{\sup_{\vartheta \in \Theta} \mathsf{R}_n(\vartheta)/(n+1) > \epsilon/3\right\} \leq 1\{M/(n+1) > \epsilon/3\}.$$

Then an application of the triangle inequality given at the beginning of the proof of Theorem 1 gives, for all $n \geq 3M/\epsilon - 1$,

$$\mathsf{P}\left\{\sup_{\vartheta} |\overline{\mathsf{R}}_\mu(\vartheta) - \mathsf{R}(\vartheta)| > \epsilon\right\} \leq \mathsf{P}\left\{\sup_{\vartheta} |\mathsf{R}_n(\vartheta) - \mathsf{R}(\vartheta)| > \epsilon/3\right\}.$$

The proof is now complete by the assumption of the uniform convergence property. □

## 4 Numerical Illustration

A simple numerical example is provided to demonstrate the recovery of the true risk by the MFGF upper risk in the Dempster-Hill setup. Two sample size conditions were considered: $n = 20$ and $200$. We simulated data from a standard normal distribution truncated to the interval $[-3, 3]$ and focused on the squared error loss function $\ell(\vartheta, y) = (y - \vartheta)^2$. A summary of simulation results can be found in Fig. 2. It is easy to see that the true risk minimizer is zero, the mean of the truncated normal distribution.

The median, the 5th, and the 95th percentiles of upper previsions across 1000 replications are displayed across a wide range of $\theta$ values in the left panel; as a benchmark, the true risk function is also superimposed. It is observed that the upper prevision overestimates the risk on average, with the discrepancy narrowing as the sample size grows. In the right panel of Fig. 2, we contrast the histograms of upper prevision minimizers obtained under the two sample size conditions. We note that the histograms of upper prevision minimizers center around the true risk minimizer (i.e., 0) and become more concentrated in larger samples.

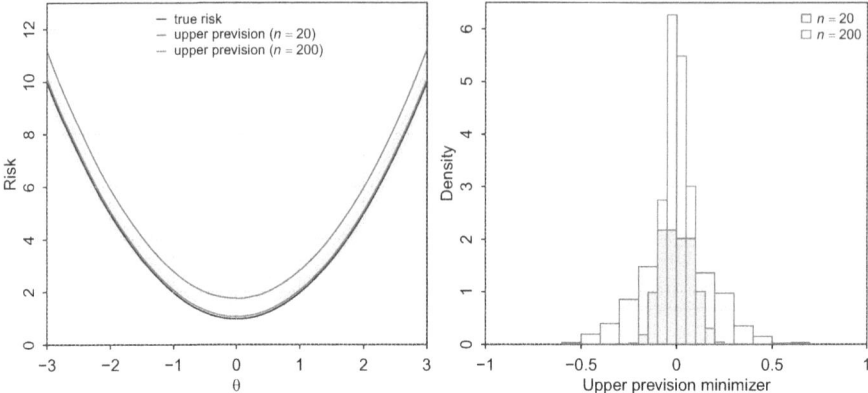

**Fig. 2.** Summary of numerical illustration in the Dempster-Hill setup. The solid, colored curves are median MFGF upper previsions pooled across 1000 replications at two sample sizes ($n = 20$ and 200), and the shaded areas indicate the middle 95% upper prevision values across replications. Right: Histograms of MFGF upper prevision minimizers across 1000 replications.

## 5 Concluding Remarks

With the foundation for the MFGF decision theory framework constructed in this paper, the most pressing next steps are to consider criteria for decision making with MFGF upper and lower risk functions. Our emphasis has been on the future formalization of decision-theoretic criteria with finite-sample validity properties, e.g., that the upper prevision/risk should not be much smaller than the true risk very often. One formalism of such a validity notion is pursued in Martin (2021), but not in a manner that directly relates the upper risk to the true risk.

A rather exhaustive collection of criteria for decision making with belief functions are surveyed in Denœux (2019), organized by criteria that induce a *complete preference* versus those that allow for imprecision due to lack of information. The former are extensions of classical precise-probabilistic-based criteria based on averaging some utility function weighted by the mass of each focal set, e.g., $\mu\{A_n(j)\}$; whereas the latter are formalizations from an imprecise-probabilistic perspective, e.g., relying on the interval produced by lower and upper expectations such as $[\underline{R}_\mu, \overline{R}_\mu]$ with $\underline{R}_\mu(\vartheta) := \inf\{R_\Delta(\vartheta) : \Delta \in \mathscr{C}(\mu)\}$. It is of interest in our future work to study MFGF decision theory incorporating criteria from both perspectives. In particular, the *pignistic* criterion for precise-probabilistic-based decision making with protection from *sure loss* is relevant, given the focus on pignistic transformations in Williams (2023).

# References

Angelopoulos, A.N., Bates, S., Fisch, A., Lei, L., Schuster, T.: Conformal risk control. In: The Twelfth International Conference on Learning Representations (2023)

Augustin, T., Walter, G., Coolen, F.P.: Statistical inference. In: Introduction to Imprecise Probabilities, pp. 135–189 (2014)

Balch, M.S., Martin, R., Ferson, S.: Satellite conjunction analysis and the false confidence theorem. Proc. Roy. Soc. A **475**, 20180565 (2019)

Carmichael, I., Williams, J.P.: An exposition of the false confidence theorem. Stat **7**(1), e201 (2018)

Coolen, F.P.: Low structure imprecise predictive inference for Bayes' problem. Stat. Probab. Lett. **36**(4), 349–357 (1998)

Denœux, T.: Decision-making with belief functions: a review. Int. J. Approximate Reasoning **109**, 87–110 (2019)

Dey, N., Williams, J.P.: Valid inference for machine learning model parameters. arXiv preprint arXiv:2302.10840 (2023)

Fisher, R.A.: The fiducial argument in statistical inference. Ann. Eugen. **6**(4), 391–398 (1935)

Hannig, J., Iyer, H., Lai, R.C., Lee, T.C.: Generalized fiducial inference: a review and new results. J. Am. Stat. Assoc. **111**(515), 1346–1361 (2016)

Hill, B.M.: Posterior distribution of percentiles: Bayes' theorem for sampling from a population. J. Am. Stat. Assoc. **63**(322), 677–691 (1968)

Huntley, N., Hable, R., Troffaes, M.C.: Decision making. In: Introduction to Imprecise Probabilities, pp. 190–206 (2014)

Martin, R.: False confidence, non-additive beliefs, and valid statistical inference. Int. J. Approximate Reasoning **113**, 39–73 (2019)

Martin, R.: Inferential models and the decision-theoretic implications of the validity property. arXiv preprint arXiv:2112.13247 (2021)

Shalev-Shwartz, S., Ben-David, S.: Understanding Machine Learning: From Theory to Algorithms. Cambridge University Press, Cambridge (2014)

Taraldsen, G., Lindqvist, B.H.: Fiducial inference and decision theory. In: Handbook of Bayesian, Fiducial, and Frequentist Inference, pp. 381–396. Chapman and Hall/CRC (2024)

Vovk, V.: Conformal predictive distributions: an approach to nonparametric fiducial prediction. In: Handbook of Bayesian, Fiducial, and Frequentist Inference, pp. 364–380. Chapman and Hall/CRC (2024)

Vovk, V., Gammerman, A., Shafer, G.: Algorithmic Learning in a Random World. Springer, Cham (2022)

Walley, P.: Statistical Reasoning with Imprecise Probabilities. Chapman & Hall (1991)

Williams, J.P.: Model-free generalized fiducial inference. arXiv preprint arXiv:2307.12472 (2023)

# Which Statistical Hypotheses are Afflicted with False Confidence?

Ryan Martin[✉]

Department of Statistics, North Carolina State University, Raleigh, NC 27695, USA
rgmarti3@ncsu.edu

**Abstract.** The false confidence theorem establishes that, for any data-driven, precise-probabilistic method for uncertainty quantification, there exists (both trivial and non-trivial) false hypotheses to which the method tends to assign high confidence. This raises concerns about the reliability of these widely-used methods, and shines promising light on the consonant belief function-based methods that are provably immune to false confidence. But an existence result alone leaves much to be desired. Towards an answer to the title question, I show that, roughly, complements of convex hypotheses are afflicted by false confidence.

**Keywords:** Bayesian · consonant beliefs · convexity · inferential model · fiducial inference · possibility theory · validity

## 1 Introduction

In *Logic of Statistical Inference*, Hacking (1976) writes: "Statisticians want numerical measures of the degree to which data support hypotheses." One such measure is a Bayesian posterior probability, but imprecise probabilists—the belief function community specifically—are well aware that precise probability theory is not the only mode of uncertainty quantification. Indeed, in a statistical inference problem, where prior information is at best incomplete and data speaks only indirectly through a model, there's good reason to question the appropriateness and/or reliability of a precise probability as statisticians' go-to quantitative expression of the degree to which data supports hypotheses.

Balch et al. (2019) expressed this concern in terms of *false confidence*. Roughly, in the context described in Sect. 2.1, false confidence corresponds to the existence of false hypotheses to which, say, a default-prior Bayesian posterior distribution tends to assign high probability, support, or confidence. Their result also applies to (generalized) fiducial inference (Dawid 2020; Fisher 1935; Hannig et al. 2016) confidence distributions (Xie and Singh 2013), etc., so it highlights a risk of unreliability inherent in *all* precise-probabilistic approaches to statistical uncertainty quantification. Since reliability is obviously a top priority, there's an exciting opportunity for imprecise probability theory to make a

---

Partially supported by the U.S. National Science Foundation, SES–205122.

fundamental contribution to statistics, a domain in which imprecise-probabilistic methods are greatly under-appreciated and largely unused. Along these lines, I've recently shown (Martin 2019, 2021, 2022a, 2022b) that a suitable *possibilistic*, or *consonant belief* framework for statistical inference is immune to false confidence; that is, this framework is reliable in the sense that it provably doesn't tend to assign high support to any false hypotheses!

Unfortunately, the *false confidence theorem*, as stated in Balch et al. (2019), is only an existence result. In a certain sense, the existence of hypotheses that are afflicted with false confidence is "obvious," and it's partly for this reason that statisticians largely haven't taken this too seriously (Carmichael and Williams 2018; Cunen et al. 2020; Martin et al. 2021). But the extent of false confidence affliction goes well beyond the hypotheses for which it's obvious: this has been demonstrated empirically in a number of specific examples, but no theoretical characterizations have been put forward. To my knowledge, all that's currently known is, for location-scale and other group transformation models, linear hypotheses about the uncertain $\Theta$ of the form "$a^\top \Theta \leq b$" are safe from false confidence (Martin 2023a). So, theoretically, we currently know effectively nothing about which hypotheses are afflicted with false confidence, but the present makes some progress in this direction. In particular, under a simple model that (approximately) represents most practical cases, I show that a class of (non-linear) hypotheses which includes those that are *co-convex*, i.e., complements of convex sets, are afflicted with false confidence. This is not a complete characterization, but it provides some insight as to what structure breeds false confidence.

Admittedly, the present paper says very little about belief functions and imprecise probability, but I still expect this investigation to make a significant contribution. Indeed, once the extent and implications of false confidence are understood, statisticians who care about reliable uncertainty quantification will have no choice but to use certain imprecise-probabilistic ideas and methods.

## 2 Background

### 2.1 Problem Setup

Let $X$ denote the data taking values in a general sample space $\mathbb{X}$. A statistical model consists of a family of probability distributions $\{\mathsf{P}_\theta : \theta \in \mathbb{T}\}$ on $\mathbb{X}$ indexed by a general parameter space $\mathbb{T}$. As is commonly assumed, suppose there is an uncertain "true" parameter value $\Theta$ such that $X$ has distribution $\mathsf{P}_\Theta$. I'll assume throughout that prior information about $\Theta$ is vacuous. The high-level goal is to quantify uncertainty about $\Theta$, given $X = x$, à la Hacking.

### 2.2 Inferential Models

Following Martin (2021), an inferential model (IM) is a map from data $x \in \mathbb{X}$ to a lower probability $\underline{\Pi}_x$ supported on subsets of $\mathbb{T}$, which depends implicitly on the statistical model and perhaps other things, e.g., prior information about $\Theta$. The interpretation is that $\underline{\Pi}_x(H)$ measures the degree of support for or belief/confidence in the truthfulness of the hypothesis "$\Theta \in H$" given data $X = x$. Examples of precise IMs include Bayes's posterior probabilities, Fisher's fiducial distributions, and Hannig's generalized fiducial distributions; examples of imprecise IMs include Dempster's seminal proposal (Dempster 1966, 1968, 2008), Walley's generalized Bayes (Walley 1991), consonant likelihood-based belief functions (Denœux 2014; Shafer 1982; Wasserman 1990), and what's briefly described in Sect. 2.4 below.

Bayesian- and fiducial-like frameworks quantify uncertainty about $\Theta$, given $X = x$, with a precise (countably additive) "posterior distribution"

$$\Pi_x(H) = \frac{\int_H L_x(\theta)\,\Pi(d\theta)}{\int_\mathbb{T} L_x(\theta)\,\Pi(d\theta)}, \quad H \subseteq \mathbb{T}, \tag{1}$$

where $\Pi$ is like a "prior distribution" for the uncertain parameter $\Theta$ and $\theta \mapsto L_x(\theta)$ is the model's likelihood function given $X = x$. The quotation marks are intended to highlight the point that, since prior information is assumed vacuous, these are "prior" and "posterior" distributions only in a formal sense. As Jeffreys (1946) explains, $\Pi$ is a default measure that gets updated to $\Pi_x$ by formally following Bayes's rule when $X = x$. In certain contexts (e.g., Hannig et al. 2016), $\Pi$ itself might depend on data, hence can't represent genuine prior information. In any case, the map $H \mapsto \Pi_x(H)$ is often used in applications to quantify uncertainty about $\Theta$, given $X = x$. But "[Bayes's rule] does not create real probabilities from hypothetical probabilities" (Fraser 2014), so a practically and theoretically important question is if this brand of (precise) probabilistic uncertainty quantification is reliable.

## 2.3 False Confidence

The false confidence theorem (Balch et al. 2019; Martin 2019) says that, for *any* precise IM, i.e., a mapping $x \mapsto \Pi_x$, with $\Pi_x$ a probability measure on $\mathbb{T}$, there exists a hypothesis–threshold pair $(H, \alpha)$ such that

$$H \not\ni \Theta \quad \text{and} \quad \mathsf{P}_\Theta\{\Pi_X(H) \geq 1 - \alpha\} > \alpha. \tag{2}$$

That is, there exists false hypotheses $H$ to which the posterior tends to assign a relatively large probability/confidence, shedding light on a lurking unreliability. That is, the statistician would tend to be confident in a hypothesis based on data $X$ if its $\Pi_X$-probability is relatively large, but this is unreliable if $\Pi_X(H)$ tends to be relatively large even if $H$ is false.

Martin (2023b) shows that false confidence implies an incoherence-like risk of monetary loss to statisticians who quantify uncertainty using precise IMs. To see this, consider the following class of (contingent) gambles

$$f_\theta^{H,\alpha}(x) = \begin{cases} 1(\theta \in H) - (1 - \alpha) & \text{if } \Pi_x(H) > 1 - \alpha \\ 0 & \text{otherwise,} \end{cases}$$

where $1(\cdot)$ is the indicator function. For every $(\alpha, H, x, \theta)$, this gamble would be acceptable to the statistician who quantifies his uncertainty with $\Pi_x$ when he observes $X = x$; that is, the expected value of $f_\Theta^{H,\alpha}(x)$, with respect to $\Pi_x$ for any fixed $x$, is positive. Now imagine another agent, a scrutinizer, who doubts the reliability of the statistician's claims. False confidence creates an opportunity for this scrutinizer—through careful considerations, background knowledge, or simply luck—to force the statistician into a systematic loss. If $(H, \alpha)$ is one of the hypothesis–threshold pairs that satisfies (2), then, as a function of $X$ for fixed $\Theta \notin H$, the statistician's winnings $f_\Theta^{H,\alpha}(X)$ are either negative (with probability $\alpha$) or zero. Therefore, his "long-run" earnings are negative, hence a systematic loss. Note that "long-run" doesn't require replications of a given experiment under the same settings or even by the same statistician. If groups of statisticians quantify their uncertainty using a precise IM, then scrutinizers can, in principle, make the statisticians collectively systematic losers. Also, the scrutinizers don't need to *know* the unknown $\Theta$ to force this systematic loss, they only need to find, even just by luck, hypotheses afflicted by false confidence. If, as I claim, hypotheses afflicted by false confidence aren't uncommon, then the above points ought to raise concern.

### 2.4 Consonant Beliefs to the Rescue

Fisher (1930) writes: "…[the likelihood function] does not obey the laws of probability; it involves no differential element." The default prior $\Pi$ also has no meaningful differential element "$d\theta$"—with vacuous prior information, there's no reason to think that measure-theoretically larger hypotheses are "more likely" than smaller ones. However, if neither the likelihood nor the prior have a meaningful differential element, then there's no sense in which the differential element on the right-hand side of (1) could be meaningful. Indeed, it's easy to find (measure-theoretically) large hypotheses that are false, hence the trivial cases of false confidence. More generally, I claim that the meaningless differential element is at the heart of false confidence (Martin 2023c).

If there exists an IM that protects all hypotheses from false confidence, then it must be imprecise; the remarks in the previous paragraph suggest that it should also be differential element free. The simplest example of this is a consonant belief function (Shafer 1976), one whose conjugate plausibility function is a maxitive possibility measure (Dubois and Prade 1988; Hose 2022). To my knowledge, the first IMs shown to be *valid*, i.e.,

$$\sup_{\Theta \notin H} \mathsf{P}_\Theta\{\underline{\Pi}_X(H) \geq 1 - \alpha\} \leq \alpha, \quad \text{for all } (H, \alpha), \qquad (3)$$

were those put forward in Martin and Liu (2015) based on nested random sets; see, also, Balch (2012) and Denœux and Li (2018). The condition (3) implies, among other things, that there's no false confidence. The valid IM construction has been generalized and streamlined in Martin (2015, 2018, 2022b), but those specific details won't be needed in what follows.

## 3 Co-convexity Breeds False Confidence

Without much loss of generality, I'll focus here on the $D$-dimensional Gaussian case $X \sim \mathsf{N}_D(\Theta, \Sigma)$, where $\Theta \in \mathbb{T} = \mathbb{R}^D$ is the uncertain parameter and the $D \times D$ covariance matrix $\Sigma$ is fixed and known. Then the likelihood is

$$L_X(\theta) \propto \exp\{-\tfrac{1}{2}(X-\theta)^\top \Sigma^{-1}(X-\theta)\}, \quad \theta \in \mathbb{T}.$$

I say "without much loss of generality" because, in most of the statistical models used in practical applications, there's a corresponding Gaussian limit experiment (e.g., van der Vaart 1998, Chapter 9). That is, if the sample size is large, then the maximum likelihood estimator (say) is an approximately minimal sufficient statistic whose sampling distribution is approximately Gaussian with mean $\Theta$ and covariance matrix a multiple of the inverse Fisher information. In this case, with vacuous prior information, the go-to precise IM for $\Theta$ is

$$\Pi_X = \mathsf{N}_D(X, \Sigma). \tag{4}$$

The precise IM in (4) has a number of desirable properties, e.g., highest posterior density credible sets are minimum volume confidence sets. But it still suffers from the inherent unreliability exposed by the false confidence theorem.

To develop some intuition, consider a function $\phi : \mathbb{T} \to \mathbb{R}$, and define

$$H_\phi = \{\theta \in \mathbb{T} : \phi(\theta) > \phi(\Theta)\}. \tag{5}$$

Clearly, hypothesis $H_\phi$ is *false*, i.e., $H_\phi \not\ni \Theta$. If $\phi$ is (non-linear) convex, which makes $H_\phi$ *co-convex*—the complement of a convex set—then Jensen's inequality immediately gives the bound $\mathsf{E}_\Theta\{\phi(X)\} > \phi(\Theta)$. Consequently, there must be non-negligible probability that $X$, the $\Pi_X$-posterior mean, is contained in the false $H_\phi$; and, if the posterior mean is in $H_\phi$, then the corresponding posterior probability, $\Pi_X(H_\phi)$, can't be small, hence an ample opportunity for false confidence in $H_\phi$. Interestingly, this apparently has little to do with the size/measure of $H_\phi$ or of $H_\phi^c$: something else is driving false confidence.

The more general, albeit less intuitive, result is presented next. I'll start with a definition. A set $G \subset \mathbb{T}$ will be called *non-linear, locally convex at $\vartheta$*, or *$\vartheta$-noloco*, if it satisfies the following three properties:

- if $G$ contains $\vartheta$ on its boundary,
- if it has a supporting hyperplane at $\vartheta$, and
- if the intersection of $G^c$ with the half-space determined by the supporting hyperplane that contains $G$ has non-zero Lebesgue measure.

To connect this to the more intuitive discussion above, if $\phi$ is a non-linear convex function, then the complement of $H_\phi$ in (5) is $\Theta$-noloco. More generally, if $G$ is convex, then a supporting hyperplane exists at each of its boundary points (e.g., Boyd and Vandenberghe 2004, Sec. 2.5.2). But $G$ could be non-convex and have a supporting hyperplane at some of its boundary points—Fig. 1 shows a

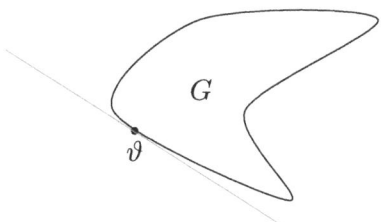

**Fig. 1.** A non-convex $G$ that's $\vartheta$-noloco; gray line defines the supporting hyperplane.

non-convex $G$ that's still $\vartheta$-noloco. The aforementioned supporting hyperplane at $\vartheta$ is determined by a vector $g_\vartheta$, i.e., that hyperplane is

$$\{\theta \in \mathbb{T} : g_\vartheta^\top (\theta - \vartheta) = 0\},$$

and the half-space that contains $G$ is

$$\text{halfsp}_\vartheta(G) = \{\theta \in \mathbb{T} : g_\vartheta^\top (\theta - \vartheta) \leq 0\}.$$

If the boundary of $G$ was linear, then its boundary would coincide with the boundary of the half-space defined above. The analysis below requires that there is some room (having non-zero Lebesgue measure) between the boundary of $G$ and the boundary of the half-space, which is enforced by the third condition above. This non-linearity and boundary separation is shown in Fig. 1.

**Proposition 1.** *For any $\Theta \in \mathbb{T}$, if $G$ is $\Theta$-noloco, then the hypothesis $H = G^c$ is afflicted by false confidence. In particular, the random variable $\Pi_X(H)$, as a function of $X \sim \mathsf{N}_D(\Theta, \Sigma)$, is stochastically larger than $\mathsf{Unif}(0,1)$.*

*Proof.* Let $g_\Theta$ denote the vector that defines the supporting hyperplane of $G$ at $\Theta$. Since $G$ is contained in the half-space $\text{halfsp}_\Theta(G)$, we get

$$H \supset H_{\text{lin}} := \{\theta \in \mathbb{T} : g_\Theta^\top (\theta - \Theta) > 0\},$$

and, consequently, $\Pi_X(H) > \Pi_X(H_{\text{lin}})$. The last inequality is strict because $\Pi_X$ is absolutely continuous with respect to Lebesgue measure and, by assumption, $H \setminus H_{\text{lin}}$ has positive Lebesgue measure. The lower bound, $\Pi_X(H_{\text{lin}})$, satisfies

$$\Pi_X(H_{\text{lin}}) = 1 - F\left(-\frac{g_\Theta^\top (X - \Theta)}{\{g_\Theta^\top \Sigma g_\Theta\}^{1/2}}\right), \tag{6}$$

where $F$ is the standard normal distribution function. As a function of $X \sim \mathsf{N}_D(\Theta, \Sigma)$, the right-hand side of (6) is $\mathsf{Unif}(0,1)$. Therefore, $\Pi_X(H)$ is (strictly) lower-bounded by a $\mathsf{Unif}(0,1)$ random variable, completing the proof.

By no means is this a complete characterization of false confidence. For one thing, it's absolutely not necessary for $\Theta$ to sit on the boundary of the hypothesis—I imposed this constraint just to make the analysis tractable. Similar results are expected for hypotheses that miss $\Theta$ but not by too much. More generally, I don't believe that noloco is fundamental to false confidence. My conjecture is that all *non-linear* hypotheses about $\Theta$, e.g., "$\phi(\Theta) \leq b$" for a non-linear map $\phi$, have at least a mild case of false confidence—the reason being that non-linear mapping can warp the parameter space in such a way that probability assignments get pushed in one direction or another systematically. Precisely diagnosing the existence and severity of affliction remains an open question.

## 4 Illustrations

In this section, I present two examples showing the existence and severity of false confidence. Both illustrations are rather simple, but they're still forceful. Indeed, if the manifestation of false confidence is relatively easy to spot in these simple examples, then we can be sure that it's present in complex, modern applications too. It's for precisely this reason that the statistical community shouldn't ignore these warnings, assume that false confidence is too rare to be concerned about, and stick with the (Bayesian) status quo.

*Example 1. Non-linear hypotheses.* Inference on the squared length of a normal mean vector is a classically challenging statistical problem, originating in Stein (1959) and appearing as the late D. R. Cox's *Challenge Question E* (Fraser et al. 2018). It's also closely related to the motivating satellite collision example in Balch et al. (2019). In the present context, if $\phi(\theta) = \|\theta\|^2$, then the set $H_\phi$ defined in (5) determines a (false) hypothesis about the mean vector's squared length; this set is also co-convex and, therefore, by Proposition 1, is afflicted with false confidence. To see the extent of affliction, the CDF $\pi \mapsto \mathsf{P}_\Theta\{\Pi_X(H_\phi) \leq \pi\}$ is shown in Fig. 2(a), where the dimension is $D=2$ and $\Theta$ is length 1. Note that, in this case, $\Pi_X(H_\phi)$ is always greater than 0.6, even though $H_\phi$ is false. For comparison, Fig. 2(a) also displays the CDF of the valid IM's lower probability $\underline{\Pi}_X(H_\phi)$ and, since it lies above the diagonal line corresponding to the Unif$(0,1)$ CDF, there's clearly no false confidence.

*Example 2. Non-linear parameter space.* Fraser (2011) considers a normal mean model $X \sim \mathsf{N}(\Theta, 1)$ but with the side information that $\Theta$ has a *known* lower bound, which I take to be 0 without loss of generality. This is motivated by relevant high-energy particle physics applications, but I'll simply point the reader to, e.g., Mandelkern (2002) for these details. Consider the (false) hypothesis $H = (\Theta, \infty)$—but note that the parameter constraint makes $H^c$ bounded. What's interesting about this example is that, apparently, the bounded and, hence, non-linear parameter space forces non-linearity in an otherwise linear hypothesis, which induces false confidence. So, this example offers a glimpse into the breadth and diversity of cases where false confidence can emerge, perhaps unexpectedly. The CDFs of the (flat-prior) Bayesian posterior probability, $\Pi_X(H)$, and of the valid IM's lower probability, $\underline{\Pi}_X(H)$, are shown in Fig. 2(b).

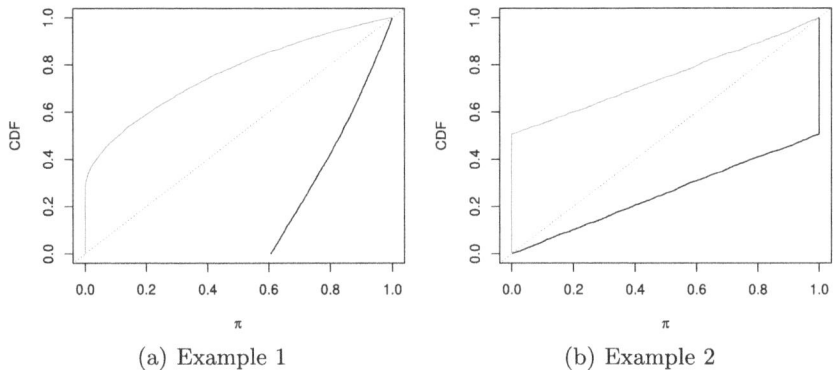

(a) Example 1        (b) Example 2

**Fig. 2.** Black lines are CDFs for the (flat-prior) Bayes posterior probabilities and red lines are the CDFs for the corresponding valid IM's lower probabilities. (Color figure online)

Note that the Bayes posterior assigns probability 1 to the false hypothesis 50% of the time, while the valid IM does the polar opposite, rightfully assigning 0 (or small) support to the false hypothesis most of the time.

## 5 Conclusion

The present paper is concerned with the following question: *which statistical hypotheses are afflicted by false confidence?* My collaborators and I have had intuition about how to answer this question for some time, but only now have I been able to formulate this intuition in a way that's conducive to mathematical analysis. The result that I proved here is quite simple, perhaps unremarkable, but I'd argue that simplicity is a virtue. After all, false confidence is the rule, rather than the exception, so it should be easy to identify hypotheses that are afflicted. What's interesting is that a property slightly more general than co-convexity is what makes the hypothesis vulnerable to false confidence.

The result presented here provides a sufficient condition for false confidence, but I seriously doubt that the same condition is necessary. As above, my conjecture is that non-linearity is enough to create at least a susceptibility to false confidence. Non-linearity alone may not be severe enough to cause false confidence-level problems as defined here; but maybe to cause the milder but still concerning "fluke confidence" that my friend and collaborator, Michael Balch, has been telling me about recently. In any case, the advancements made in the present paper make me optimistic that we'll soon be able to settle these questions.

**Acknowledgments.** Thanks to the reviewers for their helpful comments. The author is partially supported by the U.S. National Science Foundation, SES–2051225.

**Disclosure of Interests.** The author has no competing interests to declare that are relevant to the content of this article.

# References

Balch, M.S.: Mathematical foundations for a theory of confidence structures. Int. J. Approximate Reasoning **53**(7), 1003–1019 (2012)

Balch, M.S., Martin, R., Ferson, S.: Satellite conjunction analysis and the false confidence theorem. Proc. Royal Soc. A **475**(2227), 20180565 (2019)

Boyd, S., Vandenberghe, L.: Convex Optimization. Cambridge University Press, Cambridge (2004)

Carmichael, I., Williams, J.P.: An exposition of the false confidence theorem. Stat **7**(1), e201 (2018)

Cunen, C., Hjort, N.L., Schweder, T.: Confidence in confidence distributions! Proc. Roy. Soc. A **476**, 20190781 (2020)

Dawid, A.P.: Fiducial inference then and now. In: Berger, J., Meng, X., Reid, N., Xie, M. (eds.) Handbook of Bayesian, Fiducial, and Frequentist Inference, pp. 83–105. CRC Press, Boca Raton (2024)

Dempster, A.P.: New methods for reasoning towards posterior distributions based on sample data. Ann. Math. Stat. **37**, 355–374 (1966)

Dempster, A.P.: A generalization of Bayesian inference. (With discussion). J. Roy. Stat. Soc. Ser. B **30**, 205–247 (1968)

Dempster, A.P.: The Dempster-Shafer calculus for statisticians. Int. J. Approximate Reasoning **48**(2), 365–377 (2008)

Denœux, T.: Likelihood-based belief function: justification and some extensions to low-quality data. Int. J. Approximate Reasoning **55**(7), 1535–1547 (2014)

Denœux, T., Li, S.: Frequency-calibrated belief functions: review and new insights. Int. J. Approximate Reasoning **92**, 232–254 (2018)

Dubois, D., Prade, H.: Possibility Theory. Plenum Press, New York (1988)

Fisher, R.A.: Inverse probability. Proc. Camb. Philos. Soc. **26**, 528–535 (1930)

Fisher, R.A.: The fiducial argument in statistical inference. Ann. Eugenics **6**, 391–398 (1935)

Fraser, D.A.S.: Is Bayes posterior just quick and dirty confidence? Stat. Sci. **26**(3), 299–316 (2011)

Fraser, D.A.S.: Why does statistics have two theories? In: Lin, X., Genest, C., Banks, D.L., Molenberghs, G., Scott, D.W., Wang, J.-L. (eds.) Past, Present, and Future of Statistical Science. Chapman & Hall/CRC Press (2014)

Fraser, D.A.S., Reid, N., Lin, W.: When should modes of inference disagree? Some simple but challenging examples. Ann. Appl. Stat. **12**(2), 750–770 (2018)

Hacking, I.: Logic of Statistical Inference. Cambridge University Press, Cambridge (1976)

Hannig, J., Iyer, H., Lai, R.C.S., Lee, T.C.M.: Generalized fiducial inference: a review and new results. J. Am. Stat. Assoc. **111**(515), 1346–1361 (2016)

Hose, D.: Possibilistic reasoning with imprecise probabilities: statistical inference and dynamic filtering. Ph.D. thesis, University of Stuttgart (2022)

Jeffreys, H.: An invariant form for the prior probability in estimation problems. Proc. Roy. Soc. Lond. Ser. A **186**, 453–461 (1946)

Mandelkern, M.: Setting confidence intervals for bounded parameters. Stat. Sci. **17**(2), 149–172 (2002). With comments

Martin, R.: Plausibility functions and exact frequentist inference. J. Am. Stat. Assoc. **110**(512), 1552–1561 (2015)

Martin, R.: On an inferential model construction using generalized associations. J. Stat. Plann. Inference **195**, 105–115 (2018)

Martin, R.: False confidence, non-additive beliefs, and valid statistical inference. Int. J. Approximate Reasoning **113**, 39–73 (2019)

Martin, R.: An imprecise-probabilistic characterization of frequentist statistical inference. arXiv:2112.10904 (2021)

Martin, R.: Valid and efficient imprecise-probabilistic inference with partial priors, I. First results. arXiv:2203.06703 (2022a)

Martin, R.: Valid and efficient imprecise-probabilistic inference with partial priors, II. General framework. arXiv:2211.14567 (2022b)

Martin, R.: Fiducial inference viewed through a possibility-theoretic inferential model lens. In: Miranda, E., Montes, I., Quaeghebeur, E., Vantaggi, B. (eds.) Proceedings of the Thirteenth International Symposium on Imprecise Probability: Theories and Applications, Volume 215 of Proceedings of Machine Learning Research, pp. 299–310. PMLR (2023a)

Martin, R.: Fisher's underworld and the behavioral–statistical reliability balance in scientific inference. arXiv:2312.14912 (2023b)

Martin, R.: A possibility-theoretic solution to Basu's Bayesian–frequentist via media. *Sankhya A*, to appear, arXiv:2303.17425 (2023c)

Martin, R., Balch, M., Ferson, S.: Response to the comment 'confidence in confidence distributions!'. Proc. R. Soc. A. **477**, 20200579 (2021)

Martin, R., Liu, C.: Inferential Models, Volume 147 of Monographs on Statistics and Applied Probability. CRC Press, Boca Raton (2015)

Shafer, G.: A Mathematical Theory of Evidence. Princeton University Press, Princeton (1976)

Shafer, G.: Belief functions and parametric models. J. Roy. Stat. Soc. Ser. B **44**(3), 322–352 (1982). With discussion

Stein, C.: An example of wide discrepancy between fiducial and confidence intervals. Ann. Math. Stat. **30**, 877–880 (1959)

van der Vaart, A.W.: Asymptotic Statistics. Cambridge University Press, Cambridge (1998)

Walley, P.: Statistical Reasoning with Imprecise Probabilities, Volume 42 of Monographs on Statistics and Applied Probability. Chapman & Hall Ltd., London (1991)

Wasserman, L.A.: Belief functions and statistical inference. Can. J. Stat. **18**(3), 183–196 (1990)

Xie, M., Singh, K.: Confidence distribution, the frequentist distribution estimator of a parameter: a review. Int. Stat. Rev. **81**(1), 3–39 (2013)

# Algebraic Expression for the Relative Likelihood-Based Evidential Prediction of an Ordinal Variable

Frédéric Pichon[(✉)] and Sébastien Ramel

Univ. Artois, UR 3926, Laboratoire de Génie Informatique et d'Automatique de l'Artois (LGI2A), 62400 Béthune, France
{frederic.pichon,sebastien.ramel}@univ-artois.fr

**Abstract.** Given past observations of an ordinal variable, we want to predict a future observation. This paper provides the solution, according to the relative likelihood-based evidential method for statistical inference and prediction, of this problem, in an algebraic form. This result is obtained after establishing that the prediction of an ordinal variable can be computed, under some conditions, by integrating the marginals of the possibility distribution representing the estimation uncertainty in this method.

**Keywords:** Prediction · Belief function · Ordinal variable · Likelihood

## 1 Introduction

Consider observed counts of $K$ ordered categories and that we wish to predict a future observation, as illustrated by Example 1.

*Example 1.* The data in Table 1 are January precipitations, in inches, recorded during the period 1895–2004 in Arizona and categorized, as in [2, Example 7], into $K = 6$ ordered categories. The problem is to predict the precipitation category of the following January (2005).

The main result of this paper is that we provide the solution, according to the method for statistical inference and prediction proposed in [4,5], of this basic, yet arguably important, problem, in an algebraic form.

This method[1], framed in Dempster-Shafer theory (DST) [1,8], assumes a parametric model, with parameter $\theta$, for the random variable to be predicted. It is based on three steps. First, estimation uncertainty on $\theta$ is quantified by a possibility distribution; in the example above, this possibility distribution is nothing but the relative likelihood of $\theta$ given the observed counts. Next, the

---

[1] We refer the reader to [3] for a recent comparison of this method to other DST-based statistical inference approaches, as well as for a recent refinement of this method, based on random fuzzy sets, which needs not to be considered for our developments.

**Table 1.** Categorized Arizona January precipitation data, with observed count for each category.

| precipitation | category | count |
|---|---|---|
| $[0, 0.75)$ | '1' | 48 |
| $[0.75, 1.25)$ | '2' | 17 |
| $[1.25, 1.75)$ | '3' | 19 |
| $[1.75, 2.25)$ | '4' | 11 |
| $[2.25, 2.75)$ | '5' | 6 |
| $\geq 2.75$ | '6' | 9 |

random variable to be predicted is expressed as a function of a pivotal random variable and $\boldsymbol{\theta}$. Finally, the possibility distribution is combined with the pivotal distribution to yield a predictive belief function (PBF) that quantifies the uncertainty about the future observation.

This approach has been used for the prediction of quantitative variables (see [4,5]) as well as qualitative variables. The simplest qualitative case, *i.e.*, the binary case, which involves only a parameter $\boldsymbol{\theta} \in [0, 1]$, has been addressed for possibility distributions on $\boldsymbol{\theta}$ obtained in various contexts: observation of a binomial variable [5], calibration of binary classifiers [6,9], and binary classification through logistic regression [7]. Of particular interest with the binary case is that the PBF can be computed by integrating the possibility distribution on $\boldsymbol{\theta}$, under the mild condition that this distribution is unimodal and continuous. The PBF even admits an algebraic expression in the case where the possibility distribution is the relative likelihood of $\boldsymbol{\theta}$ given observed data having a binomial distribution with proportion $\boldsymbol{\theta}$ [9].

The nominal case, which involves a "structure of the second kind" [1], was considered recently in [3] for a possibility distribution on $\boldsymbol{\theta}$ obtained in the context of multinomial logistic regression. However, contrarily to the binary case, no condition simplifying the expression of the PBF has been identified and thus the computation of the PBF was only carried out approximately, through Monte Carlo simulation.

The last qualitative case, *i.e.*, the ordinal case, which relies on a "structure of the first kind" [1], is at play in [10] with a possibility distribution on $\boldsymbol{\theta}$ obtained in a context of calibrating multi-class classifiers. In this case also, the PBF was computed only approximately using Monte Carlo simulation.

In this paper, we focus on the ordinal case. First (Sect. 3), we bring to light that the PBF can be computed by integrating the possibility distribution on $\boldsymbol{\theta}$, under some conditions on this distribution, which basically extend those of the binary case. Second (Sect. 4), we show that the PBF even admits an algebraic expression, generalizing that of the binary case, when the possibility distribution is the relative likelihood of $\boldsymbol{\theta}$ given observed data having non empty categories and following a multinomial distribution whose underlying categorical distribution has parameter $\boldsymbol{\theta}$. The paper starts (Sect. 2) by recalling the prediction approach introduced in [4,5] and, in particular, the results concerning the computation of the PBF in the binary case. Proofs are omitted due to lack of space.

## 2 Relative Likelihood-Based Evidential Prediction

Let $Z$ be a random variable, with probability function $g_\theta$ where $\boldsymbol{\theta} \in \Theta$ is the unknown parameter. Consider the problem of predicting a future observation $z \in \mathcal{Z}$ of $Z$, having observed a realization $\mathbf{y} \in \mathcal{Y}$ of a random vector $\mathbf{Y}$ with probability function $f_\theta$. This section summarizes first the necessary elements of the method introduced in [4,5] for this general problem and then recalls results with respect to the specific case of predicting a binary variable $Z$, using this approach.

### 2.1 Method

Estimation uncertainty about $\boldsymbol{\theta}$ given the observation $\mathbf{y}$ is quantified by a possibility distribution $pl^\Theta$ on $\Theta$, interpreted as the contour function of a consonant belief function $Bel^\Theta$ and defined as the relative likelihood of any value $\theta$ of $\boldsymbol{\theta}$ after observing $\mathbf{Y} = \mathbf{y}$, i.e., $pl^\Theta(\theta) = L_\mathbf{y}(\theta)/L_\mathbf{y}(\hat{\theta})$, for all $\theta \in \Theta$, where $L_\mathbf{y}(\theta) = cf_\theta(\mathbf{y})$ with $c > 0$ an arbitrary constant and where $\hat{\theta}$ is a maximum likelihood estimator (MLE) of $\boldsymbol{\theta}$.

We recall that $Bel^\Theta$, being consonant, is characterized by $pl^\Theta$ and its focal sets are the sets $\Gamma(w) = \{\theta \in \Theta \mid pl^\Theta(\theta) \geq w\}$, for all $w \in [0,1]$. Moreover, let $W$ be a random variable with distribution $\lambda$, where $\lambda$ is the uniform probability measure on $[0,1]$. Then $Bel^\Theta$ is induced by the random set $\Gamma(W)$, meaning that we have $Bel^\Theta(A) = \lambda(w \in [0,1] \mid \Gamma(w) \subseteq A)$ for all $A \subseteq \Theta$.

Now, $Z$ can always be expressed as a function, called a $\varphi$-equation, of $\theta$ and a pivotal variable $V$ whose distribution $\mu$ does not depend on $\theta$:

$$Z = \varphi(\theta, V).$$

In this article, we consider only $\varphi$-equations where $V \sim \mathcal{U}([0,1])$, hence this case is assumed hereafter.

Combining the estimation uncertainty about $\boldsymbol{\theta}$ with the $\varphi$-equation yields prediction uncertainty about a future realization of $Z$, quantified by a PBF noted $Bel^\mathcal{Z}$ and induced by the random set $\varphi(\Gamma(W), V)$, i.e., for all $A \subseteq \mathcal{Z}$:

$$Bel^\mathcal{Z}(A) = \lambda \otimes \mu(\{(w,v) \in [0,1]^2 \mid \varphi(\Gamma(w), v) \subseteq A\}), \quad (1)$$

where $\lambda \otimes \mu$ is the uniform probability measure on $[0,1]^2$.

### 2.2 Prediction of a Binary Variable

Let $Z \in \mathcal{Z} = \{1,2\}$ be a binary random variable with (unknown) parameter $P_1 \in \mathcal{P}_1 = [0,1]$, where $P_1 := \mathbb{P}_Z(Z=1)$. $Z$ can be expressed as follows:

$$Z = \varphi(P_1, V) = \begin{cases} 1 & \text{if } V \leq P_1, \\ 2 & \text{otherwise.} \end{cases} \quad (2)$$

Assume estimation uncertainty about $P_1$ quantified by a possibility distribution $pl^{\mathcal{P}_1}$. Let $Bel^{\mathcal{P}_1}$ be the consonant belief function with contour function $pl^{\mathcal{P}_1}$. If $pl^{\mathcal{P}_1}$ is unimodal (with mode $0 < \hat{P}_1 < 1$) and continuous, the focal sets $\Gamma(w) = \{P_1 \in \mathcal{P}_1 \,|\, pl^{\mathcal{P}_1}(P_1) \geq w\}$ of $Bel^{\mathcal{P}_1}$ form closed intervals $\Gamma(w) = [L_1(w), U_1(w)]$ for all $w \in [0,1]$, where $L_1(w)$ and $U_1(w)$ are the two roots of the equation $pl^{\mathcal{P}_1}(P_1) = w$. Hence, $Bel^{\mathcal{P}_1}$ is induced by the random interval $[L(W), U(W)]$ with $W \sim \mathcal{U}([0,1])$. In this case, the PBF $Bel^{\mathcal{Z}}$ about a future realization of $Z$ is given by (1) with $\varphi$ defined by (2). It satisfies

$$Bel^{\mathcal{Z}}(\{1\}) = \hat{P}_1 - \int_0^{\hat{P}_1} pl^{\mathcal{P}_1}(P_1) dP_1, \qquad (3)$$

$$Bel^{\mathcal{Z}}(\{2\}) = 1 - \hat{P}_1 - \int_{\hat{P}_1}^1 pl^{\mathcal{P}_1}(P_1) dP_1. \qquad (4)$$

A particular situation where these mild conditions are respected by the possibility distribution $pl^{\mathcal{P}_1}$, is when it is defined as the relative likelihood of $P_1$ given an observation $y_1$ of a random variable $Y_1$ following a binomial distribution with parameters $n$ known and $P_1$ unknown, and with $y_1$ the number of successes out of the $n$ experiments. We have then

$$pl^{\mathcal{P}_1}(P_1) = \frac{P_1^{y_1}(1-P_1)^{n-y_1}}{\hat{P}_1^{y_1}(1-\hat{P}_1)^{n-y_1}}, \quad \forall P_1 \in [0,1],$$

with $\hat{P}_1 = y_1/n$ the MLE of $P_1$. In this case, it has even been shown [9] that Eqs. (3) and (4) admit algebraic expressions; for, e.g., $Bel^{\mathcal{Z}}(\{1\})$ we have

$$Bel^{\mathcal{Z}}(\{1\}) = \hat{P}_1 - \frac{\underline{B}(\hat{P}_1; y_1+1, n-y_1+1)}{\hat{P}_1^{y_1}(1-\hat{P}_1)^{n-y_1}},$$

with $\underline{B}(z; a, b) = \sum_{j=a}^{a+b-1} \frac{(a-1)!(b-1)!}{j!(a+b-1-j)!} z^j (1-z)^{a+b-1-j}$ the lower incomplete beta function.

## 3 Prediction of an Ordinal Variable

Let $Z$ be an ordinal random variable taking its value in $\mathcal{Z} = \{1, \ldots, K\}$, where the $K$ categories are, without lack of generality, denoted by the integers from 1 to $K$ and ordered according to the natural order between integers. The probability measure of $Z$ is characterized by the vector $\mathbf{P} = (P_1, \ldots, P_{K-1})$ of (unknown) parameters $P_j := \sum_{i=1}^{j} \mathbb{P}(Z = i)$, $1 \leq j < K$. We have $\mathbf{P} \in \mathcal{P}$, where

$$\mathcal{P} = \{(P_1, \ldots, P_{K-1}) \in [0,1]^{K-1} \,|\, P_1 \leq \cdots \leq P_{K-1}\}.$$

Following [1], $Z$ can be expressed as, with $P_K := 1$:

$$Z = \varphi(\mathbf{P}, V) = \begin{cases} j, & \text{if } P_{j-1} < V \leq P_j \text{ for some } j \text{ such that } 1 < j \leq K, \\ 1, & \text{otherwise } (i.e., V \leq P_1). \end{cases} \qquad (5)$$

It can easily be verified that $\mathbb{P}(\varphi(\mathbf{P}, V) = j) = \mathbb{P}(Z = j)$, for all $1 \leq j \leq K$.

Consider estimation uncertainty about $\mathbf{P}$ quantified by a possibility distribution $pl^{\mathcal{P}}$. Let $Bel^{\mathcal{P}}$ be the consonant belief function with contour function $pl^{\mathcal{P}}$. Its focal sets are $\Gamma(w) = \{\mathbf{P} \in \mathcal{P} \mid pl^{\mathcal{P}}(\mathbf{P}) \geq w\}$, with $w \in [0, 1]$. Under such estimation uncertainty about $\mathbf{P}$, the PBF $Bel^{\mathcal{Z}}$ about a future realization of $Z$ is given by (1) with $\varphi$ defined by (5).

Let us introduce two conditions (Assumptions 1 and 2) about $Bel^{\mathcal{P}}$:

**Assumption 1.** $pl^{\mathcal{P}}(\mathbf{P}) > 0$ if and only if $\mathbf{P} \in \mathcal{P}^* := \{(P_1, \ldots, P_{K-1}) \in (0,1)^{K-1} \mid P_1 < \cdots < P_{K-1}\}$.

**Assumption 2.** $\Gamma(w)$, for all $w \in (0, 1]$, is convex.

Under such conditions, the focal sets of the PBF $Bel^{\mathcal{Z}}$, which are $\Gamma'(w, v) := \varphi(\Gamma(w), v)$ for all $(w, v) \in [0, 1]^2$, are intervals:

**Proposition 1.** Under Assumptions 1 and 2 we have, for all $(w, v) \in [0,1]^2$:

$$\varphi(\Gamma(w), v) = [\![\ell(w, v), u(w, v)]\!], \tag{6}$$

for some $1 \leq \ell(w, v) \leq u(w, v) \leq K$.

Let $\mathcal{P}_j = [0, 1]$, for all $1 \leq j < K$, and $pl^{\mathcal{P}_j}$ be the marginal contour function of $pl^{\mathcal{P}}$ for the $j$-th component of $\mathbf{P}$, defined by [4, Eq. (76)]:

$$pl^{\mathcal{P}_j}(P_j) = \sup_{\mathbf{P}_{-j}} pl^{\mathcal{P}}(\mathbf{P}), \tag{7}$$

where $\mathbf{P}_{-j}$ is the subvector of $\mathbf{P}$ with component $j$ removed.

Let us introduce three conditions (Assumptions 3, 4 and 5) about the marginal contour functions of $Bel^{\mathcal{P}}$:

**Assumption 3.** $pl^{\mathcal{P}_j}$, $1 \leq j < K$, is continuous and unimodal (with mode $\hat{P}_j$ such that $0 < \hat{P}_j < 1$).

**Assumption 4.** $\hat{P}_i < \hat{P}_j$, for all $1 \leq i < j < K$.

**Assumption 5.** For all $1 \leq i < j < K$, there is a value $P_{i,j} \in (0,1)$ such that $pl^{\mathcal{P}_i}(P) = pl^{\mathcal{P}_j}(P)$, for $P \in (0,1)$, if and only if $P = P_{i,j}$.

Under such additional conditions, the bounds of the interval focal set $\Gamma'(w, v)$ of $Bel^{\mathcal{Z}}$ are obtained for two particular elements of $\mathcal{P}$ (illustrated by Example 2):

**Proposition 2.** Under Assumptions 1–5 we have, for all $(w, v) \in [0, 1]^2$:

$$\varphi(\Gamma(w), v) = [\![\varphi(\mathbf{U}(w), v), \varphi(\mathbf{L}(w), v)]\!],$$

where $\mathbf{U}(w) = (U_1(w), \ldots, U_{K-1}(w)) \in \mathcal{P}$ and $\mathbf{L}(w) = (L_1(w), \ldots, L_{K-1}(w)) \in \mathcal{P}$, with $L_j(w)$ and $U_j(w)$ the two roots of the equation $pl^{\mathcal{P}_j}(P_j) = w$, $1 \leq j < K$.

Algebraic Expression for the Evidential Prediction of an Ordinal Variable    155

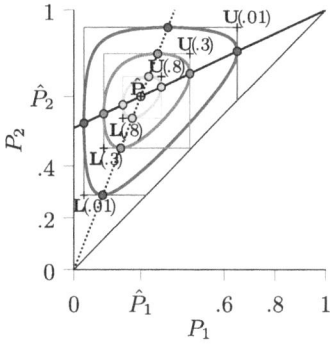
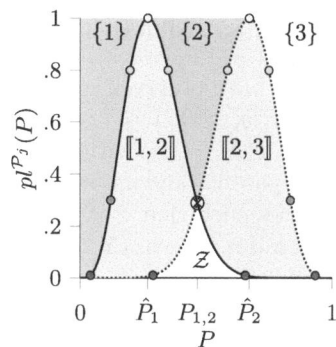

(a) Contour plot of $pl^{\mathcal{P}}$. Focal sets $\Gamma(.01)(—)$, $\Gamma(.3)(—)$, $\Gamma(.8)(—)$, Supremums (○) associated to marginals $pl^{\mathcal{P}_1}$(-○-), $pl^{\mathcal{P}_2}$(⋯○⋯), roots (+) $\mathbf{L}(w)$, $\mathbf{U}(w)$ for $w \in \{.01, .3, .8, 1\}$.

(b) Marginals $pl^{\mathcal{P}_1}$(-○-), and $pl^{\mathcal{P}_2}$(⋯○⋯), "intersection" point $P_{1,2}(\otimes)$, mass function $m^{\mathcal{Z}}$ associated to PBF $Bel^{\mathcal{Z}}$: ■ $m^{\mathcal{Z}}(\{1\})$, ■ $m^{\mathcal{Z}}(\{2\})$, ■ $m^{\mathcal{Z}}(\{3\})$, ■ $m^{\mathcal{Z}}(\llbracket 1, 2 \rrbracket)$, ■ $m^{\mathcal{Z}}(\llbracket 2, 3 \rrbracket)$, □ $m^{\mathcal{Z}}(\mathcal{Z})$.

**Fig. 1.** $Bel^{\mathcal{P}}$ respecting Assumptions 1–5 (Fig. 1a), associated $Bel^{\mathcal{Z}}$ (Fig. 1b).

*Example 2.* Figure 1a shows for a case where $K = 3$ and for $w \in \{.01, .3, .8, 1\}$, the points $\mathbf{U}(w)$ and $\mathbf{L}(w)$ as well as the focal sets $\Gamma(w)$ of a $Bel^{\mathcal{P}}$, with contour function $pl^{\mathcal{P}}$, known[2] to respect Assumptions 1-5. The marginals of $pl^{\mathcal{P}}$ (Eq. (7)) are shown in Fig. 1b, along with the point $P_{1,2}$ at play in Assumption 5.

We are now ready to provide our first main result concerning $Bel^{\mathcal{Z}}$, which is that under the preceding conditions, it can be computed by integrating the marginal contour functions (which is a generalization of Eqs. (3) and (4)):

**Theorem 1.** *Under Assumptions 1-5, $Bel^{\mathcal{Z}}$ is characterized by $Bel^{\mathcal{Z}}(\llbracket i,j \rrbracket)$ for all $1 \leq i \leq j \leq K$, with $Bel^{\mathcal{Z}}(\llbracket i,j \rrbracket)$ equal to*

$$\begin{cases} 1, & \text{if } i=1, j=K, \\ \hat{P}_j - \int_0^{\hat{P}_j} pl^{\mathcal{P}_j}(P)dP, & \text{if } i=1, j<K, \\ 1 - \hat{P}_{i-1} - \int_{\hat{P}_{i-1}}^1 pl^{\mathcal{P}_{i-1}}(P)dP, & \text{if } 1<i, j=K, \\ \hat{P}_j - \hat{P}_{i-1} - \int_{\hat{P}_{i-1}}^{P_{i-1,j}} pl^{\mathcal{P}_{i-1}}(P)dP - \int_{P_{i-1,j}}^{\hat{P}_j} pl^{\mathcal{P}_j}(P)dP, & \text{otherwise.} \end{cases} \quad (8)$$

We note that thanks to [2, Eq. (29)], the mass function associated to $Bel^{\mathcal{Z}}$ can also be expressed in terms of integration of the marginal contour functions. Due to lack of space, this is merely illustrated graphically by Fig. 1b.

The next section studies a standard setting, which happens to yield a possibility distribution $pl^{\mathcal{P}}$ respecting the conditions considered above.

---

[2] Because of Proposition 3 and its contour being of the form (9), with $\mathbf{y} = (4, 6, 5)$.

## 4 The Case of Multinomial Data

Assume we have observed a realisation $\mathbf{y} = (y_1, \ldots, y_K)$, with non empty categories (i.e., $y_i > 0$, $1 \le i \le K$), of a random vector $\mathbf{Y} = (Y_1, \ldots, Y_K)$, where $Y_j = \sum_{i=1}^{n} I(X_i = j)$, with $I(\cdot)$ the indicator function and $X_1, \ldots, X_n$ an iid sample of parent random variable $X$ following a categorical distribution with $K$ categories such that $\mathbb{P}(X \le j) = P_j$, $1 \le j < K$, with $(P_1, \ldots, P_{K-1}) \in \mathcal{P}$ unknown and $n$ known. ($\mathbf{Y}$ follows thus a multinomial distribution with parameters $n$ and $p_j$, $1 \le j \le K$, where $p_j := P_j - P_{j-1}$, with $P_0 := 0$ and $P_K := 1$.)

It is straightforward to obtain that the relative likelihood of any $\mathbf{P} \in \mathcal{P}$ given $\mathbf{y}$ is $pl_{\mathbf{y}}^{\mathcal{P}}(\mathbf{P})$ with $pl_{\mathbf{y}}^{\mathcal{P}}$ the possibility distribution such that

$$pl_{\mathbf{y}}^{\mathcal{P}}(\mathbf{P}) = \left(\frac{P_1}{\hat{P}_1}\right)^{y_1} \left(\frac{1-P_{K-1}}{1-\hat{P}_{K-1}}\right)^{y_K} \prod_{j=2}^{K-1} \left(\frac{P_j - P_{j-1}}{\hat{P}_j - \hat{P}_{j-1}}\right)^{y_j}, \quad \forall \mathbf{P} \in \mathcal{P}, \quad (9)$$

with $\hat{P}_j = \sum_{i=1}^{j} y_i/n$ the MLE of $P_j$.

**Proposition 3.** $pl_{\mathbf{y}}^{\mathcal{P}}$ defined by (9) satisfies Assumptions 1-5.

Our second main result is that, when estimation uncertainty is characterized by (9), then Eq. (8) admits an algebraic expression:

**Theorem 2.** Under estimation uncertainty given by Eq. (9), $Bel^{\mathcal{Z}}(\llbracket i, j \rrbracket) =$

$$\begin{cases} 1, & \text{if } i = 1, j = K, \\ \hat{P}_j - \frac{B(\hat{P}_j; n_j+1, n-n_j+1)}{c_j}, & \text{if } i = 1, j < K, \\ 1 - \hat{P}_{i-1} - \frac{\overline{B}(\hat{P}_{i-1}; n_{i-1}+1, n-n_{i-1}+1)}{c_{i-1}}, & \text{if } 1 < i, j = K, \\ \hat{P}_j - \hat{P}_{i-1} - \frac{B(P_{i-1,j}; n_{i-1}+1, n-n_{i-1}+1) - B(\hat{P}_{i-1}; n_{i-1}+1, n-n_{i-1}+1)}{c_{i-1}} & \text{otherwise,} \\ \quad - \frac{B(\hat{P}_j; n_j+1, n-n_j+1) - B(P_{i-1,j}; n_j+1, n-n_j+1)}{c_j}, \end{cases}$$

(10)

where $c_j = \hat{P}_j^{n_j}(1-\hat{P}_j)^{n-n_j}$, $\overline{B}(z; a, b) = \underline{B}(1-z; b, a)$ and

$$P_{i,j} = \left(\left(\left(\frac{\hat{P}_j}{1-\hat{P}_j}\right)^{\hat{P}_j}\left(\frac{1-\hat{P}_i}{\hat{P}_i}\right)^{\hat{P}_i}\left(\frac{1-\hat{P}_j}{1-\hat{P}_i}\right)\right)^{-1/(\hat{P}_j - \hat{P}_i)} + 1\right)^{-1}, \quad \forall 1 \le i < j < K.$$

(11)

Theorem 2 is illustrated by Example 3.

*Example 3.* Example 1 corresponds to having observed $\mathbf{y} = (48, 17, 19, 11, 6, 9)$. The marginal contour functions $pl^{\mathcal{P}_j}$ and the mass function $m^{\mathcal{Z}}$ associated to the PBF $Bel^{\mathcal{Z}}$ are illustrated by Figs. 2a and 2b, respectively. The degrees of belief and of plausibility that the precipitation will be, e.g., below 1.25 inches, amounts to computing the quantities $Bel^{\mathcal{Z}}(\llbracket 1, 2 \rrbracket)$ and $Pl^{\mathcal{Z}}(\llbracket 1, 2 \rrbracket)$, where $Bel^{\mathcal{Z}}(\llbracket 1, 2 \rrbracket)$ is obtained easily using the second case of (10), and $Pl^{\mathcal{Z}}(\llbracket 1, 2 \rrbracket)$ is computed using the third case of (10) and the property $Pl^{\mathcal{Z}}(\llbracket 1, j \rrbracket) = 1 - Bel^{\mathcal{Z}}(\llbracket j+1, K \rrbracket)$ for all $1 \le j \le K$. We find $Bel^{\mathcal{Z}}(\llbracket 1, 2 \rrbracket) \approx 0.53$ and , $Pl^{\mathcal{Z}}(\llbracket 1, 2 \rrbracket) \approx 0.65$.

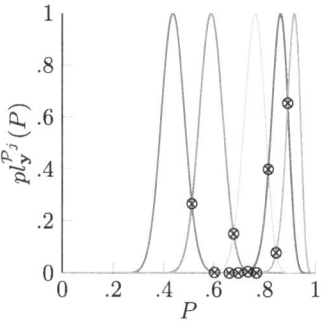
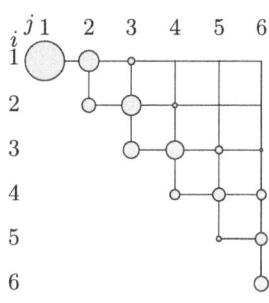

(a) Maginal contour functions: $pl_{\mathbf{y}}^{\mathcal{P}_1}(-)$, $pl_{\mathbf{y}}^{\mathcal{P}_2}(-)$, $pl_{\mathbf{y}}^{\mathcal{P}_3}(\ )$, $pl_{\mathbf{y}}^{\mathcal{P}_4}(-)$, $pl_{\mathbf{y}}^{\mathcal{P}_5}(-)$. Intersection points ($\otimes$) between marginals.

(b) Predictive mass values $m^{\mathcal{Z}}(\llbracket i,j \rrbracket)$ for all $1 \leq i \leq j \leq K$, are proportional to areas of corresponding circles.

**Fig. 2.** Marginal contour functions (Fig. 2b) and predictive mass function (Fig. 2b) for the Arizona January precipitation data.

## 5 Conclusion

In this paper, we considered the problem of the prediction of an ordinal variable, according to the relative likelihood-based evidential method for statistical inference and prediction. First, we established that this prediction can be computed, under some conditions, by integrating the marginals of the possibility distribution representing the estimation uncertainty in this method. Then, we showed that the prediction even admits an algebraic expression, when the estimation uncertainty is obtained from past observations of the variable.

Perspectives to this work include application of these results to ordinal classification problems, in the same vein as was done for binary and nominal classification problems [3,6,7,9,10], as well as extension (potentially only partially) to nominal variables.

## References

1. Dempster, A.P.: New methods for reasoning towards posterior distributions based on sample data. Ann. Math. Stat. 355–374 (1966)
2. Denœux, T.: Constructing belief functions from sample data using multinomial confidence regions. Int. J. Approx. Reason. **42**(3), 228–252 (2006)
3. Denœux, T.: Uncertainty quantification in logistic regression using random fuzzy sets and belief functions. Int. J. Approx. Reason. **168**, 109159 (2024)
4. Kanjanatarakul, O., Denœux, T., Sriboonchitta, S.: Prediction of future observations using belief functions: a likelihood-based approach. Int. J. Approx. Reason. **72**, 71–94 (2016)
5. Kanjanatarakul, O., Sriboonchitta, S., Denœux, T.: Forecasting using belief functions: an application to marketing econometrics. Int. J. Approx. Reason. **55**(5), 1113–1128 (2014)

6. Minary, P., Pichon, F., Mercier, D., Lefevre, E., Droit, B.: Evidential joint calibration of binary SVM classifiers. Soft. Comput. **23**, 4655–4671 (2019)
7. Ramel, S., Pichon, F., Delmotte, F.: A reliable version of Choquistic regression based on evidence theory. Knowl.-Based Syst. **205**, 106–252 (2020)
8. Shafer, G.: A Mathematical Theory of Evidence. Princeton University Press, Princeton (1976)
9. Xu, P., Davoine, F., Zha, H., Denœux, T.: Evidential calibration of binary SVM classifiers. Int. J. Approx. Reason. **72**, 55–70 (2016)
10. Xu, P., Davoine, F., Denœux, T.: Evidential multinomial logistic regression for multiclass classifier calibration. International Conference on Information Fusion, pp. 1106–1112 (2015)

# Information Fusion and Optimization

# Why Combining Belief Functions on Quantum Circuits?

Qianli Zhou[1], Hao Luo[1], Éloi Bossé[2,3], and Yong Deng[1,4(✉)]

[1] Institute of Fundamental and Frontier Science, University of Electronic Science and Technology of China, Chengdu 610054, China
dengentropy@uestc.edu.cn
[2] Department of Image and Information Processing, IMT-Atlantique, Brest 29238, France
[3] Expertises Parafuse, Québec City, QC G1W 4N1, Canada
[4] School of Medicine, Vanderbilt University, Nashville 37240, USA

**Abstract.** This paper aims to provide a unified Boolean algebra-based combination rule for belief functions and implement it on quantum circuits. From the perspective of operating steps, the advantages of combination rules for poss-transferable mass functions and general mass functions in quantum implementation are discussed. Through quantum implementation, the poss-transferable mass function can achieve constant-fold acceleration, while the general mass function can achieve exponential-fold acceleration.

**Keywords:** Information fusion · Dempster-Shafer theory · Quantum circuits · Boolean algebra-based combination rule · Poss-transferable mass function

## 1 Introduction

Quantum computation is implemented based on the fundamental assumptions of quantum mechanics using the quantum bit as the basic unit of information representation [7]. A $2^n$-dimensional quantum superposition state represented by $n$ quantum bits is represented by a vector under $n$ unit orthogonal bases. Compared to operations on classical bits, updating the amplitude and phase information of a quantum state utilizing unitary operators and unique entanglement operations are core information processing approaches on quantum circuits. Developing classical AI models on quantum circuits has become a hot topic in recent years [1]. The uncertainty inherent in quantum superposition states is not considered as randomness; however, it is common practice to encode probability distributions onto amplitudes in information representations [4]. Dempster-Shafer theory serve as effective tools for uncertainty reasoning, with their focal

---

This work was supported by the National Natural Science Foundation of China (Grant No.62373078).

sets being representable by binary codes. For an $n$-element frame of discernment (FoD), there are $2^n$ potential focal sets available for uncertainty modeling. Hence, there is a clear link between quantum computation and Dempster-Shafer theory in terms of information representation.

Relating quantum information to the Dempster-Shafer granule has been discussed by several scholars from various perspectives, such as lattice of subspace [12], mixed quantum states [2], complex-valued masses [13] and Dempster's rule [8]. However, the above approaches do not consider using state evolution to update the beliefs. In [15], Zhou *et al.* propose a method for encoding basic probability assignments (BPAs) onto quantum states and implementing them on quantum circuits. Utilizing the proposed encoding approach, the belief, plausibility and commonality functions can be efficiently extracted on quantum circuits by controlling not-gates (C-NOT gates). In the realm of information processing, belief function operations based on matrix calculus are executed on quantum circuits utilizing linear solvers such as HHL [15] and VQA [5]. However, these matrix calculus-based approaches have merely been transferred straightforwardly to general quantum algorithms, failing to leverage the specific advantages offered by the Dempster-Shafer structures on quantum circuits. Luo *et al.* implement common combination rules on quantum circuits from a set-theoretic perspective and develop a quantum evidential classifier, which achieves an exponential reduction in computational complexity compared to its classical counterpart [6]. In this paper, we define a type of combination rule for belief functions via Boolean algebra, and discuss its advantages of quantum implementation from the perspectives of poss-transferable mass function and general mass function, respectively.

The paper is organized as follows: Sect. 2 introduces basic notions of quantum computation and the encoding approaches of mass functions on quantum circuits. Section 3 proposes the Boolean algebra-based combination rule and implement it on quantum circuits. Section 4 demonstrates the advantages of combining belief functions on quantum circuits. In Sect. 5, we analyze further research directions for developing evidential information processing on quantum circuits.

## 2 Relation Between Mass Function and Quantum State

### 2.1 Quantum Computation

Compared to the classical bit, 0 and 1, a quantum bit can be written as a vector, $|q\rangle = a|0\rangle + b|1\rangle$, where $|0\rangle = [1,0]^T$, $|1\rangle = [0,1]^T$ and $|a|^2 + |b|^2 = 1$. For the state composed by $n$ quantum bits, the Cartesian product can combine them into a superposition state, i.e. $|q_1\rangle \otimes \cdots \otimes |q_n\rangle = |q_1 \cdots q_n\rangle$. In information updating, a unitary matrix $U$ satisfying $UU^\dagger = I$ is employed as an operator to act on the quantum state, where $\{\}^\dagger$ is the conjugate transpose operation and $I$ is the unit matrix. For the state with $n$ qubits, the corresponding dimension of evolution matrix is $2^n$. These necessary quantum gates are shown in Table 1 in [6]. This section only introduces the necessary knowledge about quantum computation for this paper, please refer to [7] for more details.

## 2.2 Dempster-Shafer Theory

For an uncertain variable $X$, its potential true value is located in the closed set $\Omega = \{\omega_1, \cdots, \omega_n\}$, denoted as frame of discernment (FoD). A mapping $m : 2^\Omega \to [0,1]$ satisfying $\sum_{F_i \subseteq \Omega} m(F_i) = 1$, is known as basic probability assignment (BPA), or mass function. When $m(F_i) > 0$, $F_i$ is a focal set, which represents the non-refineable belief mass to support $X \in F_i$ [3,10]. In an $n$-element FoD, the focal set $F_i$ can be expressed as an $n$-bit binary code, where each binary number indicates whether an element is present and $i$ is the corresponding decimal representation. For a 3-element, $F_6 = \{\omega_2 \omega_3\}$ corresponds to the binary code $\{110\}$. In addition to BPA, there are still some identical information content representations

$$b(F_i) = \sum_{F_j \subseteq F_i} m(F_j), \quad q(F_i) = \sum_{F_i \subseteq F_j} m(F_j), \quad (1)$$

$b$ function and $q$ function are dual with each other, which can implement the evidence combination rules in a simple way [11].

## 2.3 Belief Function on Quantum Circuits

By leveraging the consistency between quantum bits and quantum superposition states with binary-coded representations of focal sets, [15] presents an approach for encoding BPA onto quantum states. Consider a BPA $m$ under an $n$-element FoD, it can be written as

$$|m\rangle = \sqrt{m(F_0)}|0\cdots0\rangle + \sqrt{m(F_1)}|0\cdots1\rangle + \cdots + \sqrt{m(F_{2^n-1})}|1\cdots1\rangle. \quad (2)$$

Utilizing the correspondence between elements and qubits, the $b$ and $q$ functions can be efficiently extracted using the $C^n$-NOT gate.

$$|m\rangle = \sum_{i=0}^{2^n-1} \sqrt{m(F_i)} \left|t_i^n \cdots t_i^1\right\rangle |0\rangle \xrightarrow[C:t_i^n]{C-NOT} \sum_{\omega_n \notin F_i} \sqrt{m(F_i)} \left|t_i^n \cdots t_i^1\right\rangle |0\rangle + \sum_{\omega_n \in F_i} \sqrt{m(F_i)} \left|t_i^n \cdots t_i^1\right\rangle |1\rangle,$$

where $(t_i^n \cdots t_i^1)_2$ is the binary representation of $i$, and the amplitude of $|1\rangle$ and $|0\rangle$ are

$$\sqrt{\sum_{\omega_n \in F_i} m(F_i)} = \sqrt{q(\{\omega_n\})}, \quad \sqrt{\sum_{F_i \subseteq \{\omega_1 \cdots \omega_{n-1}\}} m(F_i)} = \sqrt{b(\{\omega_1 \cdots \omega_{n-1}\})}. \quad (3)$$

# 3 Boolean Algebra-Based Operations on Quantum Circuits

## 3.1 Boolean Algebra-Based Combination Rules

According to the Boolean algebra operations, a unified multi-source combination rule is defined, which covers the conjunctive, disjunctive combination rules, and their exclusive versions [9].

**Definition 1 (Boolean algebra-based combination rule).** *Consider BPAs $m_1 \cdots m_k$ under an $n$-element FoD from $k$ distinct sources, their Boolean algebra-based combination rule (BACR) can be written as*

$$m_1 \odot \cdots \odot m_k(F_{i_{k+1}}) = m_{k+1}(F_{i_{k+1}}) = \sum_{\mathfrak{B}(F_{i_1}, \cdots, F_{i_k}) = F_{i_{k+1}}} \prod_{j=1}^{k} m_j(F_{i_j}), \quad (4)$$

*where $F_{i_j}$ indicates the focal set of $m_j$, $i_j \in \{0, 1, \cdots, 2^n - 1\}$ and $\mathfrak{B}$ is a mapping $[F_0, F_{2^n-1}]^k \to [F_0, F_{2^n-1}]^1$, which is composed by elementary Boolean operations ($\vee$, $\wedge$ and $\neg$).*

*Remark 1.* The $\mathfrak{B}(F_{i_1}, \cdots, F_{i_k}) = F_{i_{k+1}}$ also can be written as operating on elements. Consider an $n$-element FoD, its focal set can be written as $F_i = (t_i^n, \cdots, t_i^1)_2$, where $t_i^d$ indicates whether the element $\omega_d$ in focal set $F_i$. Hence, the Boolean operation also can be implemented through their binary representation $\mathfrak{B}(F_{i_1}, \cdots, F_{i_k}) = (\mathfrak{B}(t_{i_1}^n, \cdots, t_{i_k}^n), \cdots, \mathfrak{B}(t_{i_1}^1, \cdots, t_{i_k}^1))_2$, where $F_{i_k} = (t_{i_k}^n, \cdots, t_{i_k}^1)_2$.

*Example 1.* Consider three BPAs $m_i$, $i = \{1, 2, 3\}$, where $m_1 = \{0, 0.1, 0.2, 0.7\}$, $m_2 = \{0.25, 0.06, 0.27, 0.42\}$, and $m_3 = \{0.07, 0.05, 0.16, 0.72\}$. If the Boolean operation is $\mathfrak{B}(F_{i_1}, F_{i_2}, F_{i_3}) = (F_{i_1} \wedge F_{i_2}) \vee (F_{i_1} \wedge F_{i_3}) \vee (F_{i_2} \wedge F_{i_3})$, according to the Eq. (4), the outcome of BACR is $m_4 = \{0.027, 0.046, 0.195, 0.732\}$.

**Proposition 1.** *When the sources provide certain but imprecise information, i.e. categorical mass function, $m(F_i) = 1$, BACR can be implemented using classical Boolean circuits.*

*Proof.* A categorical mass function can be seen as a classical subset of the FoD or a digital signal with binary encoded. According to Remark 1, the composition of AND gate, OR gate, and NOT gate can implement arbitrary Boolean operations on $F_i$, enabling the implementation of the BACR. □

### 3.2 BACR on Quantum Circuits

When the sources provide uncertain and imprecise information, i.e., when classical sets become random sets, classical Boolean circuits are unable to handle these logical operations. In quantum computing, superposition states represent the uncertainty of states. Although their interpretation differs from that of belief functions, as both are normalized weights, it is logical and natural to extend the use of quantum computing to implement the operations of belief functions.

*Remark 2.* Consider a Boolean operation $\mathfrak{B}(F_1, \cdots, F_k)$, it can be divided into $t$-layer operations, where $t$ denotes the minimum number of Boolean algebra operations that need to implement $\mathfrak{B}$. Boolean algebra operations that can be performed simultaneously are placed in a layer, e.g., $(\{\omega_1\omega_2\} \cap \{\omega_2\omega_3\}) \cup (\{\omega_2\omega_4\} \cap \{\omega_4\omega_5\})$, where $\{\omega_1\omega_2\} \cap \{\omega_2\omega_3\}$ and $\{\omega_2\omega_4\} \cap \{\omega_4\omega_5\}$ compose the first layer and $\{\omega_2\} \cap \{\omega_4\}$ is the second layer.

---

[1] If $i \in [0, 2^n-1], F_i \in [F_0, F_{2^n-1}]$.

**Definition 2 (BACR on quantum circuits).** *Consider $k$ BPAs under an $n$-element FoD. Their BACR's Boolean operation $\mathfrak{B}$ can be divided into t-layer operations, as shown in Remark 2. In Fig. 1, the left part shows the implementation of their BACR on the quantum circuits, where $L_i$ represents the circuit which implement Boolean operation on the i-th layer. The right part shows the implementation of elementary Boolean operations on quantum circuits, where $\wedge$, $\vee$, and $\neg$ correspond to the conjunctive combination rule, disjunctive combination rule, and negation, respectively.*

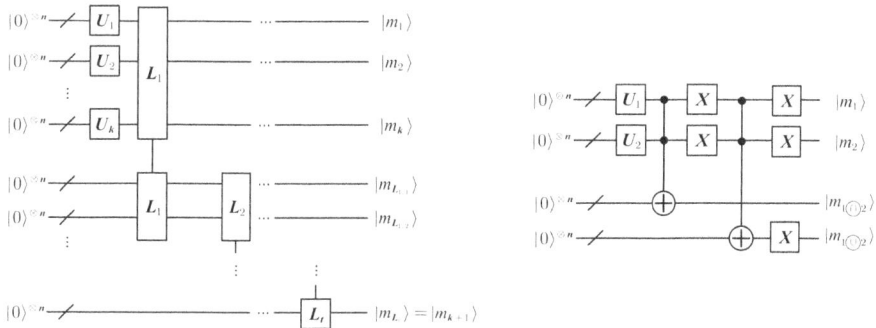

**Fig. 1.** In terms of the left part, the implementation of BACR on quantum circuits is shown. In terms of the right part, the implementation of elementary Boolean operations($\wedge$, $\vee$, and $\neg$) is shown.

**Proposition 2.** *The amplitude of $|m_{k+1}\rangle$ in Definition 2 equals the outcome of combination in the Definition 1, which satisfies*

$$|m_{k+1}\rangle = \sum_{i_{k+1}=0}^{2^n-1} \sqrt{\sum_{\mathfrak{B}(F_{i_1},\cdots,F_{i_k})=F_{i_{k+1}}} \prod_{j=1}^{k} m_j(F_{i_j})} \left|t^n_{i_{k+1}} \cdots t^1_{i_{k+1}}\right\rangle.$$

*Proof.* According to Definition 2, for multi-layer Boolean operations, quantum circuits are constructed using loops, thus allowing for generalization to any number of layers of Boolean operations by demonstrating the realizability of one layer of Boolean operations. Consider a 1-layer Boolean operation $\mathfrak{B}$ operating on $|m_1\rangle \cdots |m_k\rangle$, the initial state $|\psi_0\rangle$ is

$$|\psi_0\rangle = |m_1\rangle \cdots |m_k\rangle |0\rangle^n = \sum_{i_1=0}^{2^n-1} \cdots \sum_{i_k=0}^{2^n-1} \prod_{r=1}^{k} \sqrt{m_r(F_{i_r})} |F_{i_1}\rangle \cdots |F_{i_k}\rangle |0\rangle^n.$$

Consider to operate the $n$th qubit $t_{k+1}^n = \mathfrak{B}(t_1^n, \cdots, t_k^n)$, the $|\psi_0\rangle$ will be evolved to $|\psi_1\rangle$

$$|\psi_1\rangle = \sum_{t_{k+1}^n=0}^{2^n-1} \sum_{i_1=0}^{2^n-1} \cdots \sum_{i_k=0}^{2^n-1} \prod_{r=1}^{k} \sqrt{m_r(F_{i_r})} |F_{i_1}\rangle \cdots |F_{i_k}\rangle |0\rangle |0\rangle^{n-1}$$

$$+ \sum_{t_{k+1}^n=1}^{2^n-1} \sum_{i_1=0}^{2^n-1} \cdots \sum_{i_k=0}^{2^n-1} \prod_{r=1}^{k} \sqrt{m_r(F_{i_r})} |F_{i_1}\rangle \cdots |F_{i_k}\rangle |1\rangle |0\rangle^{n-1},$$

when all operations have been applied, the $|\psi_n\rangle$ is

$$|\psi_n\rangle = \sum_{i_{k+1}=0}^{2^n-1} \sum_{\mathfrak{B}(t_1^n,\cdots,t_k^n)=t_{k+1}^n} \cdots \sum_{\mathfrak{B}(t_1^1,\cdots,t_k^1)=t_{k+1}^1} \prod_{r=1}^{k} \sqrt{m_r(F_{i_r})} |F_{i_1}\rangle \cdots |F_{i_k}\rangle |t_{i_{k+1}}^n \cdots t_{i_{k+1}}^1\rangle,$$

and the amplitude of $\left|t_{i_{k+1}}^n \cdots t_{i_{k+1}}^1\right\rangle$ is

$$\sqrt{\sum_{\mathfrak{B}(t_1^n,\cdots,t_k^n)=t_{k+1}^n} \cdots \sum_{\mathfrak{B}(t_1^1,\cdots,t_k^1)=t_{k+1}^1} \prod_{r=1}^{k} m_r(F_{i_r})} = \sqrt{\sum_{\mathfrak{B}(F_{i_1},\cdots,F_{i_k})=F_{i_{k+1}}} \prod_{r=1}^{k} m_r(F_{i_r})}, \quad (5)$$

which is equal with the amplitude in Definition 2. □

*Example 2.* Continue to the Example 1, it can be implemented on the quantum circuits through the following steps:

1. Implement the $m_i$s on the quantum circuits:

$$|0\rangle^{\otimes 6} \xrightarrow{U_{m_1} \otimes U_{m_2} \otimes U_{m_3}} |m_1\rangle |m_2\rangle |m_3\rangle.$$

2. Execute the $C^2-$ NOT gates to implement the conjunctions:

$$|m_1 m_2 m_3\rangle |0\rangle^{\otimes 6} \to |m_i\rangle \left|m_{1 \odot 2}\right\rangle \left|m_{1 \odot 3}\right\rangle \left|m_{2 \odot 3}\right\rangle := |m_i\rangle |m_{L_{1,1}} m_{L_{1,2}} m_{L_{1,3}}\rangle$$

3. Execute the $X$ gate on $|m_{L_{1,i}}\rangle$s and $C^2-$ NOT gates on initial qubits to implement the disjunction:

$$|m_{L_{1,1}} m_{L_{1,2}} m_{L_{1,3}}\rangle |0\rangle^{\otimes 6} \to |\overline{m}_{L_{1,i}}\rangle \left|m_{\overline{L}_{1,1} \odot \overline{L}_{1,2} \odot \overline{L}_{1,3}}\right\rangle := |\overline{m}_{L_{1,i}}\rangle |m_4\rangle.$$

Through simulating the above circuit on the Qiskit platform and measuring the bits corresponding to $m_4$ with 8096 shots, the outcome probability distribution is $p(00) = 0.027, p(11) = 0.045, p(10) = 0.195, p(11) = 0.733$.

In Sect. 3, we present a unified form of the combination rules based on Boolean operations in belief functions and demonstrate its feasibility on quantum circuits. In particular, the slight error between the measurements in Example 2 and the classical version in Example 1 comes from the quantum uncertainty, not from theoretical errors.

## 4 Why Combining Belief Functions on Quantum Circuits?

For ease of analysis, we limit our discussion in this section to conjunctive combination rules. For the BACR with multi-layer Boolean operations, the complexity increase rates of the classical and quantum versions are the same.

### 4.1 Poss-Transferable Mass Function

In [14], a new type of mass function, called poss-transferable mass function is developed based on the Cartesian product. Consider an possibility distribution $\pi$ under an $n$-element FoD $\Omega$, the Cartesian product-based belief function transformation (CP-BFT) is

$$m(F_i) = \prod_{r=1:n;t^r=1} \pi(\omega_r) \prod_{r=1:n;t^r=0} (1 - \pi(\omega_r)), \quad F_i = (t^n, \cdots, t^r, \cdots, t^1)_2.$$

which also can be inverted to an $n$-dimensional $\pi$ using canonical decomposition.

*Remark 3.* Both the consonant mass function and the poss-transferable mass function can be written as possibility distributions. If a consonant mass function and a poss-transferable mass function are derived from a possibility distribution, their contour functions equal with the possibility distribution. They differ in the different triangular norms adopted in expanding from the singletons' possibility to the multi-element propositions' possibility, and their specific differences are discussed in detail in [14].

In the initial quantum state implementation, for a state implemented using only single quantum bit gates, i.e., there is no quantum gate containing control bit, it is called a separable quantum state. For separable quantum states, applying the unitary gate to each qubit separately and then generating the composite state yields the same result as combining the unitary gates first and then applying them to all the qubits simultaneously, i.e., $U_1 \left|0\right\rangle \otimes \cdots \otimes U_n \left|0\right\rangle = U_1 \otimes \cdots \otimes U_n \left|0 \cdots 0\right\rangle$.

**Proposition 3.** *Poss-transferable mass function can be implemented as a separable quantum state.*

*Proof.* For a qubit with initial state $\left|0\right\rangle$, it can be implemented as a possibility measure via a $R_Y$ gate $R_Y(\theta) = \begin{bmatrix} \cos(\theta/2) & \sin(\theta/2) \\ -\sin(\theta/2) & \cos(\theta/2) \end{bmatrix}$, i.e., $R_Y(\theta_r) \left|0\right\rangle = \cos(\theta_r/2) \left|0\right\rangle - \sin(\theta_r/2) \left|1\right\rangle$ corresponding to $\pi(\omega_r) = \sin^2(\theta_r/2)$. Extending to the $n$-element case

$$R_Y(\theta_1) \otimes \cdots \otimes R_Y(\theta_n) \left|0 \cdots 0\right\rangle = \sum_{i=0}^{2^n-1} \prod_{r=1:n;t^r=1} \sqrt{\pi(\omega_r)} \prod_{r=1:n;t^r=0} \sqrt{1 - \pi(\omega_r)} \left|t^n \cdots t^1\right\rangle,$$

where $i = (t^n \cdots t^1)_2$. Hence the amplitude of state $\left|t^n \cdots t^1\right\rangle$ equals $\sqrt{m(F_i)}$. □

In [14], the product t-norm and probability t-conorm for possibility distributions have been proven to be equivalent to the conjunctive and disjunctive combination rules for their corresponding poss-transferable BPAs, respectively.

**Proposition 4.** *For $k$ poss-transferable mass functions, it is necessary to perform $(k-1) \times n$ times multiplications to implement conjunction or disjunction, and on quantum circuits, there are $n$ $C^k$ – NOT gates are needed to implement the target state.*

*Proof.* In the realm of classical version, since the CP-BFT satisfies that the contour function $pl$ of $m$ is same with the possibility distribution $\pi$, the combination can be written as $m_1 \odot \cdots \odot m_k = \text{CP} - \text{BFT}^{-1}(pl_1 \circ \cdots \circ pl_k)$, where $\text{CP} - \text{BFT}^{-1}$ is the inverse transformation of CP-BFT. Since $pl_1 \circ \cdots \circ pl_k$ has $(k-1) \times n$ multiplications, the conjunctive combination rule of poss-transferable BPAs needs to perform $(k-1) \times n$ times multiplications. In the realm of quantum computation, according to Definition 2, since a BPA needs $n$ qubits to be represented, and a $C^k$ – NOT gate can implement an AND operator for qubits. There are $n$ $C^k$ – NOT gates are necessary for implement the fusion result. □

According to Proposition 4, for the poss-transferable mass function, from the perspective of operating steps, quantum computation achieves a constant-fold speedup. However, on existing quantum simulation platforms, gates with more than three control bits cannot be directly implemented. Therefore, the fusion of $k$ poss-transferable mass functions requires a quantum circuit with a depth of $(k-1)$. From the perspective of circuit depth, the conjunction between poss-transferable mass functions does not offer a significant advantage on quantum circuits.

### 4.2 General Mass Function

For the general mass function, the $q$ function can implement conjunction through a more simple way. Consider $k$ BPAs from distinct sources, their conjunction can be written as $q_1 \odot \cdots \odot q_k = q_1 \circ \cdots \circ q_k$. For an $n$-element FoD, the combination needs $(k-1) \times 2^n$ times multiplications. However, their conjunction on quantum circuits is same with the poss-transferable case, hence it also need $n$ operations or $(k-1) \times n$ depth quantum circuits to implement. Table 1 shows the difference between classical and quantum conjunctive rules, which is evident to show the advantages of combining belief function on quantum circuits.

**Proposition 5.** *For non-poss-transferable mass functions, implementing their quantum states requires gates that produce entanglement.*

*Proof.* The proposition also can be described as for the non-separable quantum state, its accurate simulation must need multi-quantum gates, which is a widely accepted conclusion in the field of quantum computing. □

**Table 1.** Comparison of quantum and classical conjunctive combination rules.

|  | Poss-transferable mass function | General mass function |
|---|---|---|
| Multiplication times | $(k-1) \times n$ | $(k-1) \times 2^n$ |
| Operation times | $n$ | $n$ |
| Circuit depth | $(k-1) \times n$ | $(k-1) \times n$ |

Multi-qubit gates enable entanglement operations of qubits, as unique operation in quantum computation, which cannot be realised by classical arithmetic. Continue to Proposition 2, if the amplitude of each quantum bit in $|m_{k+1}\rangle$ is measured separately, for the qubit representing $\omega_r$, based on the Eq. (3), it can be written as $|t^r\rangle = \sqrt{q(\{\omega_r\})}|1\rangle + \sqrt{b(\Omega\backslash\{\omega_r\})}|0\rangle$. They are then combined into new quantum states via the Cartesian product:

$$|t^n\rangle \cdots |t^1\rangle = \sqrt{\prod_{r=1}^{n} b(\Omega\backslash\{\omega_r\})}|F_0\rangle + \cdots + \sqrt{\prod_{r=1}^{n} q(\{\omega_r\})}|F_{2^n-1}\rangle,$$

whose amplitude of the state representing focal set is evident diffident with Eq. (5). Thus, for the general mass function on quantum circuits, the result of measuring the qubits of the quantum state separately and then compose them via Cartesian product is different from measuring entire quantum state directly. The partial information of general mass function is implicit in the quantum entanglement under the Hilbert space, and if the measurement collapses the Hilbert space midway, this part of information will be lost. The proposal is to directly connect the Hilbert space of individual mass functions using controlled bits, and then the entangled information is transferred to the target qubits. This also interprets why quantum circuits require only $n$-step operations for the $2^n$-dimensional distribution combination, whilst $2^n$-step are required for classical operations.

## 5 Conclusion

In this paper, we have proposed a unified form of combination rule and prove its feasibility on quantum circuits. More importantly, from the perspective of the number of operations, for the poss-transferable mass functions, a constant-fold speedup will be achieved on quantum circuits, and for general mass functions, an exponential-fold speedup will be achieved. It is worth noting that this advantage of processing information on quantum circuits is most evident with power set-structured information distributions, as it not only makes full use of all the states represented by the $n$ qubits, but also enables a semantic correspondences between qubits and elements. In addition, Dempster-Shafer theory, as a theory of uncertainty founded on power set representations, its information processing is proposed based on the Boolean algebra, which forms a natural tacit understanding with the quantum computing.

# References

1. Biamonte, J., Wittek, P., Pancotti, N., Rebentrost, P., Wiebe, N., Lloyd, S.: Quantum machine learning. Nature **549**(7671), 195–202 (2017)
2. Deng, X., Jiang, W.: Quantum representation of basic probability assignments based on mixed quantum states. In: 2021 IEEE 24th International Conference on Information Fusion (FUSION), pp. 1–6. IEEE (2021)
3. Denœux, T., Dubois, D., Prade, H.: Representations of uncertainty in AI: beyond probability and possibility. In: Marquis, P., Papini, O., Prade, H. (eds.) Guided Tour of Artificial Intelligence Research, pp. 119–150. Springer, Cham (2020). https://doi.org/10.1007/978-3-030-06164-7_3
4. Low, G.H., Yoder, T.J., Chuang, I.L.: Quantum inference on Bayesian networks. Phys. Rev. A **89**(6), 062315 (2014)
5. Luo, H., Zhou, Q., Li, Z., Deng, Y.: Variational quantum linear solver-based combination rules in Dempster-Shafer theory. Inf. Fusion **102**, 102070 (2024)
6. Luo, H., Zhou, Q., Pan, L., Li, Z., Deng, Y.: Attribute fusion-based evidential classifier on quantum circuits. arXiv preprint arXiv:2401.01392 (2024)
7. Nielsen, M.A., Chuang, I.L.: Quantum Computation and Quantum Information. Cambridge University Press, Cambridge (2010)
8. Pan, L., Gao, X., Deng, Y.: Quantum algorithm of dempster rule of combination. Appl. Intell. **53**(8), 8799–8808 (2023)
9. Pichon, F.: Belief functions: canonical decompositions and combination rules. HAL **2009**(0) (2009). http://dml.mathdoc.fr/item/tel-00793859
10. Shafer, G.: A Mathematical Theory of Evidence, vol. 42. Princeton University Press, Princeton (1976)
11. Smets, P.: Belief functions: the disjunctive rule of combination and the generalized Bayesian theorem. Int. J. Approximate Reasoning **9**(1), 1–35 (1993). https://doi.org/10.1016/0888-613X(93)90005-X
12. Vourdas, A.: Quantum probabilities as Dempster-Shafer probabilities in the lattice of subspaces. J. Math. Phys. **55**(8) (2014)
13. Xiao, F.: Generalized quantum evidence theory. Appl. Intell. **53**(11), 14329–14344 (2023)
14. Zhou, Q., Deng, Y., Yager, R.R.: CD-BFT: canonical decomposition-based belief functions transformation in possibility theory. IEEE Trans. Cybern. **54**(1), 611–623 (2023)
15. Zhou, Q., Tian, G., Deng, Y.: BF-QC: belief functions on quantum circuits. Expert Syst. Appl. **223**, 119885 (2023)

# SHADED: Shapley Value-Based Deceptive Evidence Detection in Belief Functions

Haifei Zhang

Université de technologie de Compiègne, UMR-CNRS 7253 Heudiasyc, Compiègne, France
`haifei.zhang@hds.utc.fr`

**Abstract.** Deceptive evidence detection is an important issue in the theory of belief functions, which can be used to solve the problem of conflicts among evidence and to assess the credibility of evidence sources. In this paper, we first define strong and weak deceptive evidence. Then, we propose a deceptive evidence detection approach that directly investigates the process of Dempster's combination rule and decision-making based on the pignistic transformation. It can distinguish between strong and weak deceptive evidence and assess the importance of each piece of evidence. Several numerical examples are used to illustrate the effectiveness and efficiency of our proposed approach.

**Keywords:** Dempster-Shafer theory · Deceptive evidence detection · Conflict management · Shapley value · Information fusion

## 1 Introduction

Dempster-Shafer theory (DST) [2,11], also known as theory of belief functions or theory of evidence, is a mathematical framework designed for combining and reasoning with uncertain information from multiple sources and has been applied in various fields, such as classification [3,4,20,21] and fault diagnosis [16,17].

In DST, combining evidence from different sources is a key issue, which provides the basis for subsequent reasoning. There are several available combination methods, among which Dempster's combination rule (DCR) is a fundamental one [2]. However, when confronted with high-conflicting evidence, using DCR may yield counter-intuitive fusion results, partly because it allocates conflicts totally to the empty set, and partly because of the unreliability of evidence sources. To address this issue, many previous works have been presented. The first strategy attempts to modify and improve the evidence combination rules by reallocating the conflict [5,7,13,18], while the second one focuses on evaluating the reliability of pieces of evidence and discounting them before the combination process [8,19].

Rather than considering all evidence as in the approaches described above, recent studies have started to investigate whether a piece of evidence should be discarded in the fusion process [6]: if a piece of evidence is determined as deceptive, then it should not be combined. In reality, deceptive evidence can arise for a variety of reasons, which may be deliberate deception (e.g., from human sources) or natural causes (e.g., failure of sensors). It will be dangerous if we trust any conclusion derived from the combination of different pieces of evidence without knowing if there are deceptive ones. Therefore, detecting deceptive evidence is crucial, not only to address the problem of conflict among evidence but also to assess the credibility of evidence sources.

As simple and obvious deceptive evidence, the "negation or the complement of true evidence" has been presented and investigated by several researchers [9,15]. However, in real-world applications, deceptive evidence is more complicated than negation. For this issue, Schubert proposed to detect deceptions using the falsity degree of each piece of evidence based on conflict and entropy, as well as their combination [10]. Schubert's approach is limited to considering only the interaction between each piece of evidence and the remaining evidence. Subsequently, Zhou et al. [22] introduced a deception detection method by calculating the Shapley value of each piece of evidence based on the distances between each pair of two subsets of evidence. Cui et al. [1] proposed a method of belief gravitational clustering for this problem, using a similarity measure based on the entropy of evidence and the conflict coefficient between two pieces of evidence. A notable limitation of these methods is their inability to connect the criteria for assessing evidence credibility with Dempster's combination rule.

Recently, Kang and Zhao [6] provided the definition of deceptive evidence and proposed to detect deceptive evidence using reinforcement learning with off-policy Q-learning: rewarding the reduction of uncertainty and the production of reasonable decisions. This approach does not assess the credibility of the evidence, nor does it distinguish between strong deceptive evidence (which can alter the decision) and weak deceptive evidence (which does not change the decision but increases the uncertainty of the decision). In addition, it may be ineffective in identifying deceptive evidence that is difficult to detect.

Considering all the issues mentioned above, we propose a new approach based on Shapley values associated with the uncertainty of combined results for deceptive evidence detection, which is designed for the following objectives:

1. to directly investigate the process of Dempster's combination rule and decision-making based on the pignistic transformation;
2. to distinguish between strong and weak deceptive evidence;
3. to provide an importance assessment for each piece of evidence, which can be further used to conduct a weighted averaging of credible evidence.

The rest of this paper is structured as follows. Section 2 briefly reviews DST and Shapley value. Our proposed method is presented in Sect. 3 and tested with several examples in Sect. 4. Finally, a conclusion is drawn in Sect. 5.

## 2 Preliminaries

### 2.1 Theory of Belief Functions

Let $\Omega = \{\omega_1, \ldots, \omega_K\}$, $K \geq 2$, be the *frame of discernment*. Given a piece of evidence, the information is represented by a *mass function*, which is a mapping $m : 2^\Omega \to [0, 1]$, such that $m(\emptyset) = 0$ and $\sum_{A \subseteq \Omega} m(A) = 1$.

For any subset $A \subseteq \Omega$, the uncertainty of the proposition that the true state lies in $A$ can be quantified by the degrees of *belief* $Bel(A)$ and *plausibility* $Pl(A)$:

$$Bel(A) = \sum_{B \subseteq A} m(B) \text{ and } Pl(A) = \sum_{B \cap A \neq \emptyset} m(B). \quad (1)$$

$Bel(A)$ and $Pl(A)$ measure the support (belief) and compatibility (plausibility), respectively, associated with the proposition that the truth lies in $A$.

Given two independent mass functions $m_1$ and $m_2$ defined on the same problem, they can be combined via *Dempster's combination rule* (DCR) [2]:

$$m(A) = (m_1 \oplus m_2)(A) = \frac{1}{1-\mathcal{K}} \sum_{B \cap C = A} m_1(B) m_2(C), \quad \forall A \subseteq \Omega, \quad (2)$$

where $\mathcal{K}$ is the degree of conflict between the two mass functions, defined as:

$$\mathcal{K} = \sum_{B \cap C = \emptyset} m_1(B) m_2(C). \quad (3)$$

Dempster's combination rule is also called the *orthogonal sum* of $m_1$ and $m_2$, and requires that the mass functions to be combined are independent and their conflict is smaller than one. This operation is commutative and associative, which makes it possible to sequentially combine a series of evidence in any order.

The combined mass function $m$ can be transformed into a probability distribution through the *pignistic transformation* [14]:

$$BetP(\omega_k, m) = \sum_{A \subseteq \Omega, \omega_k \in A} \frac{m(A)}{|A|}, \quad \forall \omega_k \in \Omega, \quad (4)$$

in which the mass of focal sets is equally assigned to their elements. Then, the state of nature that maximizes the pignistic probability is taken as the decision.

### 2.2 Shapley Values

The *Shapley value* is a concept from cooperative game theory, which offers a fair way to distribute the total gains to the players based on their individual contributions to the collective outcome of a coalitional game [12]. Formally, a coalitional game is defined as $(\mathbf{N}, v)$ where $\mathbf{N}$ is a set of $n$ players and payoff function $v : 2^{\mathbf{N}} \to \mathbb{R}$ maps the subsets of players (coalitions) to a real number. The Shapley value (marginal contribution) of player $i$ is defined as:

$$\varphi_i = \sum_{S \subseteq \mathbf{N} \setminus \{i\}} \frac{|S|!(n-|S|-1)!}{n!} \left[ v(S \cup \{i\}) - v(S) \right]. \quad (5)$$

## 3 SHADED: SHApley Value-Based Deceptive Evidence Detection

### 3.1 Definition of Deceptive Evidence

Following the definition of deceptive evidence in [6], we refine it into strong deceptive evidence and weak deceptive evidence. This refinement allows us to better understand deceptive evidence, as well as to better manage them.

**Definition 1 (Strong deceptive evidence).** For a given set of credible evidence and a certain fusion system (i.e., Dempster's combination rule ), strong deceptive evidence is a piece of or a group of evidence that can alter the decision.

**Definition 2 (Weak deceptive evidence).** For a given set of credible evidence and a certain fusion system (i.e., Dempster's combination rule), weak deceptive evidence is a piece of or a group of evidence that can not alter the decision but can increase the uncertainty associated with the decision.

In the next section, we will delve into a Shapley value-based approach designed to identify deceptive evidence, as well as to distinguish between strong and weak deceptive evidence based on the above definitions.

### 3.2 Detection of Deceptive Evidence

In this paper, for a given collection of $n$ mass functions $\mathbf{M} = \{m1, m2, \ldots, m_n\}$, we consider that the combined decision is the state of nature with the highest pignistic probability associated with the combined mass function via DCR:

$$cd = \arg\max_{\omega \in \Omega} BetP(\omega, \bigoplus_{i=1}^{n}[m_i]). \tag{6}$$

However, the combined decision may be unreasonable because of the conflict in evidence. Therefore, the first step of our method is to determine the reasonable decision by conflict resolution. We apply Murphy's approach [8], which can maintain the majority opinion, and attain the convergence of the certainty on the decision. It combines the $n$ identical averaged mass functions using DCR:

$$rd = \arg\max_{\omega \in \Omega} BetP(\omega, \bigoplus_{i=1}^{n}[m]), \text{ where } m = \frac{1}{n}\sum_{i=1}^{n} m_i. \tag{7}$$

The second step is to calculate the Shapley value of each mass function. For a coalition (set) of mass functions $S$, the payoff function $v$ is defined as the pignistic probability associated with the reasonable decision. Thus the Shapley value of the mass function $m_i, i = 1, \ldots, n$, can be calculated by

$$\varphi_i = \sum_{S \subseteq \mathbf{M}\setminus\{m_i\}} \frac{|S|!(n - |S| - 1)!}{n!} \left[ BetP(rd, \bigoplus_{m_j \in S'}[m_j]) - BetP(rd, \bigoplus_{m_j \in S}[m_j]) \right], \tag{8}$$

where $S' = S \cup \{m_i\}$. Moreover, we set $BetP(rd, \emptyset) = 1/K$ with $K$ the number of states of nature, thus $\varphi_0 = 1/K$. Here, $\varphi_0$ represents the utility (payoff) of the evidence combination (coalitional game) when there is no evidence provided or only a piece of vacuous evidence is provided, which is equivalent to the case of making a decision with total ignorance.

Based on the above setting and the properties of Shapley value, we can distinctly draw the following propositions:

1. *efficiency*: the pignistic probability associated with the combined mass function and the reasonable decision is equal to the sum of the Shapley values of the given mass functions, i.e., $BetP(rd, \underset{i=1}{\overset{n}{\oplus}}[m_i]) = \sum_{i=0}^{n} \varphi_i$;
2. *symmetry*: two identical mass functions have the same Shapley value, i.e., if $m_i = m_j$, then $\varphi_i = \varphi_j$;
3. *dummy*: the Shapley value of a vacuous mass function is zero because it brings no contribution to any coalition in the evidence combination process, i.e., if $m_i(\Omega) = 1$, then $\varphi_i = 0$.

The third step involves determining the nature of each piece of evidence. The set of credible evidence (having positive or zero contributions to the reasonable decision), denoted as $CE$, includes all evidence with non-negative Shapley values: $CE = \{m_i \mid \varphi_i \geq 0, m_i \in \mathbf{M}\}$. Consequently, the set of deceptive evidence (DE) comprises all evidence having negative contributions (with negative Shapley values) to the reasonable decision, i.e., $DE = \mathbf{M} \backslash CE$. Next, we differentiate between strong and weak deceptive evidence. If the decision derived from combining all evidence matches the reasonable decision, it implies the absence of strong deceptive evidence (*SDE*), rendering all deceptive evidence as weak (*WDE*), i.e., $WDE = DE$ and $SDE = \emptyset$. Otherwise, there must be strong deceptive evidence and it necessitates examining all subsets of $DE$ by combining them with $CE$ to determine the piece of or the group of strong deceptive evidence. In this case, $WDE = DE \backslash SDE$.

The Algorithm 1 summarises the three steps described above[1]. The Shapley values of the credible evidence can be used to calculate the weights for credible evidence:

$$w_j = \frac{\varphi_j}{\sum_{m_i \in CE} \varphi_i}, \forall\ m_j \in CE. \tag{9}$$

Then, the final fusion result is

$$p(\omega_k) = BetP(\omega_k, \underset{i=1}{\overset{|CE|}{\oplus}}[m]), \text{ where } m = \sum_{m_j \in CE} w_j \times m_j. \tag{10}$$

## 4 Experiments

In this section, we provide several numerical examples to illustrate the effectiveness and efficiency of our proposed deceptive evidence detection method.

---

[1] Our code is available on GitHub: https://github.com/Haifei-ZHANG/shaded.

**Algorithm 1:** SHApley value-based Deceptive Evidence Detection

**Input:** A set of $n$ mass functions $\mathbf{M} = \{m_1, \ldots, m_n\}$
**Output:** Set of credible evidence, strong and weak evidence
1 Calculate the combined decision $cd$ via Eq. (6)
2 Calculate the reasonable decision $rd$ via Eq. (7)
3 **for** $i = 1, \ldots, n$ **do**
4 $\quad$ Calculate Shapley value $\varphi_i$ via Eq. (8)
5 $CE = \{m_i \mid \varphi_i \geq 0, m_i \in \mathbf{M}\}$ $\quad$ // set of credible evidence
6 $DE = \mathbf{M} \setminus CE$ $\quad$ // set of deceptive evidence
7 **if** $cd = rd$ **then**
8 $\quad$ $WDE = DE$ $\quad$ // set of weak deceptive evidence
9 $\quad$ $SDE = \emptyset$. $\quad$ // set of strong deceptive evidence
10 **else**
11 $\quad$ $SDE = \{\}$
12 $\quad$ **for** all $S \subseteq DE$ **do**
13 $\quad\quad$ $S' = CE \cup S$
14 $\quad\quad$ **if** $rd \neq \arg\max_{\omega \in \Omega} BetP(\omega, \oplus_{m_j \in S'}[m_j])$ **then**
15 $\quad\quad\quad$ $SDE = SDE \cup S$
16 $\quad$ $WDE = DE \setminus SDE$
17 Return $(CE, SDE, WDE)$

### 4.1 Effectiveness Comparison

The first example is drawn from the article of Cui et al. [1] for the fault diagnosis. Five sensors can provide evidence about three kinds of fault $\Omega = \{\omega_1, \omega_2, \omega_3\}$. The set of mass functions and their Shapley values are listed in Table 1.

**Table 1.** Mass functions and Shapley values of the first example.

|       | $m(\{\omega_1\})$ | $m(\{\omega_2\})$ | $m(\{\omega_3\})$ | $m(\{\omega_1, \omega_3\})$ | $\varphi_i$ |
|-------|------|------|------|------|--------|
| $m_1$ | 0.41 | 0.29 | 0.30 | 0.00 | 0.014  |
| $m_2$ | 0.00 | 0.90 | 0.10 | 0.00 | $-0.743$ |
| $m_3$ | 0.58 | 0.07 | 0.00 | 0.35 | 0.135  |
| $m_4$ | 0.55 | 0.10 | 0.00 | 0.35 | 0.126  |
| $m_5$ | 0.60 | 0.10 | 0.00 | 0.30 | 0.136  |

For the first example, the pignistic transformation from the result of Dempster's combination rule finds the combined decision is $\omega_3$, while the reasonable decision is $\omega_1$, which means there must be strong deceptive evidence. According to the calculated Shapley values, we can determine that $m_2$ is strong deceptive evidence, which is the same as the detection made by the method of Cui et al. After deleting deceptive evidence and re-conducting combination, we find in Table 2 that our approach can reach a higher pignistic probability for the reasonable decision $\omega_1$ than other methods, i.e., the decision is more certain.

**Table 2.** Fusion results of different methods for the first example.

| | Method | $p(\omega_1)$ | $p(\omega_2)$ | $p(\omega_3)$ | Reasonable |
|---|---|---|---|---|---|
| Consider all mass functions | DCR [2] | 0.0000 | 0.1422 | 0.8578 | No |
| | Murphy [8] | 0.9636 | 0.0210 | 0.0154 | Yes |
| Delete $m2$ | Cui et al. [1] | 0.9684 | 0.0007 | 0.0309 | Yes |
| | Ours | **0.9893** | **0.0001** | **0.0106** | **Yes** |

The second example is used to compare with the reinforcement learning method [6]. The set of mass functions and their Shapley values are listed in Table 3.

**Table 3.** Mass functions and Shapley values of the second example.

| | $m(\{\omega_1\})$ | $m(\{\omega_2\})$ | $m(\{\omega_3\})$ | $m(\Omega)$ | $\varphi_i$ |
|---|---|---|---|---|---|
| $m_1$ | 0.30 | 0.60 | 0.00 | 0.10 | −0.023 |
| $m_2$ | 0.70 | 0.00 | 0.00 | 0.30 | 0.206 |
| $m_3$ | 0.65 | 0.15 | 0.00 | 0.20 | 0.170 |
| $m_4$ | 0.75 | 0.00 | 0.05 | 0.20 | 0.238 |
| $m_5$ | 0.05 | 0.45 | 0.50 | 0.00 | −0.368 |
| $m_6$ | 0.05 | 0.50 | 0.45 | 0.00 | −0.376 |

For the second example, the decision derived by Dempster's combination rule and the pignistic transformation is $\omega_2$, while the reasonable decision is $\omega_1$. According to the obtained Shapley values, our method can detect that $m_1$ is a piece of weak deceptive evidence (it can not flip the decision but reduce the certainty of the decision) and $\{m_5, m_6\}$ are strong deceptive evidence. However, Kang and Zhao's method failed to detect $m_1$ as a piece of deceptive evidence. Table 4 shows that our approach can make a reasonable decision more certain.

**Table 4.** Fusion results of different methods for the second example.

| | Method | $p(\omega_1)$ | $p(\omega_2)$ | $p(\omega_3)$ | Reasonable |
|---|---|---|---|---|---|
| Consider all mass functions | DCR [2] | 0.1814 | 0.7428 | 0.0758 | No |
| | Murphy [8] | 0.8230 | 0.1555 | 0.0215 | Yes |
| Delete $\{m_5, m_6\}$ | Kang et Zhao [6] | 0.9566 | 0.0413 | 0.0021 | Yes |
| Delete $\{m_1, m_5, m_6\}$ | Ours | **0.9762** | **0.0147** | **0.0091** | **Yes** |

## 4.2 Efficiency Comparison

In this experiment, we compared our method with Kang and Zhao's method in terms of computational efficiency. For Kang and Zhao's method, the default parameter setting in [6] was used. The experiments were conducted on a Windows 11 system, powered by an Intel i5-12450H CPU. All used mass functions, of which there are two pieces of deceptive evidence, are defined on $\Omega = \{\omega_1, \omega_2, \omega_3\}$ with one to four focal sets. Figure 1 shows that our method is much more efficient, regardless of the number of mass functions.

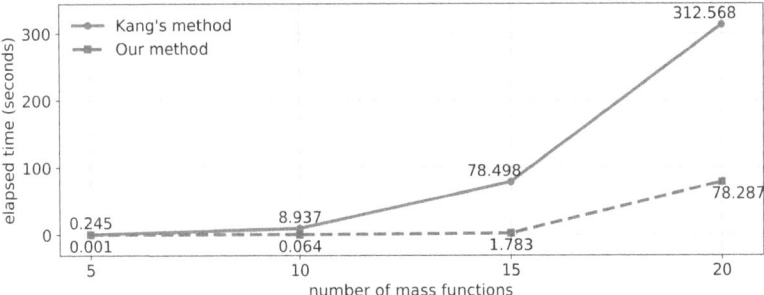

**Fig. 1.** Time required to detect deception as a function of the number of mass functions.

## 5 Conclusion

In this paper, we have defined strong and weak deceptive evidence in the information fusion of belief functions and proposed a new Shapley value-based deceptive evidence detection method, which provides more insights directly linked with the combination rule and the decision-making strategy. This method can distinguish between strong and weak deceptive evidence and assess the credibility of the evidence, which can be further used to conduct a weighted averaging process. Some numerical examples have demonstrated that our method is outperforming state-of-the-art methods in terms of detection effectiveness and efficiency.

In future works, we intend to optimize the efficiency of the proposed algorithm (especially with larger frames of discernment and a larger number of mass functions) and apply it in real-world applications. Moreover, the determination of the reasonable decision for a given set of mass functions is also an interesting open issue to investigate.

## References

1. Cui, H., Zhang, H., Chang, Y., Kang, B.: BGC: Belief gravitational clustering approach and its application in the counter-deception of belief functions. Eng. Appl. Artif. Intell. **123**, 106235 (2023)
2. Dempster, A.P.: Upper and lower probabilities induced by a multivalued mapping. Ann. Math. Stat. **38**, 325–339 (1967)

3. Denoeux, T.: A k-nearest neighbor classification rule based on Dempster-Shafer theory. IEEE Trans. Syst. Man Cybern. **25**(5), 804–813 (1995)
4. Denoeux, T.: A neural network classifier based on Dempster-Shafer theory. IEEE Trans. Syst. Man, Cybern. Part A: Syst. Hum. **30**(2), 131–150 (2000)
5. Dubois, D., Prade, H.: Representation and combination of uncertainty with belief functions and possibility measures. Comput. Intell. **4**(3), 244–264 (1988)
6. Kang, B., Zhao, C.: Deceptive evidence detection in information fusion of belief functions based on reinforcement learning. Inf. Fusion **103**, 102102 (2024)
7. Lefevre, E., Colot, O., Vannoorenberghe, P.: Belief function combination and conflict management. Inf. Fusion **3**(2), 149–162 (2002)
8. Murphy, C.K.: Combining belief functions when evidence conflicts. Decis. Support Syst. **29**(1), 1–9 (2000)
9. Pichon, F., Dubois, D., Denoeux, T.: Relevance and truthfulness in information correction and fusion. Int. J. Approximate Reasoning **53**(2), 159–175 (2012)
10. Schubert, J.: Counter-deception in information fusion. Int. J. Approximate Reasoning **91**, 152–159 (2017)
11. Shafer, G.: A Mathematical Theory of Evidence. Princeton University Press, Princeton (1976)
12. Shapley, L.S.: Notes on the n-person game-ii: The value of an n-person game (1951)
13. Smets, P.: The combination of evidence in the transferable belief model. IEEE Trans. Pattern Anal. Mach. Intell. **12**(5), 447–458 (1990)
14. Smets, P.: Decision making in the TBM: the necessity of the pignistic transformation. Int. J. Approximate Reasoning **38**(2), 133–147 (2005)
15. Smets, P.: Managing deceitful reports with the transferable belief model. In: 2005 7th International Conference on Information Fusion, vol. 2, pp. 7–pp. IEEE (2005)
16. Xiao, F., Cao, Z., Jolfaei, A.: A novel conflict measurement in decision-making and its application in fault diagnosis. IEEE Trans. Fuzzy Syst. **29**(1), 186–197 (2020)
17. Xu, Y., Li, Y., Wang, Y., Zhong, D., Zhang, G.: Improved few-shot learning method for transformer fault diagnosis based on approximation space and belief functions. Expert Syst. Appl. **167**, 114105 (2021)
18. Yager, R.R.: On the Dempster-Shafer framework and new combination rules. Inf. Sci. **41**(2), 93–137 (1987)
19. Yong, D., WenKang, S., ZhenFu, Z., Qi, L.: Combining belief functions based on distance of evidence. Decis. Support Syst. **38**(3), 489–493 (2004)
20. Zhang, H., Quost, B., Masson, M.H.: Cautious decision-making for tree ensembles. In: Bouraoui, Z., Vesic, S. (eds.) ECSQARU 2023. LNCS, vol. 14294, pp. 3–14. Springer, Cham (2023). https://doi.org/10.1007/978-3-031-45608-4_1
21. Zhang, H., Quost, B., Masson, M.H.: Cautious weighted random forests. Expert Syst. Appl. **213**, 118883 (2023)
22. Zhou, L., Cui, H., Huang, C., Kang, B., Zhang, J.: Counter deception in belief functions using Shapley value methodology. Int. J. Fuzzy Syst. **24**(1), 340–354 (2022)

# A Novel Optimization-Based Combination Rule for Dempster-Shafer Theory

Hasan Ihsan Turhan(✉) and Tugba Tanaydin

ASELSAN Inc, Ankara, Turkey
{hiturhan,tubudak}@aselsan.com

**Abstract.** In this study, a new methodology for combining probability masses from different sources is proposed for Dempster-Shafer theory. Unlike the existing works in the literature, this methodology treats the combination problem as an optimization problem and proposes an objective function that uses conflict and entropy measures to solve this problem. The proposed objective function aims to minimize the conflict and also to maximize the entropy of the combined probability masses. Thus, the difference between the combined probability masses and the masses coming from the sources is minimized while being cautious and avoiding a final certain decision. This new methodology is tested in the Matlab environment and compared with the existing methods.

**Keywords:** Dempster-Shafer theory · Belief functions · Combination rule · Decision making · Optimization

## 1 Introduction

Decision making with incomplete information is a very common problem in many fields. Dempster-Shafer theory, which is also known as "belief functions" in the literature is commonly used for this purpose. Belief functions were firstly introduced by Arthur Dempster in 1967 as "upper and lower probabilities" [1]. Later it was developed as a mathematical theory of evidence in 1976 by Glenn Shafer [2].

The theory deals with two main problems: making basic mass assignment from the available information and combining the assigned masses. Basic mass assignment can be defined as converting available information into numbers that the Dempster-Shafer theory uses. Combining assigned masses can be interpreted as combining the information about the options coming from different sources.

At the very beginning of the theory, Dempster combination rule has been used [2]. But, later it is criticized about not giving reasonable results in case of conflict. Conflict means opposite opinions of decision makers on an event. This fact is first discovered and criticized by Zadeh [3] and tried to be solved by assigning some probability masses to the complete uncertainty [2]. However, in applications it is observed that probability masses assigned to uncertainty drop quickly to zero after few combinations. Later many rules are proposed to overcome this difficulty.

In 1987, Yager proposed to assign the conflicting masses to the universal set [4]. After Yager, in 1988 Dubois and Parade introduced a new combination rule that assigns every partial conflict mass to union of conflicting sets [5], In 1994, Smets proposed the transferable belief model [6]. In 2003, Daniel proposed to redistribute the total conflicting masses to all subsets of $\Theta$ without considering whether they cause conflict or not [7]. In 2004, Smarandache and Dezert examined all combination rules proposed up to that time and revealed their drawbacks. Then, they proposed new partial conflict redistribution (PCR) rules [8, 9]. In 2006, Martin and Oswald generalized the PCR rules and proposed PCR6 combination rule [10]. In 2009, Florea proposed a new method, which considers reliability of sources [11]. Jousselme et al. devised a method to quantify conflicts by measuring the ratio of differences in evidence [13] and Murphy developed an averaging method for measuring sources of evidence (SoEs) [14]. Deng et al. utilized the weighted average method to calculate weights in the fusion process [15] and Tang et al. proposed a new conflict fusion method employing complex networks [16]. Xiao improved the correlation coefficient, considering the differences and non-interactivity of propositions [17]. Lin introduced the Jensen-Shannon (JS) divergence as an information-theoretic measure of deviation between probability distributions [18]. Jiang proposed a correlation coefficient [19, 20]. Xiao introduced a belief divergence measure [21] and Gao developed a generalized divergence measure [22].

In recent studies, proposed methods aimed at mitigating conflicts and refining fusion outcomes. Optimization-based approaches for de-combining belief functions [23], a novel approach for modifying the belief correlation measure [24] and multi criteria evaluation strategy [25] are proposed. Additionally, a new evidence fusion model utilizing principal component analysis (PCA) is proposed [26].

In this study, a novel combination rule is proposed. Unlike the existing works in the literature, this methodology treats the combination problem as an optimization problem and proposes an objective function that uses conflict and entropy measures to solve this problem. The proposed objective function aims to minimize the conflict and at the same time maximize the entropy of the combined probability masses. Thus, the proposed objective function distributes the conflict optimally only among the subsets that cause the conflict, while at the same time tries to keep the entropy of the new probability masses at the highest level in order not to reach a wrong and certain decision after a small number of combinations, which are the most criticized aspect of the Dempster combination rule.

The organization of the rest of this paper is as follows. The preliminaries of the theory are given in Sect. 2. The proposed method is described in Sect. 3. Effectiveness of the proposed method is investigated and compared with other methods in Sect. 4. Conclusions are made in Sect. 5.

## 2 Dempster-Shafer Theory

The theory is offered for the finite element universal sets. Universal set is defined as the set that contains all options and it is indicated with $\Theta = \{\theta_1, \theta_2, \ldots, \theta_n\}$. The basic mass assignment is made to subsets of the universal set excluding the empty set. $2^{\Theta} - 1 = \{\{\theta_1\}, \{\theta_2\}, \ldots, \{\theta_n\}, \{\theta_1, \theta_2\}, \ldots, \{\theta_1, \theta_2, \ldots, \theta_n\}\}$.

The Eq. (1) gives the definition of the basic mass assignment. It is a function that maps the subsets into [0, 1] and express the possibility of that subset.

$$m : 2^{\Theta} \rightarrow [0, 1], \; m(\emptyset) = 0, \; \sum_{X \in 2^{\Theta}} m(X) = 1 \qquad (1)$$

Conflict means opposite opinions of decision makers on an event. Let $m_1$ and $m_2$ basic mass assignments of two different decision makers. The total conflict among these mass assignments are computed as in (2).

$$k(m_1, m_2) = \sum_{\substack{X_1 \in 2^{\Theta}, X_2 \in 2^{\Theta} \\ X_1 \cap X_2 = \emptyset}} m_1(X_1) m_2(X_2) \qquad (2)$$

The partial conflict is defined as the amount of conflict caused by two subsets whose intersection is empty and formulized in (3).

$$k^{AB}(m_1, m_2) = m_1(A) m_2(B), \; A, B \in 2^{\Theta} \text{ ve } A \cap B = \emptyset \qquad (3)$$

Vacuous belief assignment (VBA) is defined as the mass assignment in the case of complete uncertainty.

$$m_v(\emptyset) = 0, \; m_v(\Theta) = 1 \qquad (4)$$

Entropy can be used to measure information level. If the entropy is high, then the information is uncertain or vice versa. One of the entropy definitions is made by Deng [12] in Dempster-Shafer theory. It given in the following Eq. (5) and $|X|$ in the equation express the cardinality of the subset X.

$$h = - \sum_{X \in 2^{\Theta}} m(X) log_2 \left( \frac{m(X)}{2^{|X|} - 1} \right) \qquad (5)$$

$m_1$, $m_2$ and $m_c$ are the masses of the first and second information source with combined probability masses and the combination rule is defined as operator in (6).

$$m_c = m_1 \oplus m_2 \qquad (6)$$

A proper combination rule satisfies the following conditions [9].

- Rule 1: Commutativity rule: $m_1 \oplus m_2 = m_2 \oplus m_1$
- Rule 2: Associativity rule: $m_1 \oplus (m_2 \oplus m_3) = (m_1 \oplus m_2) \oplus m_3$
- Rule 3: Must be ineffective to VBA. (i.e., combining VBA with another mass assignment must not change the result.)
- Rule 4: Must be coherent in every situation. (i.e., in case of conflict or correlation.)

## 3 The Proposed Combination Rule

The main idea behind the proposed combination rule is to redistribute the conflicting masses to these sets in unknown amounts. If these amounts can be known, then the proposed combination rule can be formulized as in (7).

$$m_c(Y) = \underbrace{\sum_{Z_1, Z_2 \in 2^{\Theta}} m_1(Z_1) m_2(Z_2)}_{\text{mass assignment from non-conflicting sets:} \overline{m}} + \underbrace{\sum_{X_1, X_2 \in 2^{\Theta}} t^Y_{\{X_1, X_2\}} k^{X_1, X_2}(m_1, m_2)}_{\text{mass assignment from conflicting sets}} \qquad (7)$$

In Eq. (7) $m_1$, $m_2$ and $m_c$, are the probability masses of the first and second decision makers and combined probability masses of these two respectively. Also $X_1$ and $X_2$ are the conflicting sets and $t^Y_{\{X_1,X_2\}}$ is the ratio that express how much of the partial conflict will be assigned to the set $Y \in \{X_1, X_2\}$. The ratio $t^Y_{\{X_1,X_2\}}$ is a number in $[0, 1]$. The proposed combination rule determines the ratio $t^Y_{\{X_1,X_2\}}$ by solving an optimization problem. The optimization problem here is to minimize the conflict at the same time to maximize the entropy of the combined probability masses. Thus, the difference between the combined probability masses and the masses coming from the sources is minimized while being cautious and avoiding a final certain decision. Hence the objective function that will be minimized composed of two terms. The first term is $k(m_c, m_1) + k(m_c, m_2)$ and it is used to compute new combined probability masses to have minimum conflict with the probability masses of the first and second decision makers. The second term of the objective function is the negative of the entropy function of the combined probability masses $m_c$. The reason of including this term into the objective function is not knowing the correlation between the information coming from and decision makers and prevent fast decision convergence in case of correlation. The weighted sum of these two terms composes the objective function. The objective function is given in (8). The weight $\omega$ that combines these two terms is a constant in $[0, 1]$.

$$f(t) = \omega(k(m_1, m_c) + k(m_2, m_c)) + (1 - \omega)(-h) \tag{8}$$

The constraints of the optimization problem are caused by the redistribution of the partial conflict to the conflicting sets. For example, the constraints of the partial conflict that are caused by $X_1$ and $X_2$ can be written as follows.

$$\begin{aligned} t^{X_1}_{\{X_1,X_2\}} + t^{X_2}_{\{X_1,X_2\}} &= 1 \\ t^{X_i}_{\{X_1,X_2\}} &\geq 0 \quad i = 1, 2 \end{aligned} \tag{9}$$

The optimization problem defined in (8) is convex and so the solution of this optimization problem gives the global optimum. The optimization problem here is defined in Matlab and solved by using Matlab optimization toolbox "fmincon" function. The convexity of the optimization problem is explained by Theorems 1 and 2.

*Theorem 1*: The term that corresponds to conflict in objective function is an affine function of the unknown $t$ parameters.

*Proof:* If the unknown constants are known the combined probability mass vector $m_c$ is computed as follow.

$$m_c = \overline{m} + Kt \tag{10}$$

In Eq. (10), $\overline{m}$ represents the un-normalized combined probability masses of the non-conflicting sets. K is the matrix that are composed of the partial conflicts and t is the unknown parameter vector. According to this the total conflict is computed as in (11). In Eq. (11) H is a matrix consisting zeros and ones. Elements in H are one if the corresponding intersection of the sets of $m_1$ and $m_2$ is empty, otherwise zero.

$$\begin{aligned} k(m_1, m_2) &= K \\ k(m_1, m_2) &= \left(m_1^T H m_2\right) \end{aligned} \tag{11}$$

The term corresponding to the conflict in the objective function is similarly calculated as in (12). In Eq. (12) G and D are the matrices consisting zeros and ones like the matrix H.

$$k(m_1, m_c) + k(m_2, m_c) = \left(m_1^T G m_c\right) + \left(m_2^T D m_c\right)$$
$$k(m_1, m_c) + k(m_2, m_c) = \left[m_1^T G(\overline{m} + Kt)\right] + \left[m_2^T D(\overline{m} + Kt)\right] \quad (12)$$

**Theorem 2:** The objective function is convex.

**Proof:** The first term of the objective function is convex with Theorem 1. The second term is convex because it is a convex function of a convex function.

## 4  Testing and Analysis Methods

The proposed method is tested by comparing with the combination rules in the literature. For this purpose, the crucial cases as complete conflict, complete correlation and vacuous belief assignment are chosen. The analysis is done by evaluating the rules given in (6). The universal set used in the tests is $\theta = \{a, b, c\}$.

The first scenario is a complete conflicting case, which Zadeh [2] criticizes the Dempster's Combination Rule. Accordingly, the probability masses assigned by the first and second decision makers are as in Table 1.

Table 1. The probability masses coming from the information sources

|       | {a} | {b} | {c} | {a, b} | {a, c} | {b, c} | {a, b, c} |
|-------|-----|-----|-----|--------|--------|--------|-----------|
| $m_1$ | 0.9 | 0.1 | 0   | 0      | 0      | 0      | 0         |
| $m_2$ | 0   | 0.1 | 0.9 | 0      | 0      | 0      | 0         |

The combination results for the first scenario are given in Table 2. The Dempster rule makes an incorrect decision and this directly violates the rule 4 given in (6). Dubois and Yager combination rule make mass assignments to the union sets. It seems to obey all rules given in (6), however assigning all conflicting mass to union sets causes not coming to a decision and this contradicts with rule 4. The proposed method and the PCR6 rule, properly distribute the conflict and assign similar masses to the sets. This assignment obeys the rule 4 and other rules are already obeyed given in (6). The method in [24] assigns to higher masses to the sets {a} and {c} and lower mass to set {b}. This can be considered disobeying to rule 4, because set {b} is the only common decision of two decision maker.

The second scenario is a complete correlating case. Accordingly, the probability masses assigned by the decision makers are as in Table 3.

The combination results for the second scenario are given in Table 4. Dubois and Yager combination rule make mass assignments to the union sets. It seems to obey all rules given in (6), however assigning all conflicting mass to union sets causes not coming

Table 2. Combination test results for the first scenario

|  | Dempster | Dubois | Yager | PCR6 [10] | Method in [24] | Proposed Method |
|---|---|---|---|---|---|---|
| $m_C(\{a\})$ | 0 | 0 | 0 | 0.45 | 0.4880 | 0.4492 |
| $m_C(\{b\})$ | 1 | 0.01 | 0.01 | 0.1 | 0.0240 | 0.1015 |
| $m_C(\{c\})$ | 0 | 0 | 0 | 0.45 | 0.4880 | 0.4492 |
| $m_C(\{a,b\})$ | 0 | 0.09 | 0 | 0 | 0 | 0 |
| $m_C(\{a,c\})$ | 0 | 0.81 | 0 | 0 | 0 | 0 |
| $m_C(\{b,c\})$ | 0 | 0.09 | 0 | 0 | 0 | 0 |
| $m_C(\{a,b,c\})$ | 0 | 0 | 0.99 | 0 | 0 | 0 |

Table 3. The probability masses coming from the information sources

|  | {a} | {b} | {c} | {a, b} | {a, c} | {b, c} | {a, b, c} |
|---|---|---|---|---|---|---|---|
| $m_1$ | 0.4 | 0.4 | 0.2 | 0 | 0 | 0 | 0 |
| $m_2$ | 0.4 | 0.4 | 0.2 | 0 | 0 | 0 | 0 |
| $m_3$ | 0.4 | 0.4 | 0.2 | 0 | 0 | 0 | 0 |
| $m_4$ | 0.4 | 0.4 | 0.2 | 0 | 0 | 0 | 0 |

to a decision and this contradicts with rule 4. Dempster, the method in [24] and PCR6 rule assigned equal and higher masses to the sets {a} and {b} and kept the mass assigned to set {c} low. They seem to obey all rules, however given that all decision-makers have made the same mass assignments, it is not reasonable to reach a final certain decision by not considering the correlation between the decisions of decision-makers and this contradicts with rule 4. The proposed method assigns equal masses to the sets {a} and {b}. Unlike the others, the mass assigned to set {c} is higher than the others. Thus, the drawback of the Dempster, the method in [24] and PCR6 rules for the rule 4 is avoided.

The third scenario is constructed to analyze the case of being neutral to vacuous belief assignment, i.e. rule 3 given in (6). Accordingly, the probability masses assigned by the first and second decision makers are as in Table 5.

The combination results for the third scenario are given in Table 6. The all results except from method in [24] are the same with the first decision maker's decision. The proposed method and the all rules except from method in [24] satisfies the rule 3, which is being neutral to vacuous belief assignment. The other rules are already obeyed given in (6). The method in [24] contradicts with both rule 3 and 4.

Table 4. Combination test results for the second scenario

|  | Dempster | Dubois | Yager | PCR6 [10] | Method in [24] | Proposed Method |
|---|---|---|---|---|---|---|
| $m_C(\{a\})$ | 0.48485 | 0.2496 | 0.2176 | 0.4506 | 0.4848 | 0.4197 |
| $m_C(\{b\})$ | 0.48485 | 0.2496 | 0.2176 | 0.4506 | 0.4848 | 0.4197 |
| $m_C(\{c\})$ | 0.0303 | 0.072 | 0.072 | 0.0988 | 0.0304 | 0.1606 |
| $m_C(\{a,b\})$ | 0 | 0.2048 | 0 | 0 | 0 | 0 |
| $m_C(\{a,c\})$ | 0 | 0.08 | 0 | 0 | 0 | 0 |
| $m_C(\{b,c\})$ | 0 | 0.08 | 0 | 0 | 0 | 0 |
| $m_C(\{a,b,c\})$ | 0 | 0.064 | 0.4928 | 0 | 0 | 0 |

Table 5. The probability masses coming from the information sources

|  | {a} | {b} | {c} | {a,b} | {a,c} | {b,c} | {a,b,c} |
|---|---|---|---|---|---|---|---|
| $m_1$ | 0.4 | 0.4 | 0.2 | 0 | 0 | 0 | 0 |
| $m_2$ | 0 | 0 | 0 | 0 | 0 | 0 | 1 |

Table 6. Combination test results for the third scenario

|  | Dempster | Dubois | Yager | PCR6 [10] | Method in [24] | Proposed Method |
|---|---|---|---|---|---|---|
| $m_C(\{a\})$ | 0.4 | 0.4 | 0.4 | 0.4 | 0.2857 | 0.4 |
| $m_C(\{b\})$ | 0.4 | 0.4 | 0.4 | 0.4 | 0.2857 | 0.4 |
| $m_C(\{c\})$ | 0.2 | 0.2 | 0.2 | 0.2 | 0.1310 | 0.2 |
| $m_C(\{a,b\})$ | 0 | 0 | 0 | 0 | 0 | 0 |
| $m_C(\{a,c\})$ | 0 | 0 | 0 | 0 | 0 | 0 |
| $m_C(\{b,c\})$ | 0 | 0 | 0 | 0 | 0 | 0 |
| $m_C(\{a,b,c\})$ | 0 | 0 | 0 | 0 | 0.2976 | 0 |

## 5 Conclusion

In this study, a novel method is proposed for the combining probability masses. This method has been examined especially for complete conflict, complete correlation and vacuous belief assignment and compared with other methods in the literature. It has been observed that the proposed method gives similar results as other conflict redistribution methods in the literature, but avoids making a certain decision when the same information with low probability values comes. Furthermore, it is neutral to vacuous belief assignment, on the contrary to method in [24]. This is our interpretation of a proper combination rule "must give reasonable results in all cases" and the proposed method provides this feature.

Since the proposed combination rule is based on optimization technique, it might be anticipated to demand significant computation time. However, a detailed computational time analysis, conducted using C + + implementation, reveals that the method operates efficiently at millisecond level. The proposed method has been integrated into our classification system, where a complete cycle of the system takes less than 200 ms. Further information regarding to computational time can be found in technical report available at [27].

## References

1. Dempster, A.P.: Upper and lower probabilities induced by a multivalued mapping. Ann. Math. Statist. **38**(2), 325–339 (1967)
2. Shafer, G.: A Mathematical Theory of Evidence. Princeton University Press, Princeton (1976)
3. Zadeh, L.A.: On the validity of Dempster's rule of combination of evidence. IEEE Trans. Syst. Man Cybern. Syst. **9**(11), 774–782 (1979)
4. Yager, R.R.: On the Dempster-Shafer framework and new combination rules. Inf. Sci. **41**, 93–138 (1987)
5. Dubois, D., Prade, H.: Representation and combination of uncertainty with belief functions and possibility measures. Comput. Intell. **4**(3), 244–264 (1988)
6. Smets, P., Kennes, R.: The transferable belief model. Artif. Intel. **66**(2), 191–234 (1994)
7. Daniel, M.: Associativity in combination of belief functions; a derivation of minC combination. Soft. Comput. **7**(5), 288–296 (2003)
8. Smarandache, F., Dezert, J.: A Simple proportional conflict redistribution rule. arXiv Archives, Los Alamos National Laboratory (2004)
9. Smarandache, F., Dezert, J.: Proportional conflict redistribution rules for information fusion. arXiv Archives, Los Alamos National Laboratory (2005)
10. Martin, A., Osswald, C.: A new generalization of the proportional conflict redistribution rule stable in terms of decision. In: Advances and Applications of DSmT for Information Fusion. Collected Works, vol. 2, pp. 69–88 (2006)
11. Florea, M.C., Jousselme, A.-L., Bossé, É., Grenier, D.: Robust combination rules for evidence theory. Inf. Fusion **10**(2), 183–197 (2009)
12. Deng, Y.: Deng entropy: a generalized Shannon entropy to measure uncertainty (2015). http://vixra.org/pdf/1502.0222v1.pdf
13. Jousselme, A.-L., Grenier, D., Bossé, É.: A new distance between two bodies of evidence. Inf. Fusion **2**(2), 91–101 (2001)
14. Murphy, C.K.: Combining belief functions when evidence conflicts. Decis. Support. Syst. **29**(1), 1–9 (2000)
15. Yang, W., Li, X., Deng, Y.: A clustering-based method to complete frame of discernment. China J. Aeronaut. **36**(4), 400–408 (2023)
16. Tang, Y., Dai, G., Zhou, Y., Huang, Y., Zhou, D.: Conflicting evidence fusion using a correlation coefficient-based approach in complex network. Chaos Solitons Fractals **176**, 114087 (2023)
17. Xiao, F., Cao, Z., Jolfaei, A.: A novel conflict measurement in decision-making and its application in fault diagnosis. IEEE Trans. Fuzzy Syst. **29**(1), 186–197 (2020)
18. Lin, J.: Divergence measures based on the Shannon entropy. IEEE Trans. Inf. Theory **37**(1), 145–151 (1991)

19. Jiang, X., Li, X., Wang, Q., Song, Q., Liu, J., Zhu, Z.: Multi-sensor data fusion-enabled semi-supervised optimal temperature-guided PCL framework for machinery fault diagnosis. Inf. Fusion **101**, 102005 (2024)
20. Jiang, W.: A correlation coefficient for belief functions. Int. J. Approx. Reason. **103**, 94–106 (2018)
21. Xiao, F.: Multi-sensor data fusion based on the belief divergence measure of evidences and the belief entropy. Inf. Fusion **46**, 23–32 (2019)
22. Gao, X., Pan, L., Deng, Y.: A generalized divergence of information volume and its applications. Eng. Appl. Artif. Intell. **108**, 104584 (2022)
23. Fan, X., Han, D., Yang, Y., Dezert, J.: De-combination of belief function based on optimization. Chin. J. Aeronaut. **35**(5), 179–193 (2022)
24. Zhang, Z., Wang, H., Zhang, J., Jiang, W.: A new correlation measure for belief functions and their application in data fusion. Entropy **25**(6), 925 (2023)
25. Dong, Y., et al.: A novel multi-criteria conflict evidence combination method and its application to pattern recognition. Inf. Fusion **108**, 102346 (2024)
26. Su, X., Shang, S., Xiong, L., Hong, Z., Zhong, J.: Research on dependent evidence combination based on principal component analysis. Math. Biosci. Eng. **21**(4), 4853–4873 (2024)
27. Turhan, H.I., Tanaydın, T.: A Novel Optimization Based Combination Rule Computation Time Results. https://sites.google.com/view/tecnical-report-for-belief2024

# Fusing Independent Inferential Models in a Black-Box Manner

Leonardo Cella[✉]

Department of Statistical Sciences, Wake Forest University,
Winston-Salem, NC 27109, USA
cellal@wfu.edu

**Abstract.** Inferential models (IMs) represent a novel possibilistic approach for achieving provably valid statistical inference. This paper introduces a general framework for fusing independent IMs in a "black-box" manner, requiring no knowledge of the original IMs construction details. The underlying logic of this framework mirrors that of the IMs approach. First, a fusing function for the initial IMs' possibility contours is selected. Given the possible lack of guarantee regarding the calibration of this function for valid inferences, a "validification" step is performed. Subsequently, a straightforward normalization step is executed to ensure that the final output conforms to a possibility contour.

**Keywords:** Inferential models · Possibility measures · Combination · p-values · Validity · Black-box

## 1 Introduction

The art of combining different sources of evidence to distinguish signal from noise remains highly relevant in today's world, where data is more abundant than ever and manifests in various forms. Take for example independent studies that were conducted to answer a common question. Each study may yield a measure of evidence for the quantity of interest. The objective is to integrate these specific measures of evidence into a composite measure that offers greater insight. Note that, in these scenarios, "data" refers to the measures of evidence themselves.

Along these lines, the present paper is primarily focused on exploring new developments in the combination/fusion of independent *inferential models* (IMs) (Martin and Liu 2015). IMs constitute a contemporary statistical inference framework that outputs provably valid possibility and necessity measures to any claim of interest concerning the inferential target. The possibilistic nature of IMs entails the existence of a possibility contour that is the base for all IMs uncertainty quantification. In what follows, the possibility contours from independent IMs will constitute the "data" to be combined.

Specifically, the goal is to fuse the independent IMs' possibility contours in a way that the resulting function maintains the calibrated possibilistic characteristic of an IM while exhibiting favorable efficiency properties. Moreover, this fusion should not require prior knowledge of how the original contours were constructed.

There exists a vast literature addressing the fusion of possibility contours. An extensive overview can be found in Dubois et al. (1999), but see also Dubois et al. (2016); Dubois and Prade (1988; 2001). However, these fusing rules typically do not take calibration assurances into consideration. Given that validity is the primary pillar of IMs, an alternative fusion strategy is needed.

After providing a brief background on the IMs approach in Sect. 2, I outline the proposed methodology in Sect. 3. I begin by examining an intuitive fusion alternative in Sect. 3.1, namely taking the minimum of the IMs' possibility contours. However, this solution turns out to be unsatisfactory for two reasons: it lacks calibration and does not conform to a possibility contour. In Sect. 3.2, I propose remedial measures to address these shortcomings. These remedial measures are not specific to the initial solution. In fact, they align well with the overall logic of the IMs approach. Consequently, in Sect. 3.3, I propose a general solution to the fusion of independent IMs. It is noteworthy that this general solution not only shares connections with the fusion methods in possibility theory mentioned above, but also exhibits similarities to methods proposed for merging independent p-values in the statistics literature (e.g., Cousins 2008; Oosterhoff 1976; Owen 2009). A concise summary and some important remarks are provided in Sect. 4.

## 2 Background on IMs

Let $Y^n = (Y_1, \ldots, Y_n)$ represent $n$ independent random draws from a distribution $\mathbb{P}_\theta$ that is associated with some unknown quantity of interest $\theta \in \Theta$. This setup is supposed to be general, encompassing scenarios where $\mathbb{P}_\theta$ conforms to a parametric model, as well as instances where the distribution of $Y^n$ remains unspecified, with $\theta$ serving to characterize a feature of this distribution. The goal is to perform probabilistic inferences on $\theta$ in the sense that, after observing some data $Y^n = y^n$, degrees of belief can be assigned to claims of interest about $\theta$.

IMs constitute a relatively novel framework for probabilistic inference, employing necessity measures to quantify degrees of belief. What sets IMs apart from other probabilistic methodologies, whether grounded in precise or imprecise probabilities, is the verifiable calibration assurance of its degrees of belief. Specifically, if $\mathcal{N}_{y^n}(A)$ represents the IMs' necessity measure attributed to the claim "$\theta \in A$" after observing data $Y^n = y^n$, the following *validity property*

$$\sup_{\theta \notin A} \mathbb{P}_\theta^n \{\mathcal{N}_{Y^n}(A) \geq 1 - \alpha\} \leq \alpha, \quad \text{for all } \alpha \in [0, 1] \text{ and all } A \subset \Theta \quad (1)$$

is satisfied. In words, (1) states that the assignment of high degree of belief to false claims about $\theta$ is a rare event with respect to $Y^n \sim \mathbb{P}_\theta^n$.

The possibilistic nature of IMs entails the existence of a *possibility contour*, a function $\pi_{y^n}(\vartheta)$ on $(\mathbb{Y}^n \times \Theta) \to [0, 1]$ satisfying

$$\sup_{\vartheta \in \Theta} \pi_{y^n}(\vartheta) = 1 \quad \text{for all } y^n. \quad (2)$$

This possibility contour serves as the foundation for computing necessity measures for claims of interest within IMs. Moreover, a necessary and sufficient condition for the IMs validity in (1) is that the possibility contour is stochastically no smaller than a $\mathsf{Unif}(0,1)$ distribution. This can be stated as

$$\mathbb{P}_\theta^n \{\pi_{Y^n}(\theta) \leq \alpha\} \leq \alpha, \quad \text{for all } \alpha \in [0,1]. \tag{3}$$

An IM is termed *exact* valid when $\pi_{Y^n}(\theta) \sim \mathsf{Unif}(0,1)$.

The "stochastically no smaller than uniform" outlined in (3) is a well known characteristic of p-values. However, it is important to emphasize that the IMs' possibility contours, besides satisfying such criteria, also satisfy (2), a characteristic that p-values don't always satisfy. As studied in detail in Martin (2021), the reason to constrain IMs to possibility measures is twofold. First, for any probabilistic approach with valid degrees of belief in the sense of (1), there exists a possibilistic IM whose necessity measure is no less efficient. Secondly, possibility and necessity measures stand out among the most computationally straightforward forms of imprecise probabilities.

But how are IMs constructed? Here I'll focus on a modern construction presented in Martin (2022a; 2022b). Suppose one has access to a measurable function $\gamma_{y^n}(\vartheta)$ on $(\mathbb{Y}^n \times \Theta) \to [0,1]$ that orders candidate values for $\theta$ in a "plausibilistic" manner given the observed data $Y^n = y^n$. That is, this function distinguishes between candidate values for $\theta$ that are more or less plausible given $y^n$, where values closer to zero and one indicate disagreement and agreement, respectively. It is usually the case that the chosen $\gamma_{y^n}$ satisfies $\sup_{\vartheta \in \Theta} \gamma_{y^n}(\vartheta) = 1$ for all $y^n$, as seen in the natural choice of the likelihood ratio for $\gamma_{y^n}$ when $\mathbb{P}_\theta$ is a parametric model. The properties of a possibility contour are satisfied for such choices, suggesting that the IM construction is complete and enabling the computation of necessity measures for claims of interest concerning $\theta$ to proceed. However, there is no guarantee that $\gamma_{y^n}$ satisfies the validity criteria in (3) so it undergoes a *validification* step:

$$\pi_{y^n}(\vartheta) = \mathbb{P}_\vartheta^n\{\gamma_{Y^n}(\vartheta) \leq \gamma_{y^n}(\vartheta)\}, \quad \vartheta \in \Theta, \tag{4}$$

giving rise to a new possibility contour that is guaranteed to satisfy (3). To see this, note that $\pi_{y^n} = G(\gamma_{y^n})$, where $G$ is the distribution function of $\gamma_{Y^n}$, so $\pi_{Y^n}(\theta) = G(\gamma_{Y^n}(\theta)) \sim \mathsf{Unif}(0,1)$.

## 3 Fusing IMs

### 3.1 Setup, Objectives and a First Idea

Let $\pi_1, \ldots, \pi_k$ represent $k$ independent IMs' possibility contours, each constructed to make valid probabilistic inferences on an unknown quantity $\theta$. While each IM may be based on a different sample size, I will assume here that this information is not available, so the notation for the IMs' contours suppresses sample sizes for the remainder of the paper. It will be clear below that the sample sizes are not even necessary for the proposed methodology, but see Sect. 4. The goal is to fuse the $k$ possibility contours in a way that

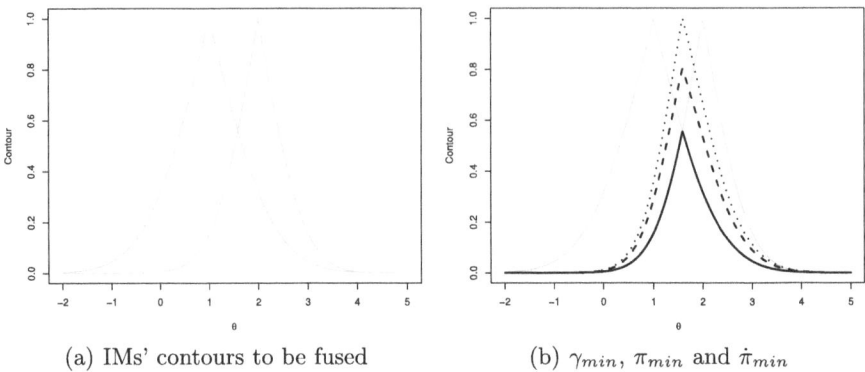

(a) IMs' contours to be fused  (b) $\gamma_{min}$, $\pi_{min}$ and $\dot\pi_{min}$

**Fig. 1.** Details from the example in Sects. 3.1 and 3.2. The solid, dashed and dotted black lines in Panel (b) represent $\gamma$, $\pi$ and $\dot\pi$, respectively.

1. the resulting function is also a valid possibility contour;
2. favorable efficiency properties are exhibited;
3. no knowledge of how each contour was constructed is required.

The first requirement is straightforward, meaning that the solution must satisfy properties (2) and (3), so that it can be used later on to make valid probabilistic inferences about $\theta$. The second requirement is intuitively reasonable. Since we are combining $k$ independent sources of information about $\theta$ we ideally want the solution to show efficiency gains. The third requirement means that the fusion of IMs will be done in a black-box manner

I start by considering a very intuitive approach to fusing independent IMs. With a focus on ensuring the efficiency of the solution, it involves taking the minimum' of the $k$ contours:

$$\gamma_{min}(\vartheta) = \min\{\pi_1(\vartheta),\ldots,\pi_k(\vartheta)\}, \quad \vartheta \in \Theta.$$

While $\gamma_{min}$ won't exceed any of the $k$ original contours in size, its validity is uncertain. Moreover, there's no guarantee that $\gamma_{min}$ satisfies (2), meaning that $\gamma_{min}$ may not necessarily qualify as a possibility contour. This can be seen Fig. 1(b), where $\gamma_{min}$ is displayed for the contours depicted in Fig. 1(a).

### 3.2 Validification and Normalization

Considering $\gamma_{\min}$ by itself is not an entirely satisfactory solution for fusing independent IMs. However, I will argue here that there is still potential in this solutions. More specifically, there is an opportunity to refine and make it desirable. And the means to do so align well with the general logic of the IMs framework, presented in Sect. 2. For the rest of this section, $\pi_i, i = 1,\ldots k$, and $\gamma_{min}$ will be denoted by $\Pi_i$ and $\Gamma_{min}$, respectively, when representing random variables.

Moreover, I will assume $\pi_1, \ldots, \pi_k$ are contours of exact valid IMs, so $\Pi_1(\theta), \ldots, \Pi_k(\theta)$ are independent and identically distributed (iid) $\mathsf{Unif}(0,1)$.

The issues with $\gamma_{min}$ were twofold: it isn't guaranteed to satisfy the validity property outlined in (3), and it doesn't necessarily qualify as a possibility contour. For the lack of validity, remember from Sect. 2 that, in the IMs construction, the selected function to plausibilistically rank candidate values of the unknown parameter given observed data might lack calibration, despite its intuitive appeal. This is precisely why the validification step in (4) is performed. Here, the concept remains the same, and $\gamma_{min}$ will undergo validification.

Towards this, recall the well known result in probability theory that if $V$ is a random variable defined as the minimum of $k$ independent $\mathsf{Unif}(0,1)$ random variables, then $V \sim \mathsf{Beta}(1,k)$, a distribution that is stochastically smaller than $\mathsf{Unif}(0,1)$ for every $k$. This confirms the lack of validity of $\gamma_{min}$ since $\Gamma_{min}(\theta) \sim \mathsf{Beta}(1,k)$. However, by following the structure of (4), $\gamma_{min}$ can be validified:

$$\pi_{min}(\vartheta) = F(\gamma_{min}(\vartheta)), \quad \vartheta \in \Theta,$$

where $F$ is the distribution function of $\mathsf{Beta}(1,k)$. Validity of $\pi_{min}$ comes from the fact that $F(\Gamma_{min}(\theta)) \sim \mathsf{Unif}(0,1)$.

Unfortunately, validification alone doesn't automatically convert $\gamma_{min}$ into a possibility contour. This is depicted in Fig. 1(b), where both $\pi_{min}$ and $\gamma_{min}$ for the contours in Fig. 1(a) are displayed. But this issue can be resolved with ease by normalizing $\pi_{min}$:

$$\dot{\pi}_{min}(\vartheta) = \frac{\pi_{min}(\vartheta)}{\max_{t \in \Theta} \pi_{min}(t)}, \quad \vartheta \in \Theta.$$

Note that this normalization step does not compromise the validity of $\pi_{min}$ since $\dot{\pi}_{min}$ is an inflated version of it. This is illustrated in Fig. 1(b).

### 3.3 A General Solution

The solution proposed above, in terms of minimum of the $k$ IMs possibility contours, while intuitive, isn't the exclusive option available. In fact, any continuous function from $[0,1]^k$ to $\mathbb{R}$ that is monotonic in each coordinate can be used to merge the $k$ contours. This chosen function can then be: i) validified, by leveraging the fact that the $k$ contours are iid $\mathsf{Unif}(0,1)$; and ii) normalized, if the validified function doesn't reach the value of one. The outcome is a fused IMs possibility contour that is provably valid. I formally present this general solution next, in the form of a theorem.

**Theorem 1.** *Assume $\pi_1, \ldots, \pi_k$ are $k$ independent and exact valid IMs possibility contours for a common unknown quantity of interest $\theta$. Let $\gamma_c$ be any continuous function from $[0,1]^k$ to $\mathbb{R}$ that is monotonic in each coordinate and $F$ be the cumulative distribution function of $\gamma_c(U^k)$ for $U^k = (U_1, \ldots, U_k)$ iid $\mathsf{Unif}(0,1)$. Then*

$$\dot{\pi}_c(\vartheta) = \frac{\pi_c(\vartheta)}{\max_{t \in \Theta} \pi_c(t)} \quad \vartheta \in \Theta, \tag{5}$$

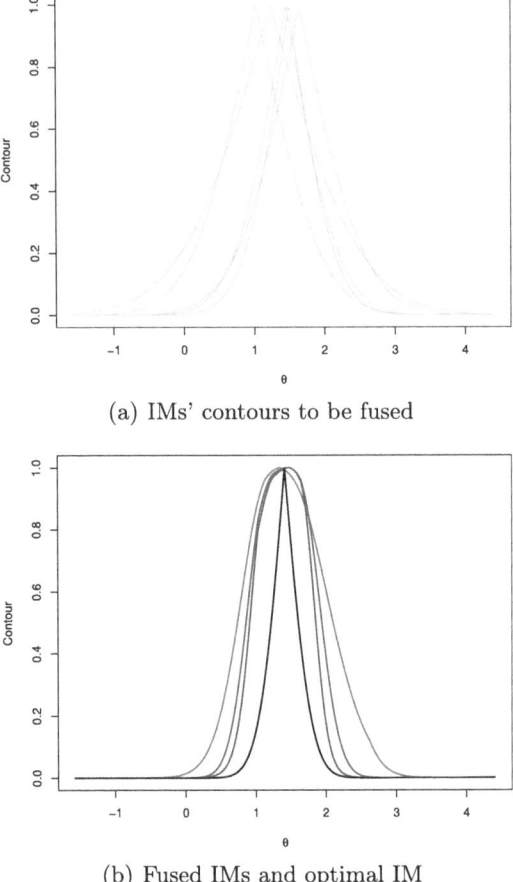

(a) IMs' contours to be fused

(b) Fused IMs and optimal IM

**Fig. 2.** Details for the example in Sect. 3.3.

where
$$\pi_c(\vartheta) = F\left(\gamma_c(\pi_1(\vartheta), \ldots, \pi_k(\vartheta))\right),$$
is a valid fused IMs' possibility contour in the sense that
$$\mathbb{P}_{U^k}(\dot{\Pi}_c(\theta) \leq \alpha) \leq \alpha, \quad \text{for all } \alpha \in [0,1]. \tag{6}$$

*Proof.* Since $\pi_1, \ldots, \pi_k$ are independent and exact valid IMs' possibility contours, $\Pi_1(\theta), \ldots, \Pi_k(\theta)$ are iid $\mathsf{Unif}(0,1)$ and $\Pi_c(\theta) \sim \mathsf{Unif}(0,1)$. The normalization in (5) leads to $\dot{\pi}_c(\vartheta) \geq \pi_c(\vartheta)$ for all $\vartheta \in \Theta$, thereby yielding (6). □

Two remarks are in order regarding the choice of $\gamma_c$. First, despite Theorem 1 stating that any monotone function can be used to fuse IMs, as all $k$ contours provide reliable information about $\theta$ I follow Dubois and Prade (2001) and recommend conjunctive or statistical combinations. Second, from a computational standpoint, the normalization in (5) would probably be non-trivial if $\theta$ is vector-valued. Consequently, it may be preferable to opt for a $\gamma_c$ whose validified form already constitutes a possibility contour. Note that such choice would also make the fused IM exact valid.

As a final illustration, consider the five independent contours depicted in Fig. 2(a), constructed for making inferences on some population mean $\theta$. Figure 2(b) displays three fused IMs: one based on $\gamma_{min}$ (red), another based on $\gamma_c$ obtained by multiplying the initial contours (blue), aligning with Fisher's suggestion for combining p-values (Fisher 1932), and a third that takes $\gamma_c$ as the average of the initial contours (green). The average-based fused IM happens to be the most efficient among the three in this case. For additional comparison, the IM that takes into account the knowledge that data comes from a normal distribution and also assumes access to the original data is displayed in black. This is the best possible IM for inferences on $\theta$ (Martin and Syring 2019). It is, of course, much more efficient than the three fused IMs considered. However, it's important to recall that the fused IMs lack knowledge about normality, original data or even the sample sizes, but are still valid solutions.

## 4 Conclusion and Remarks

This paper introduces a general framework for fusing independent IMs when no information regarding the construction of each individual IM is available. The underlying logic of this framework mirrors the core rationale of the IMs approach, making it akin to an application of IMs to problems where data consists of IMs' possibility contours themselves. This framework has close connections with methodologies to combine p-values in the statistics literature and methodologies to combine possibility measures in the imprecise probability literature. In summary, all three frameworks begin with a fusing function $\gamma_c$, but while p-values fusing rules simply validify $\gamma_c$ and contour fusing rules solely normalize $\gamma_c$, IMs fusing rules first validifies $\gamma_c$ and then normalizes it.

This section concludes with several remarks and some future directions. First, even though the proposed framework does not require knowledge of how the individual IMs were constructed, it is not uncommon for information regarding the sample sizes to be available. There is a compelling rationale to integrate weights for the different contours in such instances, in the hope that the solution is more efficient. This will be explored elsewhere. Second, the normalization in (5), despite intuitive, is not the only option to guarantee the solution is a possibility measure. For alternatives and the reasoning behind them, see Cella and Martin (2022). Third, it has been suggested that the validification step should precede the normalization step. But one may wonder if normalization could be conducted first. After all, this is the approach typically taken when selecting a function like the likelihood ratio in traditional IMs construction. However, in the present context, if normalization is done first, one can no longer leverage the uniform nature of the individual contours to perform the validification step. Fourth, all the derivations above assumed that the IMs' contours to be combined are exact valid, but, in some applications, it is possible that some of them are just valid. Since these "just valid" contours are "larger" than necessary, when fusing the IMs' contours we obtain an inflated $\gamma_c$. Importantly, the validification step above still outputs a valid combination, albeit less efficient. Lastly, efficiency and optimality evaluations of different solutions, as well as the problem of fusing dependent IMs will be explored in future work.

**Acknowledgements.** The author thanks Professor Ryan Martin and the two anonymous reviewers for their valuable feedback on earlier versions of this manuscript.

**Disclosure of Interests.** The author has no competing interests to declare that are relevant to the content of this article.

# References

Cella, L., Martin, R.: Valid inferential models for prediction in supervised learning problems. Int. J. Approximate Reasoning **150**, 1–18 (2022)

Cousins, R.D.: Annotated bibliography of some papers on combining significances or p-values. arXiv:0705.2209 (2008)

Dubois, D., Liu, W., Ma, J., Prade, H.: The basic principles of uncertain information fusion. An organised review of merging rules in different representation frameworks. Inf. Fusion **32**, 12–39 (2016)

Dubois, D., Prade, H.: Representation and combination of uncertainty with belief functions and possibility measures. Comput. Intell. **4**(3), 244–264 (1988)

Dubois, D., Prade, H.: Possibility theory in information fusion. In: Della Riccia, G., Lenz, H.-J., Kruse, R. (eds.) Data Fusion and Perception, pp. 53–76. Springer (2001)

Dubois, D., Prade, H., Yager, R.: Merging fuzzy information. In: Bezdek, J.C., Dubois, D., Prade, H. (eds.) Fuzzy Sets in Approximate Reasoning and Information Systems. The Handbooks of Fuzzy Sets Series, vol. 5, pp. 335–401. Springer, Boston (1999). https://doi.org/10.1007/978-1-4615-5243-7_7

Fisher, R.A.: Statistical Methods for Research Workers, 4th edn. Oliver and Boyd, Edinburgh (1932)

Martin, R.: An imprecise-probabilistic characterization of frequentist statistical inference (2021). arXiv:2112.10904

Martin, R.: Valid and efficient imprecise-probabilistic inference with partial priors, i. first results. arXiv:2203.06703 (2022a)

Martin, R.: Valid and efficient imprecise-probabilistic inference with partial priors, ii. general framework. arXiv:2211.14567 (2022b)

Martin, R., Liu, C.: *Inferential Models: Reasoning with Uncertainty*. Monographs in Statistics and Applied Probability Series. Chapman & Hall/CRC Press (2015)

Martin, R., Syring, N.: Validity-preservation properties of rules for combining inferential models. In: De Bock, J., de Campos, C. P., de Cooman, G., Quaeghebeur, E., Wheeler, G., (eds.) Proceedings of the Eleventh International Symposium on Imprecise Probabilities: Theories and Applications, vol. 103 , pp. 286–294. PMLR (2019)

Oosterhoff, J.: Combination of one-sided statistical tests. MC Tracts (1976)

Owen, A.B.: Karl Pearson's meta-analysis revisited. Ann. Stat. **37**(6B), 3867–3892 (2009)

# Optimization Under Severe Uncertainty: a Generalized Minimax Regret Approach for Problems with Linear Objectives

Tuan-Anh Vu[✉], Sohaib Afifi, Éric Lefèvre, and Frédéric Pichon

Univ. Artois, UR 3926, Laboratoire de Genie Informatique et d'Automatique de l'Artois (LGI2A), 62400 Bethune, France
{tanh.vu,sohaib.afifi,eric.lefevre,frederic.pichon}@univ-artois.fr

**Abstract.** We study a general optimization problem with an uncertain linear objective. We address the uncertainty using two models: belief functions and, more generally, capacities. In the former model, we use the generalized minimax regret criterion introduced by Yager, while in the latter one, we extend this criterion, to find optimal solutions. This paper identifies some tractable cases for the resulting problem. Furthermore, when focal sets of the considered belief functions are Cartesian products of intervals, we develop a 2-approximation method that mirrors the well-known midpoint scenario method used for minimax regret optimization problems with interval data.

**Keywords:** Minimax regret · Belief functions · Capacities · Linear programming

## 1 Introduction

Uncertainty is ubiquitous in optimization problems, leading to numerous frameworks for handling it. This paper revisits the famous minimax regret criterion within robust optimization. In essence, this criterion arises from two key motivations in decision-making under uncertainty: (i) the common human tendency to regret choices especially if a better option is discovered later; and (ii) the desire for an option with the best worst-case performance. The minimax regret criterion, widely studied for optimization problems with uncertain coefficients in the objectives [8], assumes a classical setting where only a so-called scenario set of possible realizations of coefficients, is available. Under this limited information, the criterion aims to find a solution that minimizes the maximum regret across all scenarios.

However, some partial information is usually at hand in real-life situations. For example, knowing the scenario set allows us to consult experts who can assess the likelihood of each scenario occurring. In such cases, refining the minimax regret criterion to account for partial information becomes necessary to better reflect real-world situations. Interestingly, under *evidential* uncertainty, i.e., when the uncertainty is modeled by belief functions [10], a notion of generalized minimax regret has already been introduced by Yager [13]. Similarly, a recent work by Adam and Destercke [1] discussed a related notion within the possibilistic framework.

Following our recent paper on a general optimization problem with uncertain linear objective [11], we also study the same problem. Unlike [11] which considered five other criteria, this paper uses generalized minimax regret criteria to find optimal solutions. Furthermore, we address situations where the uncertainty about the objective coefficients is *severe*, i.e., we cannot identify a single probability measure to represent it. For this reason, we use more general frameworks, namely belief functions and capacities, to model this uncertainty.

The paper is organized as follows. Section 2 presents some elements about belief functions. In Sect. 3, we incorporate Yager's criterion [13] to the considered problem. For Yager's criterion, two types of belief functions, where (i) their frames are finite and (ii) their frames are infinite but their focal sets take a special form, are addressed in Sect. 4 and Sect. 5, respectively. We then extend Yager's criterion to a more general setting where uncertainty is modeled by capacities or lower probabilities [4,6] in Sect. 6. The paper ends with a conclusion.

## 2 Belief Function Theory

Let $\Omega$ be the set of all possible values of a variable of interest $\omega$. In this paper, we assume that $\Omega$ is a closed subset of $\mathbb{R}^n$. In belief function theory [10], adapting the presentation of [12], partial information about the true (unknown) value of $\omega$ is given by a mapping $m : \mathcal{C} \mapsto [0,1]$ called mass function, where $\mathcal{C}$ is a finite collection of closed subsets of $\Omega$, such that $\sum_{A \in \mathcal{C}} m(A) = 1$ and $m(\emptyset) = 0$. If $\Omega$ is finite, we usually take $\mathcal{C} = 2^\Omega$.

Mass $m(A)$ quantifies the amount of belief allocated to the fact of knowing only that $\omega \in A$. A focal set of $m$ is a subset $A \subseteq \Omega$ such that $m(A) > 0$. Let $\mathcal{F} = \{F_1, \ldots, F_K\}$ be the set of all focal sets of $m$. The mass function $m$ induces a belief function $Bel$ defined on $\mathcal{B}(\Omega)$ the Borel subsets of $\Omega$ where $Bel(A) = \sum_{B \in \mathcal{F}: B \subseteq A} m(B)$.

## 3 Problem Formulation

In this paper, we focus on a general problem with a linear objective:

$$\{\min c^T x : x \in \mathcal{X} \subseteq \mathbb{R}^n_{\geq 0}\} \tag{P}$$

where $\mathcal{X}$ is a compact set and $c \in \mathbb{R}^n$ is the coefficient vector of the objective. (P) is a linear programming problem if $\mathcal{X} = \{x \in \mathbb{R}^n_{\geq 0} : Mx \leq b\}$ where $M$ is a $q \times n$ matrix and $b$ is a $q$-vector. If $\mathcal{X} \subseteq \{0,1\}^n$, then (P) is a combinatorial problem. In this paper, we assume that the coefficient vector $c$ is uncertain and let $\Omega \subseteq \mathbb{R}^n$ be the set of possible values of $c$. Each $c \in \Omega$ is then called a scenario.

## 3.1 The Minimax Regret Criterion

The regret $R(x,c)$ of a solution $x$ under a scenario $c \in \Omega$ is defined as $R(x,c) := c^T x - val^*(c)$ where $val^*(c) := \min_{x \in \mathcal{X}} c^T x$ the optimal value of (P) under $c$. The maximum regret $R(x)$ of $x$ is defined as $R(x) := \max_{c \in \Omega} R(x,c)$: it represents the regret of $x$ in the worst case scenario in $\Omega$. The goal is to find a solution $x$ having minimum $R(x)$ by solving the problem:

$$\min_{x \in \mathcal{X}} R(x) = \min_{x \in \mathcal{X}} \max_{c \in \Omega} (c^T x - val^*(c)) \tag{MR}$$

## 3.2 The Generalized Minimax Regret Criterion Under Evidential Uncertainty

If some partial knowledge about $c$ is given by a mass function $m$, we generalize the minimax regret criterion as follows [13]. For each focal set $F$ of $m$, the maximum regret of $x$ is $R^F(x) := \max_{c \in F}(c^T x - val^*(c))$. The expected maximal regret of $x$ with respect to $m$ is then defined as

$$\overline{R}(x) := \sum_{r=1}^{K} m(F_r) R^{F_r}(x). \tag{1}$$

In this paper, we focus on addressing Problem (GMR):

$$\min_{x \in \mathcal{X}} \overline{R}(x) = \min_{x \in \mathcal{X}} \sum_{r=1}^{K} m(F) \max_{c \in F_r}(c^T x - val^*(c)). \tag{GMR}$$

Note that if $m$ is a vacuous mass function, i.e., $\Omega$ is the only focal set of $m$, then (GMR) becomes (MR).

*Remark 1.* The information about the true scenario can be given by a *possibility distribution* $\pi : \Omega \to [0,1]$ with values of $\pi$ representing possibility degrees of elements in $\Omega$, among which there exists a $c$ such that $\pi(c) = 1$. This representation of uncertainty is practical, as $\pi$ can, for instance, be constructed from expert assessments. Assume that $1 = \alpha_1 > \ldots > \alpha_K > \alpha_{K+1} = 0$ are the distinct values of $\pi$. For each $\alpha_i$, the associated $\alpha_i$ cut of $\pi$ is defined as: $F_{\alpha_i} = \{c \in \Omega : \pi(c) \geq \alpha_i\}$. Obviously, $F_{\alpha_1} \subset \ldots \subset F_{\alpha_K}$. If we construct a mass function on $\Omega$ with focal sets $F_{\alpha_i}$ and $m(F_{\alpha_i}) = \alpha_i - \alpha_{i+1} \; \forall i \in \{1, \ldots K\}$, we return to the version of generalized minimax regret criterion under possibilistic framework, introduced in [1].

## 4 When $\Omega$ is Finite

In this case, we have a mass function $m$ on a finite set of $l$ elements $\Omega = \{c^1, \ldots, c^l\} \subset \mathbb{R}^n$. For combinatorial optimization problems, the intractability of (MR) has been well-documented, see e.g., [8], thereby implying the intractability of (GMR) for such problems as well. The main result in this section, therefore, concerns a case where (GMR) is tractable.

**Proposition 1.** *Assume that (P) is a linear programming problem. Then (GMR) can be solved efficiently provided $|\mathcal{F}|$ is not large. In particular, if $|\mathcal{F}|$ is polynomially bounded in $l$ then (GMR) can be solved in polynomial time.*

*Proof.* We reformulate (GMR) as:

$$\min \sum_{F \in \mathcal{F}} m(F) z_F$$
$$z_F \geq c^T x - val^*(c) \ \forall F \in \mathcal{F}, c \in F \quad (2)$$
$$Mx \leq b, \ x \in \mathbb{R}_{\geq 0}^n.$$

Note that (2) is a linear programming problem. Moreover, for each $c \in \Omega$, $val^*(c) = \min\{c^T x : x \in \mathbb{R}_{\geq 0}^n, Mx \leq b\}$ can be computed efficiently by standard linear programming solvers. Therefore, (2) can be solved efficiently provided the number of focal sets is not large and be solvable in polynomial time if $|\mathcal{F}|$ is polynomially bounded. □

## 5 When $\Omega$ is Infinite and Focal Sets of $m$ are Cartesian Products of Intervals

In this section, we assume that each focal set $F_r$ of $m$ is a Cartesian product of intervals, i.e.,

$$F_r = \times_1^n [l_i^r, u_i^r] \ \forall r.$$

When $m$ has a unique focal set of such type, we get back to the famous interval uncertainty representation in robust optimization. We remark that under interval representation, the classical minimax regret Problem (MR) is intractable in both cases where (P) is a combinatorial or linear programming problem [8]. Fortunately, a well-known heuristic exists to obtain a 2-approximation algorithm for (MR): it uses an optimal solution of (P) under the so-called midpoint scenario [5,8]. The goal here is to adapt this heuristic for our considered uncertainty representation, for which we follow the approach in [5]. We denote

$$\bar{u}_i := \sum_{r=1}^{K} m(F_r) u_i^r \text{ and } \bar{l}_i := \sum_{r=1}^{K} m(F_r) l_i^r. \quad (3)$$

**Proposition 2.** *Let $\bar{c}$ be a vector in $\mathbb{R}^n$ such that $\bar{c}_i = \frac{\bar{u}_i + \bar{l}_i}{2}$ and $y$ be an optimal solution of (P) under $\bar{c}$, i.e., $\bar{c}^T y = \min_{x \in \mathcal{X}} \bar{c}^T x$. Let $x^*$ be any optimal solution of (GMR). Then $\overline{R}(y) \leq 2\overline{R}(x^*)$.*

To prove Proposition 2, we need some preliminary observations. First, notice that for any $F_r$, $R^{F_r}(x^*) = \max_{x \in \mathcal{X}} \max_{c \in F_r} c^T(x^* - x)$, and thus

$$R^{F_r}(x^*) \geq \max_{c \in F_r} c^T(x^* - y). \quad (4)$$

Note also that $\max_{c \in F_r} c^T(x^* - y) = \sum_{i:x_i^* > y_i} u_i^r(x_i^* - y_i) - \sum_{i:x_i^* < y_i} l_i^r(y_i - x_i^*)$. Therefore, $R^{F_r}(x^*) \geq \sum_{i:x_i^* > y_i} u_i^r(x_i^* - y_i) - \sum_{i:x_i^* < y_i} l_i^r(y_i - x_i^*)$. Using (1) and (3),

$$\overline{R}(x^*) \geq \sum_{i:x_i^* > y_i} \overline{u}_i(x_i^* - y_i) - \sum_{i:x_i^* < y_i} \overline{l}_i(y_i - x_i^*) \tag{5}$$

In the subsequent, we use the notation $\delta(y - x^*, F_r) := \max_{c \in F_r} c^T(y - x^*)$. Referring to [5, Property 2.2], we have that $R^{F_r}(y) \leq R^{F_r}(x^*) + \delta(y - x^*, F_r) \, \forall r$. From (1),

$$\overline{R}(y) \leq \overline{R}(x^*) + \sum_{r=1}^{K} m(F_r) \delta(y - x^*, F_r). \tag{6}$$

We are ready to prove Proposition 2.

*Proof (Proof of Proposition 2).* By the optimality of $y$, $\sum_{i=1}^{n}(\overline{u}_i + \overline{l}_i)x_i^* \geq \sum_{i=1}^{n}(\overline{u}_i + \overline{l}_i)y_i$. Equivalently, $\sum_{i=1}^{n} \overline{u}_i(x_i^* - y_i) \geq \sum_{i=1}^{n} \overline{l}_i(y_i - x_i^*)$. It follows that

$$\sum_{i:x_i^* > y_i} \overline{u}_i(x_i^* - y_i) - \sum_{i:x_i^* < y_i} \overline{u}_i(y_i - x_i^*) \geq \sum_{i:x_i^* < y_i} \overline{l}_i(y_i - x_i^*) - \sum_{i:x_i^* > y_i} \overline{l}_i(x_i^* - y_i) \tag{7}$$

$$\sum_{i:x_i^* > y_i} \overline{u}_i(x_i^* - y_i) - \sum_{i:x_i^* < y_i} \overline{l}_i(y_i - x_i^*) \geq \sum_{i:x_i^* < y_i} \overline{u}_i(y_i - x_i^*) - \sum_{i:x_i^* > y_i} \overline{l}_i(x_i^* - y_i). \tag{8}$$

It can be easily checked that the right hand side of (8) equals $\sum_{r=1}^{K} m(F_r) \delta(y - x^*, F_r)$. Hence, it follows from (5) that $\overline{R}(x^*) \geq \sum_{r=1}^{K} m(F_r) \delta(y - x^*, F_r)$. Finally, Proposition 2 is true because of (6). □

## 6 Beyond Belief Functions

We still consider the case of finite $\Omega = \{c^1, \ldots, c^l\}$ as in Sect. 4. However, in this context, the partial knowledge about the true coefficient vectors is modeled by non-additive measures, namely capacities which are more general than belief functions. We quickly summarize some basics elements adapted from [6].

A capacity on $\Omega$ is a set function $\mu : 2^\Omega \to [0, 1]$ such that $\mu(\Omega) = 1$, $\mu(\emptyset) = 0$ and if $A \subseteq B$, $\mu(A) \leq \mu(B)$. Note that $\mu$ is a probability measure if it is additive, i.e., $\mu(A \cup B) = \mu(A) + \mu(B) \, \forall A, B \in 2^\Omega$ with $A \cap B = \emptyset$. Furthermore, $\mu$ is a 2-monotone capacity if $\mu(A \cup B) + \mu(A \cap B) \geq \mu(A) + \mu(B) \, \forall A, B \subseteq \Omega$. A belief function, also known as a *complete monotonicity* capacity, is a special 2-monotone capacity [6, 10].

*Remark 2.* In combinatorial optimizations [7], a 2-monotone capacity $\mu$ is called a supermodular set function while its dual $\bar{\mu}$, defined as $\bar{\mu}(A) = 1 - \mu(\Omega \setminus A) \, \forall A \subseteq \Omega$, is called submodular.

In imprecise probability [4], $\mu$ is usually called a lower probability where values of $\mu$ are interpreted as lower bounds of values of the true (yet unknown) probability measure $P^*$ on $\Omega$. Under this view, the so-called *credal set* of $\mu$ consisting of all compatible probability measures with $\mu$ on $\Omega$, is defined as $\mathcal{M}(\mu) := \{P : P(A) \geq \mu(A) \; \forall A \subseteq \Omega\}$. We will henceforth view any element in $\mathcal{M}(\mu)$ as a vector $p \in [0,1]^l$, and thus $\mathcal{M}(\mu)$ is a polytope:

$$\mathcal{M}(\mu) = \{p \in [0,1]^l : \sum_{i \in A} p_i \geq \mu(\{c^i : i \in A\}) \; \forall A \subseteq \{1,\ldots,l\}, \; \sum_{j=1}^{l} p_i = 1\}. \quad (9)$$

Because explicitly listing all $2^l$ values of $\mu(A)$ is intractable, we use a typical assumption from optimizations [7, Chapter 10].

**Assumption 1.** *We have access to an evaluation oracle that returns $\mu(A)$ for each query $A \subseteq \Omega$.*

We proceed to extend the generalized minimax regret criterion, discussed in Sect. 3.2, to incorporate the notion of capacity as follows. The expected regret of a solution $x \in \mathcal{X}$ with respect to a probability measure $p \in \mathcal{M}(\mu)$ is $\sum_{i=1}^{l} p_i R(x, c^i)$. Since the only available information is that the true probability measure lies in $\mathcal{M}(\mu)$, a reasonable approach is to seek a solution that minimizes the worst-case of expected regret among all compatible probabilities. In other words, we need to solve:

$$\min_{x \in \mathcal{X}} \max_{p \in \mathcal{M}(\mu)} \sum_{i=1}^{l} p_i R(x, c^i) = \min_{x \in \mathcal{X}} \max_{p \in \mathcal{M}(\mu)} \sum_{i=1}^{l} p_i \left((c^i)^T x - val^*(c^i)\right). \quad \text{(CGMR)}$$

If $\mu$ is a belief function on $\Omega$ and $m$ is its associated mass function (see Sect. 2), a well-known result [6] states that

$$\max_{p \in \mathcal{M}(\mu)} \sum_{i=1}^{l} p_i \left((c^i)^T x - val^*(c^i)\right) = \sum_{r=1}^{K} m(F_r) \max_{c \in F_r}(c^T x - val^*(c)). \quad (10)$$

Thus, in this case, CGMR reverts to (GMR).

*Remark 3.* If $\mu$ is 2-monotone, it is well-known that a $p^*$ that maximizes the left hand side of (10) can be efficiently computed by using only $l$ accesses to the oracle, as follows [6,7]. Reindex elements of $\Omega$ such that $(c^1)^T x - val^*(c^1) \geq \ldots \geq (c^l)^T x - val^*(c^l)$ and let $A_j = \{c^j, \ldots, c^l\} \; \forall j \in \{1,\ldots l\}$ and $A_{l+1} = \emptyset$. Finally, take $p_j^* = \mu(A_j) - \mu(A_{j+1}) \; \forall j \in \{1,\ldots l\}$. Moreover, such $p^*$ is also an extreme point of $\mathcal{M}(\mu)$.

We now show that under the computational model described in Assumption 1, CGMR is tractable if (P) is a linear programming problem. Let $Ext(\mu)$ be the set of extreme points of $\mathcal{M}(\mu)$. Note that $Ext(\mu)$ is finite but can be very large, i.e., $|Ext(\mu)|$ is exponential in $l$. We first observe that

$$\max_{p \in \mathcal{M}(\mu)} \sum_{i=1}^{l} p_i \left((c^i)^T x - val^*(c^i)\right) = \max_{p \in Ext(\mu)} \sum_{i=1}^{l} p_i \left((c^i)^T x - val^*(c^i)\right). \quad (11)$$

**Proposition 3.** *If (P) is a linear programming problem and assume that $\mu$ is 2-monotone. Then (CGMR) can be solved in polynomial time.*

*Proof.* Using (11), we reformulate (CGMR) as:

$$\min t \tag{12}$$

$$t \geq \sum_{i=1}^{l} p_i \left((c^i)^T x - val^*(c^i)\right) \quad \forall p \in Ext(\mu) \tag{13}$$

$$Mx \leq b, \ x \in \mathbb{R}_{\geq 0}^n. \tag{14}$$

Problem (12–14) is a linear programming problem but it has a vast number of constraints due to (13). Because (P) is a linear programming problem, $val^*(c^i)$ is computed in polynomial time. To demonstrate the polynomial solvability of (12–14), we employ the celebrated ellipsoid method [7]. According to this method, we need to show that the separation problem associated with (12–14) can be solved in polynomial time: either confirms if given a point $(x^0, t^0) \in \mathbb{R}^n$ satisfies all the constraints (13–14) or return a constraint that it violates. Checking if $(x^0, t^0)$ satisfies (14) can be easily done in polynomial time. Furthermore, checking if $(x^0, t^0)$ satisfies (13) amounts to testing if $t^0 \geq \max_{p \in \mathcal{M}(\mu)} \sum_{i=1}^{l} p_i \left((c^i)^T x^0 - val^*(c^i)\right)$, which can be done in polynomial time because of Remark 3. We conclude that the separation problem, and thus (12–14) is polynomial solvable. □

*Remark 4.* Because of the popularity of the minimax regret criterion, similar forms to (CGMR) have already appeared in the literature of optimization under distributional uncertainty, to cite only a few [2,3]. However, to the best of our knowledge, Proposition 3 is new.

While the ellipsoid algorithm is theoretically polynomial, it is known to be slow in practice [7]. Consequently, alternative approaches are necessary. Note that the function $f(x) := \max_{p \in \mathcal{M}(\mu)} \sum_{i=1}^{l} p_i \left((c^i)^T x - val^*(c^i)\right)$ is convex in $x$ (it is a pointwise maximum of affine functions). Therefore, (CGMR) is a convex optimization problem. A standard approach to solving it is using subgradient methods [9], where a subgradient of $f$ is required at each iteration. Recall that a subgradient of $f$ at $x$ is a vector $\eta$ such that $f(y) \geq f(x) + \eta^T(y-x) \ \forall y$. The next result follows from standard calculations in convex analysis. For the completeness, we include a proof.

**Proposition 4.** *For any $x$, let $p^* \in \mathrm{argmax}_{p \in \mathcal{M}(\mu)} \sum_{i=1}^{l} p_i \left((c^i)^T x - val^*(c^i)\right)$. Then $\eta := \sum_{i=1}^{l} p_i^* c^i$ is a subgradient of $f$.*

*Proof.* For any $y$, $f(y) = \max_{p \in \mathcal{M}(\mu)} \sum_{i=1}^{l} p_i \left((c^i)^T x - val^*(c^i) + (c^i)^T(y-x)\right)$. By the optimality of $p^*$,

$$f(y) \geq \sum_{i=1}^{l} p_i^* \left((c^i)^T x - val^*(c^i)\right) + \sum_{i=1}^{l} (p_i^* c^i)^T (y-x) = f(x) + \eta^T(y-x).$$

□

Thanks to Remark 3 and Proposition 4, in case of 2-monotone capacities, a subgradient of $f$ can be computed efficiently.

# 7 Conclusion

In this paper, we have used the generalized minimax regret criteria for optimization problems with uncertain objectives, where the uncertainty is modeled by belief functions and, more generally, capacities. We have identified some tractable cases and developed a 2-approximation method when focal sets of the considered belief functions are Catersian products of intervals. Future work includes applying subgradient methods to problem (CGMR) for linear programming problems or investigating problems (GMR) and (CGMR) for some practical combinatorial problems.

## References

1. Adam, L., Destercke, S.: Possibilistic preference elicitation by minimax regret. In: Uncertainty in Artificial Intelligence, pp. 718–727. PMLR (2021)
2. Agarwal, A., Zhang, T.: Minimax regret optimization for robust machine learning under distribution shift. In: Conference on Learning Theory, pp. 2704–2729. PMLR (2022)
3. Al Taha, F., Yan, S., Bitar, E.: A distributionally robust approach to regret optimal control using the Wasserstein distance. In: 2023 62nd IEEE Conference on Decision and Control (CDC), pp. 2768–2775. IEEE (2023)
4. Augustin, T., Coolen, F.P., De Cooman, G., Troffaes, M.C.: Introduction to Imprecise Probabilities. John Wiley & Sons, Hoboken (2014)
5. Conde, E.: A 2-approximation for minmax regret problems via a mid-point scenario optimal solution. Oper. Res. Lett. **38**(4), 326–327 (2010)
6. Grabisch, M.: Set Functions, Games and Capacities in Decision Making, vol. 46. Springer, Cham (2016). https://doi.org/10.1007/978-3-319-30690-2
7. Grötschel, M., Lovász, L., Schrijver, A.: Geometric Algorithms and Combinatorial Optimization. Springer, Berlin, Heidelberg (1993). https://doi.org/10.1007/978-3-642-78240-4
8. Kasperski, A., Zieliński, P.: Robust discrete optimization under discrete and interval uncertainty: a survey. In: Doumpos, M., Zopounidis, C., Grigoroudis, E. (eds.) Robustness Analysis in Decision Aiding, Optimization, and Analytics. ISORMS, vol. 241, pp. 113–143. Springer, Cham (2016). https://doi.org/10.1007/978-3-319-33121-8_6
9. Nesterov, Y.: Lectures on Convex Optimization, vol. 137. Springer, Cham (2018). https://doi.org/10.1007/978-3-319-91578-4
10. Shafer, G.: A Mathematical Theory of Evidence. Princeton University Press, Princeton (1976)
11. Vu, T.A., Afifi, S., Lefèvre, E., Pichon, F.: Optimization problems with evidential linear objective. Int. J. Approx. Reason. **161**, 108987 (2023)
12. Wasserman, L.A.: Belief functions and statistical inference. Can. J. Stat. **18**(3), 183–196 (1990)
13. Yager, R.R.: Decision making using minimization of regret. Int. J. Approx. Reason. **36**(2), 109–128 (2004)

# Measures of Uncertainty, Conflict and Distances

# A Mean Distance Between Elements of Same Class for Rich Labels

Arthur Hoarau[✉], Constance Thierry, Jean-Christophe Dubois, and Yolande Le Gall

Univ Rennes, CNRS, IRISA, DRUID, Rennes, France
arthur.hoarau@outlook.fr

**Abstract.** The prevalence of imperfections in data, characterized by uncertainty and imprecision, prompts the need for effective modeling techniques. The theory of belief functions offers a mathematical framework to address this challenge. In this paper, we tackle the problem of calculating the mean distance between elements of the same class, especially when class membership is uncertain and imprecise. Leveraging belief functions and a notion of similarity between elements, we propose a solution and validate its efficacy through experimental evaluations. The proposed method proves effective when labels exhibit low imprecision, whereas unsupervised methods may be more effective for labels closer to complete ignorance.

**Keywords:** Belief Functions · Rich Labels · Mean Distance

## 1 Introduction

The imperfection [9] in data is now prevalent in many application domains. It may be uncertainty (lack of knowledge, *e.g.* "Tomorrow it might be sunny") or imprecision (quantitative or completeness deficiency, *e.g.* "It will be sunny tomorrow or the day after"). The possibility of representing them provides a better way of taking them into account. The theory of belief functions [1,8] allows for the mathematical modeling of this uncertainty and imprecision, and the notion of distance between multiple bodies of evidence has been extensively studied [5] in this context. In connection with the notion of distances, there are numerous applications in unsupervised machine learning, such as clustering [2], notably with the evidential c-means [7]. Additionally, a related problem, which approaches the issue we are addressing, is that of missing value imputation [6]. In all cases, the representation of imperfection is an expanding field, especially in machine learning, and several recent works have been conducted with the aim of collecting real uncertain and imprecise labels from actual users.

In this paper, we focus on the mean distance between elements of the same class, which is easily calculable when the classes are known but less straightforward when the membership of elements to a class is defined uncertainly and imprecisely. We propose to address this issue using the theory of belief functions and a notion of similarity between labels. A degenerate and informal application of this method has been practiced in an active learning setup [3] (reduced to the similarity of the element's imperfect label with the labels of other elements and not that of the true classes). Since the notions of distances and belief functions are addressed, it may be interesting to note that an impossibility has been demonstrated [10] between the conjunctive combination of beliefs and the use of distances. However, this does not concern this case directly since the classical definition of distance is used, which separates two points, such as the Euclidean distance. The theory of belief functions intervenes here on the classes and not on the explanatory variables.

The document is structured as follows: Sect. 2 introduces the problem by recalling the calculation of the average distance between elements of the same class for hard labels. Section 3 presents the method by introducing the theory of belief functions. Section 4 proposes two experiments that help understand the behavior of the proposed method as well as its performance in a practical case. Finally, Sect. 5 concludes this article.

## 2   Mean Distance Between Elements for Hard Classes

Let $\mathcal{X} = \{x^n = (x_1^n, \ldots, x_P^n) | n = 1, \ldots, N\}$ represent a $P$ features collection of $N$ samples, and $\Omega = \{q_1, \ldots, q_C\}$ a set of $C$ classes. Let $d$ be a distance over the features space, for the proofs and experiments in this paper, the Euclidean distance will be adopted; however, any other distance can be used. It is defined as follows:

$$d(x^i, x^j) = \sqrt{\sum_{k=1}^{K}(x_k^i - x_k^j)^2}, \qquad (1)$$
$$= ||x^i - x^j||,$$

with $||x||$ the Euclidean norm of $x$, for convenience we denote $d(x^i, x^j) = d^{i,j}$.

A simple way to compute the mean distance d between all elements is to sum all pairwise distances and divide by the total number of pairs (excluding the distance between an element and itself):

$$\mathsf{d} = \frac{\sum_{i=1}^{N}\sum_{j=1}^{N} d^{i,j}}{N^2 - N}. \qquad (2)$$

This equation can be simplified for complexity reasons by computing only half of the matrix (of dimension $N * N$), dividing the number of elements by 2.

The interest lies in accounting for the class of each observation. The mean distance between elements of the same class $\mathsf{d}_q$ can be calculated as follows:

$$\mathsf{d}_q = \frac{\sum_{i=1}^{N_q}\sum_{j=1}^{N_q} d^{i,j}}{N_q^2 - N_q}, \qquad (3)$$

with $q$ as the corresponding class, and $N_q$ the number of elements of class $q$ (for the sums over $N_q$, we simplify notation by implying that the summed distances $d^{i,j}$ refer to those between $x^i$ and $x^j$ belonging to class $q$).

The tackled issue arises when the class assignment of an observation lacks certainty and precision. Thus, we introduce a mean distance between elements belonging to the same class, tailored for rich (uncertain and imprecise) label representations.

## 3 Mean Distance Between Elements for Rich Labels

This section presents the proposed method of mean distance between elements of the same class when the class is known uncertainly and imprecisely. The theory of belief functions [1,8] is used to model these rich labels.

### 3.1 Mean Distance for Rich Labels (MDRL)

The goal of the proposed method is to extend Equation (3) when the labels are uncertain and imprecise (when $N_q$ is unknown). For this purpose, a similarity measure between the target class $q$ and the rich label is used to weigh the contribution of each observation in the total calculation. In this paper, we arbitrarily choose the similarity measure $1 - d_J$, where $d_J$ is the Jousselme distance [4] between two mass functions. The method is defined by the following equation:

$$\text{MDRL}_q = \frac{\sum_{i=1}^{N}\sum_{j=1}^{N}(1 - d_J^{q,i})(1 - d_J^{q,j})d^{i,j}}{[\sum_{i=1}^{N}(1 - d_J^{q,i})]^2 - \sum_{i=1}^{N}(1 - d_J^{q,i})^2}, \quad (4)$$

with $d_J^{q,i}$ the Jousselme distance between $m_q$ (the categorical mass function on class $q$) and $m_i$ the mass function defining the class of $x^i$, and with $d^{i,j}$ the Euclidean distance between $x^i$ and $x^j$.

**Proposition 1.** This equation is equal to the classical mean distance between all observations (2) when considering complete ignorance.

**Proposition 2.** Equation (4) is equal to (3) for hard labels.

**Proposition 3.** This mean distance is null for identical objects, positive if an object is distinct from others, and symmetric under permutation of elements.

Propositions 1 and 2 are proven below. For Proof 1, all labels are completely ignorant ($m_i(\Omega) = 1, \forall i \in [0, N]$), therefore $d_J$ becomes constant, let $(1 - d_J^{q,i}) = \Delta^\Omega$. For Proof 2, and thus in the case of hard labels, the Jousselme distance between two elements of the same class becomes 0, and 1 otherwise for a different class.

**Proof 1:**

$$\text{MDRL}_q = \frac{\sum_{i=1}^{N}\sum_{j=1}^{N}(1-d_J^{q,i})(1-d_J^{q,j})d^{i,j}}{[\sum_{i=1}^{N}(1-d_J^{q,i})]^2 - \sum_{i=1}^{N}(1-d_J^{q,i})^2}$$

$$= \frac{\sum_{i=1}^{N}\sum_{j=1}^{N}(\Delta^{\Omega})(\Delta^{\Omega})d^{i,j}}{[\sum_{i=1}^{N}(\Delta^{\Omega})]^2 - \sum_{i=1}^{N}(\Delta^{\Omega})^2} \quad (5)$$

$$= \frac{(\Delta^{\Omega})^2 \sum_{i=1}^{N}\sum_{j=1}^{N} d^{i,j}}{(\Delta^{\Omega})^2 [(\sum_{i=1}^{N} 1)^2 - \sum_{i=1}^{N} 1]}$$

$$= \frac{\sum_{i=1}^{N}\sum_{j=1}^{N} d^{i,j}}{N^2 - N} \iff (2)$$

**Proof 2:**

$$\text{MDRL}_q = \frac{\sum_{i=1}^{N}\sum_{j=1}^{N}(1-d_J^{q,i})(1-d_J^{q,j})d^{i,j}}{[\sum_{i=1}^{N}(1-d_J^{q,i})]^2 - \sum_{i=1}^{N}(1-d_J^{q,i})^2}$$

$$= \frac{\sum_{i=1}^{N_q}\sum_{j=1}^{N_q}(1)(1)d^{i,j}}{[\sum_{i=1}^{N_q} 1]^2 - \sum_{i=1}^{N_q}(1)^2} \quad (6)$$

$$= \frac{\sum_{i=1}^{N_q}\sum_{j=1}^{N_q} d^{i,j}}{N_q^2 - N_q} \iff (3)$$

**Example:** We consider students who have obtained grades in three subjects (they belong to class 1 or class 2: $\Omega = \{1,2\}$). The goal is to determine the homogeneity[1] of the students' level in the two classes, the mean distance between the students (on the grades) according to their class is then calculated. This intra-class inertia allows us to compare the homogeneity level of each class. Students, grades, and their true class are described in left hand part of Table 1. A numerical conversion is made (from F to $A^2$). The mean distance is calculated using Eq. (3), and the obtained values are 11.2 for students in true class 1 and 5.5 for students in true class 2. Class 2 is thus much more homogeneous than class 1. Now, suppose that the students' class is partially known, this uncertainty and imprecision are described in the right hand part of Table 1. The formula used is no longer applicable[3]. With the proposed Eq. (4), we obtain MDRL values of 10.0 for class 1 and 6.78 for class 2. This also indicates that class 2 is more homogeneous than class 1.

## 4 Experiments

In this section, we propose two experiments to demonstrate the usefulness of the proposed method on several datasets presented in Table 2. These datasets contain quantitative variables that have been processed to remove the mean and scale to unit variance. Each draw is performed 100 times (one draw corresponds to the selection of noised observations). Firstly, a preliminary experiment describes the behavior of the method with respect to the quality of the labels and to its theoretical limit between the true mean distance based on classes and a naive

---

[1] Homogeneity is represented by the mean distance between students of the same class.
[2] Grades are: $A, A^-, B^+, B, B^-, C^+, C, C^-, D^+, D, D^-, F$.
[3] For the class that maximizes the pignistic probability, the mean distances are 9.3 for class 1 and 7.8 for class 2.

**Table 1.** Students' grades for each course (on the left) with true class (in the middle) and rich labels indicating class membership (on the right).

| Student | Course 1 | Course 2 | Course 3 | True Class | Class 1 | Class 2 | $\Omega$ |
|---|---|---|---|---|---|---|---|
| Alice | 11 | 8 | 11 | 1 | 1 | 0 | 0 |
| Bob | 6 | 0 | 11 | 2 | 0 | 1 | 0 |
| Carol | 4 | 2 | 0 | 1 | 0.8 | 0 | 0.2 |
| Dave | 1 | 11 | 5 | 1 | 0 | 0.1 | 0.9 |
| Eve | 8 | 4 | 9 | 2 | 0 | 0.8 | 0.2 |
| Mallory | 10 | 8 | 7 | 2 | 0.1 | 0 | 0.9 |
| Oscar | 8 | 0 | 3 | 1 | 1 | 0 | 0 |
| Trudy | 7 | 6 | 10 | 2 | 0 | 1 | 0 |

mean distance over the entire dataset. The second experiment compares the performance of the proposed method with other methods, both supervised and unsupervised. For both experiments, the mean distances are calculated based on the noise level as follows.

*Imprecision noise:* An observation is randomly chosen and the corresponding label loses one degree of precision, with another class chosen at random in $\Omega$ (*e.g.* If an observation is labeled *Virginica* on Iris dataset, the noisy label becomes either *Virginica* $\cup$ *Setosa* or *Virginica* $\cup$ *Versicolor*). A 50% noisy dataset would mean that half of the labels have lost a degree of precision.

**Table 2.** Datasets description, with class distribution entropy.

| Dataset | Observations | Classes | Features | Entropy |
|---|---|---|---|---|
| Ecoli | 336 | 8 | 7 | 0.73 |
| Glass | 214 | 6 | 9 | 0.83 |
| Seeds | 210 | 3 | 7 | 1.00 |
| Wine | 178 | 3 | 13 | 0.99 |
| Heart | 303 | 2 | 7 | 1.00 |
| Iris | 150 | 3 | 4 | 1.00 |
| Liver | 345 | 2 | 6 | 0.98 |
| Pima | 768 | 2 | 8 | 0.93 |
| Parkinson | 195 | 2 | 22 | 0.81 |
| Balance | 625 | 3 | 4 | 0.83 |
| Post-Operative | 86 | 2 | 8 | 0.85 |
| Sonar | 208 | 2 | 60 | 1.00 |
| Ionosphere | 351 | 2 | 34 | 0.94 |
| Banana | 5300 | 2 | 2 | 0.99 |
| Breast Cancer | 569 | 2 | 30 | 0.95 |

### 4.1 Experiment 1: Average Behavior

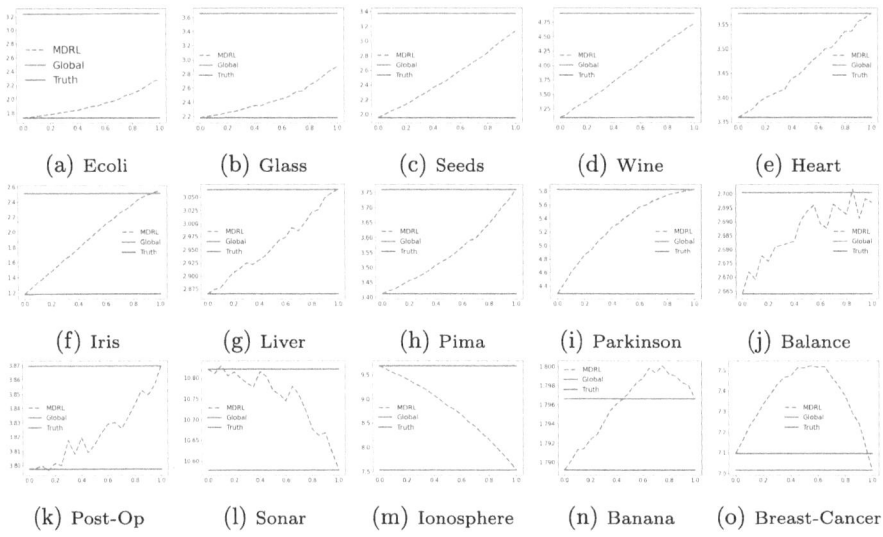

**Fig. 1.** Mean Distance for Rich Labels Vs. Noise. (Class 0)

This first experiment focuses on the evolution of the proposed Mean Distance for Rich Labels (MDRL) across multiple datasets, limited by the mean distances between elements of the same true class (3) and the global naive distance between all elements (2), these values are thus invariant to noise. Figure 1 illustrates the behavior of the proposed method with regard to the noise. It varies from 0 (unnoisy dataset) to 1 (fully noised). One class[4] is depicted for each dataset, and the *Ground Truth* line represents the true mean distance between elements of this class. The *Global* line represents the mean distance between all elements of the dataset.

For datasets with two classes (Heart, Liver, Pima, Parkinson, Post-Operative, Sonar, Ionosphere, Banana, and Breast Cancer), the proposed method starts, as theoretically expected, exactly at the true mean distance and converges to the global mean distance when the noise level reaches 100%. Indeed, the noise used translates to total ignorance for datasets limited to two classes. For datasets with a large number of classes (Ecoli and Glass), the proposed method remains closer to the true value. If the noise added total ignorance instead of a degree of imprecision, the curve would also converge to the global mean distance when the dataset is fully noisy. Only the Breast Cancer dataset makes the task very challenging for estimating the mean distance between elements of the same class with respect to noise, due to the particular distribution of observations in the

---
[4] The first class present in each dataset is always depicted.

variables space for this dataset. The method is therefore largely capable of representing a mean distance that varies between the truth and the least informative value (without using any labels at all). The second experiment then aims to determine whether this method is relevant in terms of performance.

## 4.2 Experiment 2: Performance of the Method

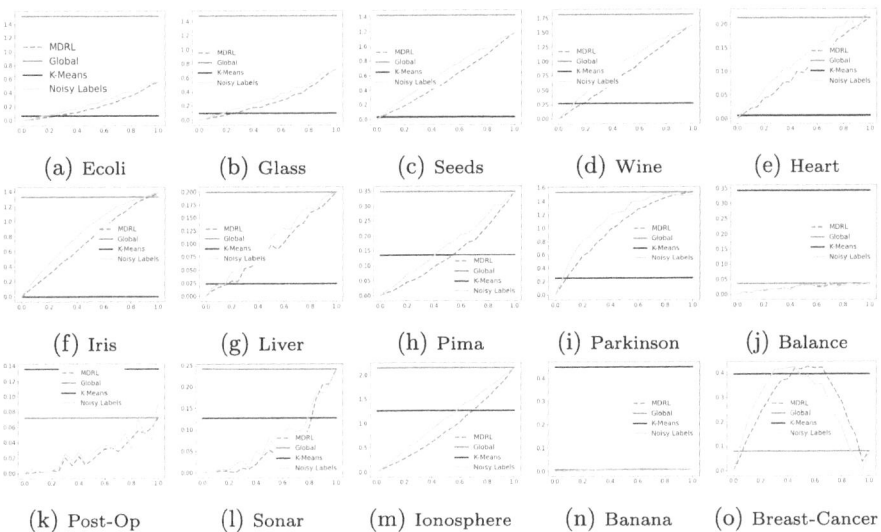

**Fig. 2.** Error across different methods Vs. Noise. (Class 0)

In this experiment, the proposed Mean Distance for Rich Labels (MDRL) method is compared with three other methods for estimating the mean distance between elements of the same class on the datasets presented earlier. Once again, only one class per dataset is being studied.

The naive *Global* method calculates the mean distance over all observations without considering the class, as mentioned earlier. Another method, *Noisy labels*, involves selecting the class that maximizes the pignistic probability for an observation and calculating the mean distance between elements of that class. The last compared method utilizes the unsupervised clustering algorithm of *K-means* to create clusters that maximize inter-class distance and minimize intra-class distance. The mean distance between the elements of this cluster is then calculated. A significant advantage given to this method is that the true number of classes is provided to the K-means algorithm to form its clusters. Moreover, the closest mean distance to the truth, among all created clusters, is chosen for comparison with the studied true class.

Figure 2 presents the difference between the estimated value and the true mean distance between elements of the studied class for each level of noise. Since the Global and K-means methods are unsupervised, it is expected for their curves to be constant, as they do not depend on labels and therefore not on noise. The lower the curve, the closer the value is to the true mean distance, indicating better performance. The least performing method is naturally the Global mean, which does not take into account the labels of the observations. The only dataset where this method is particularly effective is the Breast-Cancer dataset. The proposed MDRL method is better performing than the hard *Noisy Labels* but follows its trend. This phenomenon is theoretically expected since the proposed method aims to be an improvement over it. Finally, the K-means method is much more competitive, often close to 0. However, since the proposed method equals the true mean value when there is no imprecision, it is always better performing than K-means at least with little noise. Then the performance degrades with the addition of noise. The relevance of using the proposed method therefore depends on the noise (and more generally on the uncertainty and imprecision of the sources).

## 5 Conclusion

In this paper, we propose a mean distance between elements of the same class when classes are not known with certainty and precision but represented by a belief function. This measure is shown to be limited by the true value of the mean distance between elements of the same class when labels are known with certainty and precision and the naive measure of the mean distance between all elements in the case of complete ignorance. Two proofs and experiments are also conducted to theoretically support these properties.

A distance and a dissimilarity measure are necessary. Therefore, the Euclidean distance and the Jousselme distance have been arbitrarily chosen here, but other distances and dissimilarity measures can be used. It has been observed during our experiments that this method can be useful under moderate noise (or imprecision), but it could be more appropriate to use unsupervised methods (such as K-means) when noise is significant. Many issues can be addressed with such a measure, and its practical use is already ongoing in machine learning problems, specifically in active learning.

*Special thanks to Vincent Lemaire for conducting a pre-peer review.*

## References

1. Dempster, A.P.: Upper and lower probabilities induced by a multivalued mapping. Ann. Math. Stat. **38**(2), 325–339 (1967)
2. Denœux, T., Kanjanatarakul, O.: Evidential clustering: a review. In: Huynh, V.N., Inuiguchi, M., Le, B., Le, B.N., Denoeux, T. (eds.) Integrated Uncertainty in Knowledge Modelling and Decision Making, pp. 24–35 (2016)
3. Hoarau, A., Martin, A., Dubois, J.C., Le Gall, Y.: Imperfect labels with belief functions for active learning. In: Le Hegarat-Mascle, S., Bloch, I., Aldea, E. (eds.) BELIEF 2022. LNCS, vol. 13506, pp. 44–53. Springer, Heidelberg (2022). https://doi.org/10.1007/978-3-031-17801-6_5

4. Jousselme, A.L., Grenier, D.: Éloi Bossé: a new distance between two bodies of evidence. Inf. Fusion **2**(2), 91–101 (2001)
5. Jousselme, A.L., Maupin, P.: Distances in evidence theory: comprehensive survey and generalizations. Int. J. Appro. Reason. **53**(2), 118–145 (2012)
6. Liu, Z., Pan, Q., Dezert, J., Martin, A.: Adaptive imputation of missing values for incomplete pattern classification. Pattern Recogn. **52**, 85–95 (2016)
7. Masson, M.H., Denæux, T.: ECM: an evidential version of the fuzzy c-means algorithm. Pattern Recogn. **41**(4), 1384–1397 (2008)
8. Shafer, G.: A Mathematical Theory of Evidence. Princeton University Press, Princeton (1976)
9. Smets, P.: Imperfect information: imprecision and uncertainty. In: Motro, A., Smets, P. (eds.) Uncertainty Management in Information Systems, pp. 225–254. Springer, Boston (1997). https://doi.org/10.1007/978-1-4615-6245-0_8
10. Zhang, Y., Destercke, S., Zhang, Z., Bouadi, T., Martin, A.: On computing evidential centroid through conjunctive combination: an impossibility theorem. IEEE Trans. Artif. Intell. **43**, 487–496 (2022)

# Threshold Functions and Operations in the Theory of Evidence

Alexander Lepskiy[✉]

Higher School of Economics, 20 Myasnitskaya Ulitsa, Moscow 101000, Russia
alepskiy@hse.ru
https://www.hse.ru/en/org/persons/10586209

**Abstract.** The article introduces and discusses threshold belief and plausibility functions. When forming such functions, only focal elements that are "significant" for a given set are taken into account. The significance of focal elements is determined using a similarity measure and a threshold. Threshold functionals of uncertainty, external and internal conflicts, threshold rules of combination are introduced and considered on the basis of threshold functions of the theory of evidence. A number of examples are given to illustrate the use of threshold tools.

**Keywords:** Similarity measures · Theory of evidence

## 1 Introduction

All focal elements that intersect with a given set are taken into account in the theory of evidence when assessing the plausibility that the true alternative belongs to the given set. However, some focal elements may have been formed inaccurately. If a focal element overlaps "weakly" (relative to some similarity measure) with a given set, then the degree of confidence that this focal element is important in assessing the plausibility of membership of a true alternative in a given set will be small. Therefore, the problem of taking into account the degree of intersection or inclusion of focal elements with a given set is relevant when forming the main functions of the theory of evidence (belief, plausibility, etc.).

This problem is related to the analysis of the sensitivity of the main functions of the theory of evidence to small changes in focal elements. Traditionally, sensitivity to small changes in focal elements is analyzed using generalization-specialization procedures [6]. But these procedures are performed on the body of evidence itself and do not take into account the degree of intersection or inclusion with a given set.

A similar problem of taking into account significant (i. e., having a large degree of intersection or inclusion) focal elements is relevant when performing operations of aggregating bodies of evidence, assessing conflict, degree of uncertainty, etc.

---

The study was implemented in the framework of the Basic Research Program at the National Research University Higher School of Economics (HSE University) in 2024.

© The Author(s), under exclusive license to Springer Nature Switzerland AG 2024
Y. Bi et al. (Eds.): BELIEF 2024, LNAI 14909, pp. 216–224, 2024.
https://doi.org/10.1007/978-3-031-67977-3_23

The procedure for calculating plausibility functions taking into account only significant focal elements with respect to some similarity measure [2] will be discussed in this article. Similarly, focal elements that are "significantly" contained in a given set can be taken into account when forming a belief function. We will call such functions threshold, since they take into account focal elements for which the similarity measure exceeds a certain threshold. In general, threshold functions may not satisfy some important properties of evidence theory. The article discusses the conditions when these properties will be satisfied.

In addition, threshold functions give rise to the concepts of threshold functionals of uncertainty, external and internal conflicts of bodies of evidence, as well as threshold rules of combination. All these concepts are discussed in this article.

Taking into account the significant focal elements allows for a controlled reduction in the measure of uncertainty in the description of bodies of evidence compared to non-threshold functions.

## 2 Necessary Information from the Theory of Evidence

Let us recall the necessary information from the theory of evidence [10]. Let $X = \{x_1, ..., x_n\}$ be a finite set; $2^X$ be the set of all subsets on $X$; m: $2^X \to [0,1]$, $\sum_{A \in 2^X} m(A) = 1$, $m(\emptyset) = 0$ is a mass function; $\mathcal{A}$ is a set of all focal elements, i.e. $A \in \mathcal{A}$ if $m(A) > 0$; $F = (\mathcal{A}, m)$ is the body of evidence; $\mathcal{F}(X)$ be the set of all bodies of evidence on the $X$. The body of evidence $F = (\mathcal{A}, m)$ is symbolically convenient to represent in the form $F = \sum_{A \in \mathcal{A}} m(A) F_A$, where $F_A = (\{A\}, 1)$ is a categorical body of evidence.

The body of evidence $F = (\mathcal{A}, m)$ uniquely defines the belief function

$$Bel(A) = \sum_{B \subseteq A} m(B)$$

and its dual plausibility function

$$Pl(A) = 1 - Bel(\neg A) = \sum_{B \cap A \neq \emptyset} m(B). \qquad (1)$$

It's always true that $Bel(A) \leq Pl(A) \ \forall A \subseteq X$, and the length of the interval $[Bel(A), Pl(A)]$ determines the degree of uncertainty of the event $x \in A$ [1].

## 3 Threshold Functions of Belief and Plausibility

The summation of the mass functions in the formula (1) is carried out over all focal elements that have a non-empty intersection with the given set. This sum may include focal elements that have small (relative to some measure) intersection compared to the measures of the intersecting sets themselves. Such elements can be considered insignificant for assessing the plausibility of belonging to a given set. If we want to form a plausibility function in which only significant elements are taken into account, then this can be done using the formula

$$Pl_h(A) = \sum_{B:s(A,B)>h} m(B), \qquad (2)$$

where $h \in [0,1)$ and $s(A,B)$ is a measure (index) of similarity [2], satisfying the conditions: 1) $0 \leq s(A,B) \leq 1$; 2) $s(A,B) = 0 \Leftrightarrow A \cap B = \emptyset$; 3) $s(A,A) = 1 \ \forall A \neq \emptyset$ (or weaker condition $\max_B s(A,B) = s(A,A)$). The similarity measure must often (but not necessarily) satisfy the symmetry condition: $s(A,B) = s(B,A)$. Asymmetric similarity measures are called inclusion measures. Examples of similarity measures: a) Jaccard index $J(A,B) = \frac{|A \cap B|}{|A \cup B|}$; b) $s(A,B) = \frac{|A \cap B|}{|X|}$; c) $s(A,B) = \begin{cases} 1, A \cap B \neq \emptyset, \\ 0, A \cap B = \emptyset; \end{cases}$ d) Simpson coefficient $s(A,B) = \frac{|A \cap B|}{\min\{|A|,|B|\}}$; e) Sörensen inclusion measure $s(A,B) = \frac{|A \cap B|}{|B|}$; f) Sörensen coefficient $s(A,B) = \frac{2|A \cap B|}{|A|+|B|}$.

We will call a function of the form (2) the threshold plausibility function. The parameter $h \in [0,1)$ regulates the degree of similarity between the focal elements that are taken into account when calculating the plausibility and the given set. The larger $h$, the higher this degree of closeness.

Transformations (1), (2) can be represented in matrix form $\sum_B S(A,B)m(B) = \mathbf{Sm}$, where $\mathbf{S} = (S(A,B))_{A,B \in 2^X}$, $\mathbf{m}$ – $2^{|X|}$-dimensional column vector, the coordinates of which are the values of the mass function $m(A)$, $A \in 2^X$. For example, $Pl_h = \mathbf{S}_h \mathbf{m}$, where $\mathbf{S}_h = (S_h(A,B))_{A,B \in 2^X}$ and $S_h(A,B) = \begin{cases} 1, \text{ if } s(A,B) > h, \\ 0, \text{ otherwise.} \end{cases}$ Such matrix transformations were considered, for example, in [5].

*Remark 1.* In general, the transformation matrix can be non-binary. Its elements may depend on the measure of similarity between the focal elements and the set, the plausibility function of which is calculated. For example, $\widetilde{S}_h(A,B) = \begin{cases} s(A,B), \text{ if } s(A,B) > h, \\ 0, \qquad \text{otherwise.} \end{cases}$ Note that the plausibility function $\widetilde{Pl_0} = \widetilde{\mathbf{S}}_0 \mathbf{m}$ with subscript e) coincides with the pignistic probability [11] $\widetilde{Pl_0}(A) = \sum_B \frac{|A \cap B|}{|B|} m(B)$.

Properties of functions $Pl_h$:

1) $Pl_0 = Pl$. This follows from condition 2) for the similarity measure;
2) $Pl_{h_1}(A) \leq Pl_{h_2}(A) \ \forall A \in 2^X$ if $h_1 \geq h_2$.
3) $Pl_h(\emptyset) = 0$, but $Pl_h(X) \leq 1$.

Property 3) means that the inequality $Pl_h(X) < 1$ will be true for some similarity measures $s$ and sufficiently large $h \in [0,1)$. But, for example, equality $Pl_h(X) = 1 \ \forall h \in [0,1)$ is always true for the Simpson coefficient or index e).

4) if the similarity measure $s(A, B)$ is monotone with respect to $A$ (i.e. $s(A', B) \subseteq s(A'', B)$ implies that $A' \subseteq A''$ ), then the function $Pl_h$ will be monotonic.

The monotonicity condition for the similarity measure $s(A, B)$ with respect to the first argument is certainly satisfied by measures b), c), e).

Let us find the threshold belief function $Bel_h$ as dual to $Pl_h$. We have

$$Bel_h(A) = 1 - Pl_h(\neg A) = 1 - \sum_{B:s(\neg A, B) > h} m(B) = \sum_{B:s(\neg A, B) \leq h} m(B). \quad (3)$$

In other words, the threshold belief function is formulated taking into account focal sets, which may contain a "small" number of elements not included in the considered set.

*Remark 2.* The classical concept of a belief function (including a threshold one) assumes that the summation in (3) must be performed over sets $B$ that intersect $A$ and satisfy the condition $s(\neg A, B) \leq h$ in the threshold case. But the statement $s(\neg A, B) \leq h \Rightarrow A \cap B \neq \emptyset$ is not true in the general case. It will be true, for example, for similarity measures c), d), e). These similarity measures are preferable to use in threshold belief functions.

*Remark 3.* Note that $Bel_h(X) = 1 \; \forall h \in [0, 1)$ is true for $Bel_h$ and any similarity measure. But it may be $Bel_h(\emptyset) > 0$ for some similarity measures and some $h \in [0, 1)$. This situation corresponds to the so-called open world concept, which is considered within the framework, for example, of the Transferable Belief Model [12] or the Generalized evidence theory [4]. At the same time, $Bel_0 = Bel$.

The duality relationship and property 2) imply that:

2') $Bel_{h_1}(A) \geq Bel_{h_2}(A) \; \forall A \in 2^X$ if $h_1 \geq h_2$;

4') if the similarity measure $s(A, B)$ is monotone with respect to $A$, then the function $Bel_h$ will be monotone.

*Remark 4.* If the threshold belief and plausibility functions are defined on a finite set $X$, then they can be represented in matrix form. Let $0 = h_0 < \ldots < h_l = 1$ be an ordered set of different values of the similarity measure $s(A, B)$, $A, B \in 2^X$. It is clear that the values of $Bel_h(A)$ and $Pl_h(A)$ will not change within the intervals $h \in [h_j, h_{j+1})$, $j = 0, \ldots, l-1$. Let $\{A_i\}_{i=1}^{2^X - 1}$ be the lexicographically ordered set of all proper subsets of $X$. Then the function $Bel_h$ can be represented by the matrix $\mathbf{Bel} = (bel_{ij})$, where $bel_{ij} = Bel_{h_j}(A_i)$. The matrix $\mathbf{Pl}$ for $Pl_h$ is formed in a similar way. Note that the partition $H = \{h_j\}_{j=0}^{l}$ and the size of the matrices depend only on the similarity measure $s$ and $|X|$, but do not depend on evidence bodies.

In general, the agreement condition

$$Bel_h(A) \leq Pl_h(A) \quad \forall A \in \mathcal{A}. \quad (4)$$

may not be fulfilled.

Since for $h = 0$ the condition (4) is true, the function $Pl_h$ does not increase, and the function $Bel_h$ does not decrease with respect to $h$, then there is a value $h_0 = \sup\{h : Bel_h(A) \leq Pl_h(A), A \in \mathcal{A}\} > 0$ that the condition (4) is true on the interval $[0, h_0)$ and false on the interval $(h_0, 1)$. The equalities $Pl_h(X) = 1$ and $Bel_h(\emptyset) = 0\ \forall h \in [0, h_0]$ follow from (4) and the fact that $Bel_h(X) = 1$, $Pl_h(\emptyset) = 0\ \forall h \in [0, 1)$.

The following estimates for $h_0$ follow from (2) and (3).

**Proposition 1.** *If* $\{B \in \mathcal{A} : s(\neg A, B) \leq h\} \subseteq \{B \in \mathcal{A} : s(A, B) > h\}\ \forall A \in \mathcal{A}$, *then (4) is true.*

**Corollary 1.** *Inequality (4) is true if we have:*

a) $s(A,B) = \dfrac{|A \cap B|}{|B|}$ and $h < \dfrac{1}{2}$; b) $s(A,B) = \dfrac{|A \cap B|}{|X|}$ and $h < \dfrac{\min_{B \in \mathcal{A}} |B|}{2|X|}$;

c) $J(A,B) = \dfrac{|A \cap B|}{|A \cup B|}$ and $h < \min_{B \in \mathcal{A}} \dfrac{|B|}{|X| + |B|}$; d) $s(A,B) = \dfrac{|A \cap B|}{\min\{|A|, |B|\}}$

and $h < \dfrac{1}{2}$; e) $s(A,B) = \dfrac{2|A \cap B|}{|A| + |B|}$ and $h < \min_{B \in \mathcal{A}} \dfrac{2|B|}{|X| + 2|B|}$.

**Proposition 2.** *We have for measures a) or e) and for* $h \in (0, 1)$: $Bel_h(X) = 1$,

$$Bel_h(\{x_i\}) = \sum_{B \in \mathcal{A} : x_i \in B, |B| \leq \frac{1}{1-h}} m(B),\quad Pl_h(\{x_i\}) = \sum_{B \in \mathcal{A} : x_i \in B, |B| < \frac{1}{h}} m(B).$$

*In addition, we have*

$$Pl_h(X) = \sum_{B \in \mathcal{A} : |B| > h|X|} m(B),\quad Bel_h(\emptyset) = \sum_{B \in \mathcal{A} : |B| \leq h|X|} m(B)$$

*for measure a) and* $Pl_h(X) = 1$, $Bel_h(\emptyset) = 0$ *for measure e).*

*Example 1.* Let $F = 0.2F_{\{a,b\}} + 0.3F_{\{a,c\}} + 0.4F_{\{b\}} + 0.1F_{\{a,b,c\}}$ is the body of evidence on $X = \{a, b, c\}$. Let us find the matrices **Bel** and **Pl** for the similarity measure e). We have the partition $H = \{0, \frac{1}{3}, \frac{1}{2}, \frac{2}{3}, 1\}$ for this measure. Then

$$\mathbf{Bel} = \begin{array}{c|cccc} & [0, \tfrac{1}{3}) & [\tfrac{1}{3}, \tfrac{1}{2}) & [\tfrac{1}{2}, \tfrac{2}{3}) & [\tfrac{2}{3}, 1) \\ \hline \{a\} & 0 & 0 & 0.5 & 0.6 \\ \{b\} & 0.4 & 0.4 & 0.6 & 0.7 \\ \{c\} & 0 & 0 & 0.3 & 0.4 \\ \{a,b\} & 0.6 & 0.7 & 1 & 1 \\ \{a,c\} & 0.3 & 0.4 & 0.6 & 0.6 \\ \{b,c\} & 0.4 & 0.5 & 1 & 1 \end{array},\quad \mathbf{Pl} = \begin{array}{c|cccc} & [0, \tfrac{1}{3}) & [\tfrac{1}{3}, \tfrac{1}{2}) & [\tfrac{1}{2}, \tfrac{2}{3}) & [\tfrac{2}{3}, 1) \\ \hline \{a\} & 0.6 & 0.5 & 0 & 0 \\ \{b\} & 0.7 & 0.6 & 0.4 & 0.4 \\ \{c\} & 0.4 & 0.3 & 0 & 0 \\ \{a,b\} & 1 & 1 & 0.7 & 0.6 \\ \{a,c\} & 0.6 & 0.6 & 0.4 & 0.3 \\ \{b,c\} & 1 & 1 & 0.5 & 0.4 \end{array}.$$

The first columns of these matrices correspond to the classical functions $Bel$ and $Pl$. Condition (4) is satisfied only for the first two columns, i.e. for $h \in [0, \frac{1}{2})$.

## 4 Threshold Uncertainty and Internal Conflict

Functional $U_h : \mathcal{F}(X) \to [0,1]$

$$U_h(F) = \frac{1}{2^n - 2} \sum_A (Pl_h(A) - Bel_h(A)), \; ; \; h \in [0, h_0)$$

has the meaning of the uncertainty value [1] of the body of evidence $F = (\mathcal{A}, m)$ for a given threshold value $h \in [0, h_0)$. It is easy to see that the functional $U_h(F)$ does not increase with respect to $h \in [0, h_0)$.

The value $E_h(F) = U_0(F) - U_h(F)$ will characterize the total error in calculating the uncertainty at a given threshold value $h \in [0, h_0)$.

*Example 2.* The functionals $U_h(F)$ and $E_h(F)$ for the body of evidence from Example 1 are equal, respectively

$$U_h(F) = \begin{cases} \frac{13}{30}, \; h \in [0, \frac{1}{3}), \\ \frac{1}{3}, \; h \in [\frac{1}{3}, \frac{1}{2}). \end{cases} \quad E_h(F) = \begin{cases} 0, \; h \in [0, \frac{1}{3}), \\ \frac{1}{10}, \; h \in [\frac{1}{3}, \frac{1}{2}). \end{cases}$$

The internal conflict of the body of evidence $Con\_in : \mathcal{F}(X) \to [0,1]$ characterizes the degree of unconsolidation of focal elements [8]. There are different ways to assess internal conflict. For example, the measure of internal conflict proposed in [3]:

$$Con\_in(F) = 1 - \max_{1 \leq i \leq n} Pl(x_i).$$

But the internal conflict of the body of evidence will be zero with respect to such a measure in any case of logical consistency of the set of focal elements: $\bigcap_{A \in \mathcal{A}} A \neq \emptyset$. This "strict" requirement for the degree of internal conflict may not always be justified. For example, if the electoral college must choose several candidates from the set $\{a, b, c, d, e\}$ and half of the electors indicated candidates $\{a, b, c\}$, and the other half indicated candidates $\{c, d, e\}$, then the measure $Con\_in(F) = 0$. However, such a body of evidence must be considered internally conflicting.

If we use the threshold plausibility function $Pl_h$ instead of the function $Pl$, we obtain a threshold measure of internal conflict

$$Con\_in_h(F) = 1 - \max_{1 \leq i \leq n} Pl_h(x_i).$$

Since the function $Pl_h$ does not increase, the measure of internal conflict will not decrease with increasing $h \in [0, h_0)$. For example, the measure of internal conflict for the above example with electors and for the Jaccard similarity measure would be equal to

$$Con\_in_h(F) = 1 - \max_{1 \leq i \leq n} \sum_{B \in \mathcal{A}: x_i \in B, |B| < \frac{1}{h}} m(B) = \begin{cases} 0, \; h \in [0, \frac{1}{3}), \\ 1, \; h \in [\frac{1}{3}, \frac{1}{2}]. \end{cases}$$

*Example 3.* The functional $Con\_in_h(F)$ for the body of evidence from Example 1 is equal to $Con\_in_h(F) = 1 - Pl_h(\{b\}) = \begin{cases} 0.3, \ h \in [0, \frac{1}{3}), \\ 0.4, \ h \in [\frac{1}{3}, \frac{1}{2}). \end{cases}$

Then the optimization problem of finding a threshold $h \in [0, h_0)$ that would minimize the PP functional

$$\Phi_h(F) = U_h(F) + \lambda Con\_in_h(F) \to \min$$

can be formulated. The parameter $\lambda > 0$ adjusts the priority between uncertainty and internal conflict.

*Example 4.* The functional $\Phi_h(F)$ for the body of evidence from Example 1 is equal to

$$\Phi_h(F) = \begin{cases} \frac{13}{30} + 0.3\lambda, \ h \in [0, \frac{1}{3}), \\ \frac{1}{3} + 0.4\lambda, \ h \in [\frac{1}{3}, \frac{1}{2}), \end{cases} \quad \arg\min_h \Phi_h(F) = \begin{cases} [0, \frac{1}{3}), \ \lambda > 1, \\ [\frac{1}{3}, \frac{1}{2}), \ \lambda \in (0, 1]. \end{cases}$$

The pair $(Con\_in_h(F), U_h(F))$ characterizes the "quality" of the body of evidence. Less uncertainty and less internal conflict correspond to a higher "quality" body of evidence. We can pose the problem of finding a threshold $h \in [0, h_0]$ at which the uncertainty will be minimal, provided that the internal conflict does not exceed a given value.

## 5 Threshold Aggregation and External Conflict

Suppose that two bodies of evidence $F_1 = (\mathcal{A}_1, m_1)$ and $F_2 = (\mathcal{A}_2, m_2)$ are given on one and the same set $X$. Only strongly interacting focal elements can be taken into account when aggregating these bodies of evidence into one body of evidence $\mathcal{F}(X) \times \mathcal{F}(X) \to \mathcal{F}(X)$. Then the conjunctive threshold aggregation, similar to Dempster's rule, will take the form

$$m_h(A) = \frac{1}{k_h} \sum_{B \cap C = A, s(B,C) > h} m_1(B) m_2(C), \quad m_h(\emptyset) = 0, \tag{5}$$

where $k_h = \sum_{s(B,C) > h} m_1(B) m_2(C) \neq 0$. If $k_h = 0$, then the rule (5) is not applicable. Rule (5) is a special case of Zhang's center combination rule [13]. Note that $s(B, C) > 0 \Rightarrow A = B \cap C \neq \emptyset$ in (5).

The value

$$Con_h(F_1, F_2) = 1 - k_h = \sum_{s(B,C) \leq h} m_1(B) m_2(C) \tag{6}$$

has the meaning of a threshold measure of external conflict between bodies of evidence. Measures of conflict with weights were also considered in the [9]. It is easy to see that for any similarity measure and any $F_1, F_2 \in \mathcal{F}(X)$ the following is true:

1) $Con_0(F_1, F_2) = \sum_{B \cap C = \emptyset} m_1(B) m_2(C)$ is the most popular measure of external conflict from the Dempster rule (for a review of other measures of external conflict see [7]);
2) $Con_1(F_1, F_2) = 1$.

*Example 5.* Let three bodies of evidence be given on $X = \{a, b, c\}$: $F_1 = 0.2F_{\{a,b\}} + 0.3F_{\{a\}} + 0.4F_{\{b\}} + 0.1F_{\{a,b,c\}}$, $F_2 = 0.6F_{\{b,c\}} + 0.4F_{\{a,b,c\}}$, $F_3 = 0.4F_{\{a\}} + 0.4F_{\{b\}} + 0.2F_{\{a,b,c\}}$. Suppose we select two bodies of evidence from these three with the least external conflict for subsequent aggregation. The values of the conflict measure $Con_h$ for each pair, calculated using the formula (6) for the Jaccard similarity measure, are presented in the Table 1.

**Table 1.** The values of the conflict measure $Con_h$.

| $h$ | $(F_1, F_2)$ | $(F_1, F_3)$ | $(F_2, F_3)$ |
|---|---|---|---|
| $[0, \frac{1}{3})$ | 0.18 | 0.28 | 0.24 |
| $[\frac{1}{3}, \frac{1}{2})$ | 0.58 | 0.44 | 0.56 |
| $[\frac{1}{2}, \frac{2}{3})$ | 0.82 | 0.72 | 0.8 |
| $[\frac{2}{3}, 1)$ | 0.96 | 0.82 | 0.92 |

If we use the usual (threshold-free, i.e. $h = 0$) measure of external conflict, then the pair $F_1, F_2$ will have the least conflict. But if we want to take into account (weakly) overlapping focal elements when assessing conflict, then we can use the integral characteristic of conflict. For example,

$$ICon_w(F_1, F_2) = \int_0^1 w(h) Con_h(F_1, F_2) \, dh,$$

where the non-negative weight function $w(h)$ satisfies the normalization condition $\int_0^1 w(h) \, dh = 1$ and regulates the priority of values $h \in [0, 1)$. Small values of $h$ should have higher priority. Therefore, the function $w(h)$ must be non-increasing.

We will get for evidence bodies from Example 5 and $w(h) = 1$: $ICon_w(F_1, F_2) \approx 0.613$, $ICon_w(F_1, F_3) = 0.56$, $ICon_w(F_2, F_3) \approx 0.613$. In this case, choosing the pair $F_1, F_3$ will be preferable. If we use the weight $w(h) = \frac{3}{2} - h$, we obtain $ICon_w(F_1, F_2) \approx 0.523$, $ICon_w(F_1, F_3) \approx 0.607$, $ICon_w(F_2, F_3) \approx 0.534$. In this case, it would be preferable to choose the pair $F_1, F_2$ or $F_2, F_3$.

Integral conjunctive aggregation can be defined analogously: $im_w(A) = \int_0^1 w(h) m_h(A) \, dh$, $A \in 2^X$, where $m_h$ are calculated, for example, by the formula (5) (if $A \in 2^X$ is not a focal element for $h$, then we assume $m_h(A) = 0$).

## 6 Conclusion

The main advantage of describing bodies of evidence using threshold functions is that we can control the degree of uncertainty and conflict in such a description. In addition, the set of all represents with different thresholds gives us a more complete description of bodies of evidence and their aggregation. The problems of finding the optimal threshold at which a compromise is achieved between the accuracy of the description and uncertainty, between uncertainty and internal conflict, etc. can be posed. Some examples of such problems have been considered.

## References

1. Bronevich, A., Lepskiy, A.: Imprecision indices: axiomatic, properties and applications. Int. J. Gen. Syst. **44**(7–8), 812–832 (2015)
2. Coletti, G., Bouchon-Meunier, B.: A study of similarity measures through the paradigm of measurement theory: the classic case. Soft Comput. **23**, 6827–6845 (2019)
3. Daniel, M.: Conflicts within and between belief functions. In: Hüllermeier, E., Kruse, R., Hoffmann, F. (eds.) IPMU 2010. LNCS (LNAI), vol. 6178, pp. 696–705. Springer, Heidelberg (2010). https://doi.org/10.1007/978-3-642-14049-5_71
4. Deng, Y.: Generalized evidence theory. Appl. Intell. **43**, 530–543 (2015)
5. Denneberg, D., Grabisch, M.: Interaction transform of set functions over a finite set. Inf. Sci. **121**, 149–170 (1999)
6. Dubois, D., Prade, H.: A set-theoretic view on belief functions: logical operations and approximations by fuzzy sets. Int. J. Gen. Syst. **12**, 193–226 (1986)
7. Lepskiy, A.: Analysis of information inconsistency in belief function theory. Part I: external conflict. Control Sci. **5**, 2–16 (2021)
8. Lepskiy, A.: Analysis of information inconsistency in belief function theory. Part II: internal conflict. Control Sci. **6**, 2–12 (2021)
9. Lepskiy, A.: Conflict measure of belief functions with blurred focal elements on the real line. In: Denœux, T., Lefèvre, E., Liu, Z., Pichon, F. (eds.) BELIEF 2021. LNCS (LNAI), vol. 12915, pp. 197–206. Springer, Cham (2021). https://doi.org/10.1007/978-3-030-88601-1_20
10. Shafer, G.: A Mathematical Theory of Evidence. Princeton Univ. Press, Princeton (1976)
11. Smets, P.: Decision making in TBM: the necessity of the pignistic transformation. Int. J. Approx. Res. **38**, 133–147 (2005)
12. Smets, P., Kennes, R.: The transferable belief model. Artif. Intell. **66**, 191–243 (1994)
13. Zhang, L.: Representation, independence and combination of evidence in the Dempster-Shafer theory. In: Yager, R.R., et al. (eds.) Advances in the Dempster-Shafer Theory of Evidence, pp. 51–69. John Wiley & Sons, New York (1994)

# Mutual Information and Kullback-Leibler Divergence in the Dempster-Shafer Theory

Prakash P. Shenoy(✉)

School of Business, University of Kansas, Lawrence, KS 66045, USA
pshenoy@ku.edu
https://pshenoy.ku.edu/

**Abstract.** In probability theory, the mutual information between two discrete random variables, $X$ and $Y$, measures the average reduction in uncertainty about $Y$ when we learn the value of $X$. It is defined using the Shannon entropy of probability distributions. This paper defines a corresponding concept of mutual information between two variables in the Dempster-Shafer (D-S) belief function theory using the decomposable entropy defined by Jirousek and Shenoy. We also define the Kullback-Liebler (KL) divergence for the D-S theory as similar to the KL divergence for probability theory.

**Keywords:** Shannon's entropy · mutual information · Kullback-Leibler divergence · Dempster-Shafer theory of belief functions · decomposable entropy of belief functions

## 1 Introduction

The main goal of this paper is to define mutual information (MI) between two variables in the D-S belief function theory [2,13]. Our definition is based on the decomposable entropy for belief functions defined in [5], which satisfies the compound distributions property analogous to the one that characterizes Shannon's definitions of entropy and conditional entropy for probability mass functions [15]. We also define a generalization of the KL divergence between two belief functions defined for the same set of variables and express mutual information in terms of KL divergence, similar to probability theory.

The definition of MI between two variables in a belief-function graphical model is analogous to the definition between two variables in a probabilistic graphical model. It satisfies many of the properties of MI in the probabilistic case. An exception is that we are unable to prove that MI in the belief-function case is always non-negative. We also define a generalization of the KL divergence for the case of DS belief functions analogous to the probabilistic case. Unlike the probabilistic case, the KL divergence between two belief functions is not always non-negative. This is not a fatal flaw. We conjecture that if $Q_{X,Y}$ is a commonality function for $\{X,Y\}$ with marginals $Q_X$ for $X$, and $Q_Y$ for $Y$, then the KL divergence between $Q_{X,Y}$ and $Q_X \oplus Q_Y$ ($\oplus$ denotes Dempster's combination rule) is always non-negative. If this conjecture is true, then it would follow that MI between two variables (in a belief function graphical model) is always non-negative.

The concepts of MI and KL divergence in probability theory are widely used to construct probabilistic graphical models [8]. We believe the MI and KL divergence concepts defined in this paper will be equally useful for constructing belief function graphical models [1].

An outline of the remainder of the paper is as follows. In Sect. 2, we briefly review the definition of Shannon's entropy of a probability mass function, conditional entropy, and their properties. We also review the definition and properties of mutual information and KL divergence in probability theory. Most of this material is taken from [3,10,15]. In Sect. 3, we review the representations, operators, and conditional belief functions in the D-S theory of belief functions. In Sect. 4, we review the definitions of decomposable entropy and conditional decomposable entropy for the D-S theory and state some of their properties [5,7]. In Sect. 5, we define mutual information of a variable given another for a joint belief function for the two variables. Also, we define the KL divergence between two belief functions for the same set of variables. As in the probabilistic case, we express mutual information in terms of the KL divergence of two joint belief functions. Finally, in Sect. 6, we summarize, discuss future research, and conclude.

## 2 Shannon's Entropy, MI, and KL Divergence

This section briefly reviews Shannon's definitions of entropy of probability mass functions (PMFs) and conditional entropy of conditional probability tables (CPTs) and their properties. We also review the definitions of mutual information between two variables and the KL divergence between two probability mass functions defined for the same set of variables. Most of the material in this section is taken from [3,10,15]. We use some notation (such as probabilistic combination, $\otimes$) from [16].

**Definition 1 (Shannon's entropy [15]).** *Suppose $P_X$ is a PMF of a discrete variable $X$ with state space $\Omega_X$. Shannon's entropy of $P_X$, denoted by $H_s(P_X)$, is defined as:*

$$H_s(P_X) = - \sum_{x \in \Omega_X : P_X(x) > 0} P_X(x) \log_2(P_X(x)). \tag{1}$$

**Definition 2 (Shannon's conditional entropy [15]).** *Suppose $P_{Y|X}$ is a CPT for $Y$ given $X$ for all $x \in \Omega_X$ such that $P_X(x) > 0$. Shannon's conditional entropy of $P_{Y|X}$, denoted by $H_s(P_{Y|X})$, is defined as:*

$$H_s(P_{Y|X}) = \sum_{x \in \Omega_X : P_X(x) > 0} P_X(x) \, H_s(P_{Y|x}). \tag{2}$$

Thus, Shannon's conditional entropy of $P_{Y|X}$ is the average of Shannon's entropy $H_s(P_{Y|x})$ for each value $x \in \Omega_X$ weighted by $P_X(x)$.

Some important properties of Shannon's entropy are as follows [10]:

1. $H_s(P_X) \geq 0$. $H_s(P_X) = 0$ if and only if there is an $x \in \Omega_X$ such that $P_X(x) = 1$.
2. Shannon's entropy is decomposable, i.e., if $P_{X,Y} = P_X \otimes P_{Y|X}$, then $H_s(P_{X,Y}) = H_s(P_X) + H_s(P_{Y|X})$.
3. It is shown in [15] that $H_s(P_{X,Y}) \leq H_s(P_X) + H_s(P_Y)$, where $P_X$ and $P_Y$ are marginal PMFs of $X$ and $Y$ computed from joint PMF $P_{X,Y}$, with equality only if $X$ and $Y$ are independent with respect to $P_{X,Y}$.
4. $H_s(P_{Y|X}) \leq H_s(P_Y)$. Thus, the entropy of $P_Y$ is never increased by knowledge of $X$. It will be decreased unless $X$ and $Y$ are independent, in which case it stays the same.

The concept of mutual information between two random variables is introduced in [15].

**Definition 3 (Mutual information).** *Consider a joint PMF $P_{X,Y}(x,y) = P_X(x) P_{Y|X}(x,y)$ defined in terms of marginal PMF $P_X$ and CPT $P_{Y|X}$. Let $P_Y = (P_{X,Y})^{\downarrow Y}$ denote the marginal of $P_{X,Y}$ for $Y$. The mutual information of $Y$ with respect to $X$, denoted by $I(Y;X)$, is defined as:*

$$I(Y;X) = H_s(P_Y) - H_s(P_{Y|X}) \qquad (3)$$

Mutual information $I(Y;X)$ can be interpreted as a measure of $Y$'s dependence on $X$, where the measure is the reduction of Shannon's entropy of $Y$ after observation of $X$. Some properties of $I(Y;X)$ are as follows [3,10].

1. $I(Y;X) \geq 0$. $I(Y;X) = 0$ if and only if $Y$ is independent of $X$ with respect to the joint PMF $P_{X,Y}$.
2. $I(X;Y) = I(Y;X)$.
3. $I(Y;X) \leq H_s(P_Y)$ and $I(X;Y) \leq H_s(P_X)$.
4. $I(Y;X) = H_s(P_X) + H_s(P_Y) - H_s(P_{X,Y})$.

**Definition 4 (KL divergence [9]).** *Suppose $P$ and $Q$ are two PMFs for $X$ defined on the state space $\Omega_X$ such that if $Q(x) = 0$ for some $x \in \Omega$, then $P(x) = 0$. The KL divergence between $P$ and $Q$, denoted by $D_{KL}(P||Q)$, is defined as:*

$$D_{KL}(P||Q) = \sum_{x \in \Omega_X} P(x) \log \left( \frac{P(x)}{Q(x)} \right) \qquad (4)$$

*If the condition that $Q(x) = 0$ implies $P(x) = 0$ is not satisfied, then $D_{KL}(P||Q)$ is considered as $+\infty$.*

KL divergence satisfies Gibb's inequality, i..e., $D_{KL}(P||Q) \geq 0$, with equality if and only if $P = Q$ [10].

Mutual information $I(Y;X)$ with respect to joint PMF $P_{X,Y}$ can be expressed in terms of KL divergence as follows. Suppose $P_{X,Y}$ is a joint PMF for $\{X,Y\}$ with marginals $P_X$ and $P_Y$ for $X$ and $Y$, respectively. Then,

$$I(Y;X) = D_{KL}(P_{X,Y}||P_X \otimes P_Y) \tag{5}$$

Thus, it follows from the properties of KL divergence that $I(Y;X) \geq 0$, and $I(Y;X) = 0$ if and only if $P_{X,Y} = P_X \otimes P_Y$, i.e., $X$ and $Y$ are independent with respect to $P_{X,Y}$. Also, as $D_{KL}(P_{X,Y}||P_X \otimes P_Y)$ is symmetric in $X$ and $Y$, i.e., $P_{X,Y} = P_{Y,X}$ and $P_X \otimes P_Y = P_Y \otimes P_X$, it follows that $I(X;Y) = I(Y;X)$.

## 3 Basic Definitions in the D-S Theory

*Notation.* Let $\mathcal{V}$ denote a finite set of variables. Elements of $\mathcal{V}$ are denoted by upper-case Roman letters, $X$, $Y$, $Z$, etc. Subsets of $\mathcal{V}$ are denoted by lower-case Roman alphabets $r$, $s$, $t$, etc. Each variable $X$ is associated with a finite state space $\Omega_X$ that contains all possible values of $X$. For subset $r \subseteq \mathcal{V}$, let $\Omega_r = \times_{X \in r} \Omega_X$ denote the state space of $r$. Let $2^{\Omega_r}$ denote the set of all subsets of $\Omega_r$.

*Basic Probability Assignment.* A *basic probability assignment* (BPA) $m$ for $r$ is a function $m: 2^{\Omega_r} \to [0,1]$ such that:

$$m(\emptyset) = 0, \text{ and } \sum_{\mathsf{a} \subseteq \Omega_r} m(\mathsf{a}) = 1. \tag{6}$$

$m$ represents some knowledge about variables in $r$, and we say the *domain* of $m$ is $r$. Subsets $\mathsf{a}$ such that $m(\mathsf{a}) > 0$ are called *focal elements* of $m$. If $m$ has only one focal element (with probability 1), we say $m$ is *deterministic*. If the focal element of a deterministic BPA is $\Omega_r$, we say $m$ is *vacuous*. A vacuous BPA for $r$ is denoted by $\iota_r$. If all focal elements of $m$ are singleton subsets, we say $m$ is *Bayesian*. We say $m$ is *consonant* if the focal elements of $m$ are nested. We say $m$ is *quasi-consonant* if the intersection of all focal elements of $m$ is non-empty. A BPA that is consonant is also quasi-consonant, but not vice-versa.

*Commonality Function.* The information in a BPA $m$ for $r$ can also be represented by a corresponding commonality function (CF) $Q_m$ for $r$ that is defined as follows:

$$Q_m(\mathsf{a}) = \sum_{\mathsf{b} \in 2^{\Omega_r}: \mathsf{b} \supseteq \mathsf{a}} m(\mathsf{b}) \quad \text{for all} \quad \mathsf{a} \in 2^{\Omega_r}. \tag{7}$$

For the vacuous BPA $\iota_r$ for $r$, the corresponding CF $Q_{\iota_r}$ is given by $Q_{\iota_r}(\mathsf{a}) = 1$ for all $\mathsf{a} \in 2^{\Omega_r}$. If $m$ is a Bayesian BPA for $r$, then the corresponding CF $Q_m$ is such that $Q_m(\mathsf{a}) = m(\mathsf{a})$ if $|\mathsf{a}| = 1$, and $Q_m(\mathsf{a}) = 0$ if $|\mathsf{a}| > 1$.

*Operations in the D-S Theory.* The D-S theory has two main operations: Dempster's combination rule and marginalization.

*Notation.* Projection of states simply means dropping extra coordinates; for example, if $(x, y)$ is a state of $(X, Y)$, then the projection of $(x, y)$ to $X$, denoted by $(x, y)^{\downarrow X}$, is simply $x$, which is a state of $X$.

The projection of subsets of states is achieved by projecting every state in the subset. Suppose $\mathsf{b} \in 2^{\Omega_{X,Y}}$. Then $\mathsf{b}^{\downarrow X} = \{x \in \Omega_X : (x, y) \in \mathsf{b} \text{ for some } y \in \Omega_Y\}$. Notice that $\mathsf{b}^{\downarrow X} \in 2^{\Omega_X}$.

*Dempster's Combination Rule.* In the D-S theory, we combine two BPAs $m_1$ and $m_2$ representing distinct pieces of evidence by Dempster's rule [2] and obtain the BPA $m_1 \oplus m_2$, which represents the combined evidence. Dempster referred to this rule as the product-intersection rule, as the product of the BPA values is assigned to the intersection of the focal elements, followed by normalization. Normalization consists of discarding the value assigned to $\emptyset$ and normalizing the remaining values so that they add to 1.

In terms of CFs, Dempster's rule is pointwise multiplication of CFs followed by normalization, which is similar to the probabilistic combination rule of pointwise multiplication of probability potentials followed by normalization. This similarity with probability theory is one of the motivations behind our definitions of entropy and conditional entropy.

*Marginalization.* Suppose $m$ is a BPA for $\{X, Y\}$. Then, the marginal of $m$ for $X$, denoted by $m^{\downarrow X}$, is a BPA for $X$ such that for each $\mathsf{a} \in 2^{\Omega_X}$,

$$m^{\downarrow X}(\mathsf{a}) = \sum_{\mathsf{b} \in 2^{\Omega_{X,Y}} \,:\, \mathsf{b}^{\downarrow X} = \mathsf{a}} m(\mathsf{b}). \tag{8}$$

*Conditional Belief Functions.* Consider a BPA $m_X$ for $X$ and $x \in \Omega_X$ such that $m_X(\{x\}) > 0$. Suppose that there is a BPA for $Y$ expressing our belief about $Y$ if we know that $X = x$, and denote it by $m_{Y_x}$. Notice that $m_{Y_x}$ is a BPA for $Y$. We can embed this BPA for $Y$ into a conditional BPA for $\{X, Y\}$, which is denoted by $m_{Y|x}$, such that the following two conditions hold:

1. $m_{Y|x}$ tells us nothing about $X$, i.e., $m_{Y|x}^{\downarrow X}(\Omega_X) = 1$.
2. If we combine $m_{Y|x}$ with the deterministic BPA $m_{X=x}$ for $X$ such that $m_{X=x}(\{x\}) = 1$ using Dempster's rule, and marginalize the result to $Y$ we obtain $m_{Y_x}$, i.e., $(m_{Y|x} \oplus m_{X=x})^{\downarrow Y} = m_{Y_x}$.

Henceforth, we refer to BPA $m_{Y|x}$ as a BPA for $Y$ given $x \in \Omega_X$. Conditional BPAs are studied further in [4].

Smets suggests one way to obtain such an embedding [19] (see also [14]), called *conditional embedding*. It consists of taking each focal element $\mathsf{b} \in 2^{\Omega_Y}$ of $m_{Y_x}$, and converting it to the corresponding focal element

$$(\{x\} \times \mathsf{b}) \cup ((\Omega_X \setminus \{x\}) \times \Omega_Y) \in 2^{\Omega_{X,Y}} \tag{9}$$

of $m_{Y|x}$ with the same mass. It is easy to confirm that this embedding method satisfies the two conditions described in the previous paragraph.

This completes our brief review of the D-S belief function theory. For further details, the reader is referred to [13].

## 4 The Decomposable Entropy for the D-S Theory

The D-S theory has numerous definitions of entropy (see a review in [6]). In this section, we focus on decomposable entropy (d-entropy) of belief functions in the D-S theory [5] and describe its properties [7]. The definition of d-entropy is designed to satisfy a compound distribution property analogous to the compound distribution property that characterizes Shannon's entropy of PMFs.

**Definition 5 (d-entropy of a CF).** *Suppose $Q_X$ is a CF for $X$ with state space $\Omega_X$. Then, the d-entropy of $Q_X$, denoted by $H_d(Q_X)$, is defined as*

$$H_d(Q_X) = \sum_{\mathsf{a} \in 2^{\Omega_X}} (-1)^{|\mathsf{a}|} Q_X(\mathsf{a}) \log(Q_X(\mathsf{a})). \tag{10}$$

**Definition 6 (Conditional d-entropy).** *Suppose $Q_X$ is a CF for $X$, and suppose $Q_{Y|X}$ is a conditional CF for $Y$ given $X$. Then, the conditional d-entropy of $Q_{Y|X}$, denoted by $H_d(Q_{Y|X})$, is defined as follows:*

$$H_d(Q_{Y|X}) = \sum_{\mathsf{a} \in 2^{\Omega_{X,Y}}\, :\, Q_X(\mathsf{a}^{\downarrow X}) > 0} (-1)^{|\mathsf{a}|} Q_X(\mathsf{a}^{\downarrow X}) Q_{Y|X}(\mathsf{a}) \log(Q_{Y|X}(\mathsf{a})). \tag{11}$$

Using the definition of expectation for belief functions in [17], the conditional d-entropy in Eq. (11) can be considered as an expectation of $H_d(Q_{Y|x})$ as in the probabilistic case.

Some important properties of our definitions in Eqs. (10) and (11) are as follows: (proofs of all properties can be found in [7]).

1. (*Compound distributions*) Suppose $Q_X$ is a CF for $X$, and suppose $Q_{Y|X}$ is a conditional CF for $Y$ given $X$. Let $Q_{X,Y} = Q_X \oplus Q_{Y|X}$. Then, $H_d(Q_{X,Y}) = H_d(Q_X) + H_d(Q_{Y|X})$.
2. (*Non-negativity*) Suppose $m$ is a BPA for $X$ and suppose $|\Omega_X| = 2$. Then, $H_d(m) \geq 0$. For $|\Omega_X| > 2$, $H_d(m)$ does *not* satisfy the non-negativity property as shown in an example in [7]. Lack of non-negativity is not a drawback. Shannon's definition of entropy for continuous random variables characterized by probability density functions can be negative [15].

## 5 MI and KL Divergence for Belief Functions

We will define mutual information for two variables whose behavior is defined by a joint BPA $m_{X,Y}$ for $\{X,Y\}$. The exposition will mirror the definition of mutual information in probability theory in Sect. 2.

**Definition 7 (MI for the D-S theory).** *Consider a joint BPA $m_{X,Y} = m_X \oplus m_{Y|X}$ for $\{X,Y\}$ defined in terms of a marginal BPA $m_X$ for $X$ and a conditional BPA $m_{Y|X}$ for $Y$ given $X$. Let $m_Y$ denote the marginal BPA $m_{X,Y}^{\downarrow Y}$ for $Y$. The mutual information of $Y$ with respect to $X$, denoted by $I_d(Y;X)$, is defined as follows:*

$$I_d(Y;X) = H_d(m_Y) - H_d(m_{Y|X}) \tag{12}$$

Some comments/properties of Definition 7:

1. The definition of MI $I_d(Y;X)$ is similar to the probabilistic MI. The subscript $d$ in $I_d(X;Y)$ is to differentiate MI for the D-S theory from the corresponding probabilistic definition.
2. Unlike Shannon's entropy, $d$-entropy is not non-negative. But, MI $I_d(Y;X)$ is the difference of two $d$-entropies of $Y$. We conjecture that $I_d(Y;X) \geq 0$.
3. If our conjecture in Property 2 is true, $H_d(m_{Y|X}) \leq H_d(m_Y)$. Thus, the $d$-entropy of $m_Y$ is never increased by knowledge of $X$. It will be decreased unless $X$ and $Y$ are independent, in which case it stays the same.
4. If our conjecture in Property 2 is true, then $H_d(m_{X,Y}) \leq H_d(m_X) + H_d(m_Y)$ with equality iff $X$ and $Y$ are independent with respect to $m_{X,Y}$.
5. For probabilistic mutual information $I(X;Y) = I(Y;X)$. For the D-S case, if we have $m_{X,Y} = m_X \oplus m_{Y|X}$, it is not always the case that there exists a conditional $m_{X|Y}$ for $X$ given $Y$ such that $m_{X,Y} = (m_{X,Y})^{\downarrow Y} \oplus m_{X|Y}$. If there does exist a conditional $m_{X|Y}$ for $X$ given $Y$, then $I_d(X;Y) = I_d(Y;X)$. See proof in [18].

*Example 1.* Consider two binary variables $X$ and $Y$ with states $\Omega_X = \{x, \bar{x}\}$ and $\Omega_Y = \{y, \bar{y}\}$. Suppose BPA $m_X$ for $X$ is as follows: $m_X(\{x\}) = 0.3$, $m_X(\{\bar{x}\}) = 0.3$, $m_X(\{x, \bar{x}\}) = 0.4$. Suppose $Y$ is a deterministic function of $X$: $m_{Y_x}(\{y\}) = 1$, and $m_{Y_{\bar{x}}}(\{\bar{y}\}) = 1$. After conditional embedding and Dempster combination, conditional $m_{Y|X}$ for $Y$ given $X$ is as follows: $m_{Y|X}(\{(x,y),(\bar{x},\bar{y})\}) = 1$.

Finally, the joint BPA $m_{X,Y} = m_X \oplus m_{Y|X}$ is as follows: $m_{X,Y}(\{(x,y)\}) = 0.3$, $m_{X,Y}(\{(\bar{x},\bar{y})\}) = 0.3$, $m_{X,Y}(\{(x,y),(\bar{x},\bar{y})\}) = 0.4$. It follows from the definitions of $d$-entropy and conditional $d$-entropy that $H_d(m_X) \approx 0.19$, $H_d(m_{Y|X}) = 0$ (as it is deterministic), $H_d(m_{X,Y}) \approx 0.19$. Notice that the marginal of the joint for $Y$, $m_Y = (m_{X,Y})^{\downarrow Y}$ is as follows: $m_Y(\{y\}) = 0.3$, $m_Y(\{\bar{y}\}) = 0.3$, $m_Y(\{y, \bar{y}\}) = 0.4$. Thus, $H_d(m_Y) \approx 0.19$. This example illustrates the following results:

1. $H_d(m_X) + H_d(m_{Y|X}) \approx 0.19 + 0 = 0.19 = H_d(m_{X,Y})$ (chain rule of entropy).
2. $I(Y;X) = H_d(m_Y) - H_d(m_{Y|X}) \approx 0.19 - 0 = 0.19 \geq 0$ (Property 2 of Definition 7).
3. For this example, the joint $m_{X,Y}$ can also be factored into $m_Y \oplus m_{X|Y}$, where $m_{X|Y} = m_{Y|X}$. Thus, $I(X;Y) = H_d(m_X) - H_d(m_{X|Y}) \approx 0.19 - 0 = 0.19 = I(X;Y)$ (Property 5 of Definition 7).
4. $H_d(m_{X,Y}) \approx 0.19 \leq H_d(m_X) + H_d(m_Y) \approx 0.19 + 0.19 = 0.38$ (Property 4 of Definition 7).

*KL Divergence for the D-S Theory.* Next, we will define KL divergence for the D-S theory and express mutual information in terms of KL divergence, similar to probability theory.

**Definition 8.** *Suppose $Q_1$ and $Q_2$ are CFs for $X$ with state space $\Omega_X$ such that if $Q_2(\mathsf{a}) = 0$, then $Q_1(\mathsf{a}) = 0$. The KL divergence between $Q_1$ and $Q_2$, denoted by $D_{KL}(Q_1||Q_2)$, is defined as:*

$$D_{KL}(Q_1||Q_2) = \sum_{a \in 2^{\Omega_X}} (-1)^{|a|+1} Q_1(a) \log\left(\frac{Q_1(a)}{Q_2(a)}\right) \tag{13}$$

*If the condition $Q_2(a) = 0$ implies $Q_1(a) = 0$ is not satisfied, then $D_{KL}(Q_1||Q_2)$ is considered to be $+\infty$.*

Some comments about KL divergence:

1. Using the definition of expectation for belief functions in [17], Definition 8 can be interpreted as an expectation of $\log(Q_1/Q_2)$ with respect to CF $Q_1$, analogous to the definition of KL divergence for probability theory.
2. If $Q_1$ and $Q_2$ are both Bayesian CFs, i.e., $Q_i(a) = 0$ if $|a| > 1$, then $D_{KL}(Q_1||Q_2)$ reduces to the probabilistic definition.
3. The KL divergence $D_{KL}(Q_1||Q_2)$ does not satisfy the non-negativity property of probabilistic KL divergence. See a counter-example in [18].
4. If $Q_{X,Y} = Q_X \oplus Q_{Y|X}$ is a joint CF for $\{X,Y\}$, then $I(Y;X)$ can be expressed as in the probabilistic case, i.e., $I(Y;X) = H_d(Q_Y) - H_d(Q_{Y|X}) = D_{KL}(Q_{X,Y}||Q_X \oplus Q_Y)$. See proof in [18].

## 6 Summary and Conclusion

We have generalized the concepts of mutual information [15] and KL divergence [9] in probability theory to the D-S theory using $d$-entropy defined in [5]. What makes this possible is the decomposability property of $d$-entropy.

We need to resolve the issue of non-negativity of MI $I_d(Y;X)$. As far as we know, there is no prior literature on mutual information for the D-S theory. There are several definitions of KL divergence for the D-S theory, e.g., [11,12,20,21]. A comparison of these definitions with the definition in this paper is yet to be done. This paper is a condensed version of [18].

**Acknowledgments.** The author is grateful to Radim Jiroušek and Václav Kratochvíl for their comments and encouragement. Thanks to Radim for pointing out that the definition of KL divergence for the D-S theory in this paper is not always non-negative.

## References

1. Almond, R.G.: Graphical Belief Modeling. Chapman & Hall, London (1995)
2. Dempster, A.P.: Upper and lower probabilities induced by a multivalued mapping. Ann. Math. Stat. **38**(2), 325–339 (1967)
3. Fano, R.M.: Transmission of Information: A Statistical Theory of Communications. MIT Press, Cambridge (1961)
4. Jiroušek, R., Kratochvíl, V., Shenoy, P.P.: On conditional belief functions in directed graphical models in the Dempster-Shafer theory. Int. J. Approx. Reason. **160**(7), 108976 (2023)
5. Jiroušek, R., Shenoy, P.P.: A decomposable entropy of belief functions in the Dempster-Shafer Theory. In: Destercke, S., Denoeux, T., Cuzzolin, F., Martin, A. (eds.) BELIEF 2018. LNCS (LNAI), vol. 11069, pp. 146–154. Springer, Cham (2018). https://doi.org/10.1007/978-3-319-99383-6_19

6. Jiroušek, R., Shenoy, P.P.: A new definition of entropy of belief functions in the Dempster-Shafer theory. Int. J. Approx. Reason. **92**(1), 49–65 (2018)
7. Jiroušek, R., Shenoy, P.P.: On properties of a new decomposable entropy of Dempster-Shafer belief functions. Int. J. Approx. Reason. **119**(4), 260–279 (2020)
8. Koller, D., Friedman, N.: Probabilistic Graphical Models: Principles and Techniques. MIT Press, Cambridge (2009)
9. Kullback, S., Leibler, R.A.: On information and sufficiency. Ann. Math. Stat. **22**(1), 79–86 (1951)
10. MacKay, D.J.C.: Information Theory, Inference, and Learning Algorithms. Cambridge University Press, Cambridge (2003)
11. Ramasso, E.: Inference and learning in evidential discrete latent Markov models. IEEE Trans. Fuzzy Syst. **25**(5), 1102–1114 (2017)
12. Ramasso, E., Rombaut, M., Pellerin, D.: Forward-backward-viterbi procedures in the transferable belief model for state sequence analysis using belief functions. In: Mellouli, K. (ed.) ECSQARU 2007. LNCS (LNAI), vol. 4724, pp. 405–417. Springer, Heidelberg (2007). https://doi.org/10.1007/978-3-540-75256-1_37
13. Shafer, G.: A Mathematical Theory of Evidence. Princeton University Press, Princeton (1976)
14. Shafer, G.: Belief functions and parametric models. J. Roy. Stat. Soc. B **44**(3), 322–352 (1982)
15. Shannon, C.E.: A mathematical theory of communication. Bell Syst. Tech. J. **27**, 379–423 (1948)
16. Shenoy, P.P.: Conditional independence in valuation-based systems. Int. J. Approx. Reason. **10**(3), 203–234 (1994)
17. Shenoy, P.P.: An expectation operator for belief functions in the Dempster-Shafer theory. Int. J. Gen. Syst. **49**(1), 112–141 (2020)
18. Shenoy, P.P.: Mutual information and Kullback-Leibler divergence in the Dempster-Shafer theory of belief functions. Working Paper 345, University of Kansas School of Business, Lawrence, KS 66045 (2024). https://pshenoy.ku.edu/Papers/WP345.pdf
19. Smets, P.: Un modele mathematico-statistique simulant le processus du diagnostic medical. Ph.D. thesis, Free University of Brussels (1978)
20. Soubaras, H.: Towards an axiomatization for the generalization of the Kullback-Leibler divergence to belief functions. In: Proceedings of the 7th Conference of the European Society for Fuzzy Logic and Technology (EUSFLAT-11), pp. 1090–1097. Advances in Intelligent Systems Research, Atlantis Press (2011)
21. Wang, H., Deng, X., Jiang, W., Geng, J.: A new belief divergence measure for Dempster-Shafer theory based on belief and plausibility function and its application in multi-source data fusion. Eng. Appl. Artif. Intell. **97**(1), 104030 (2021)

# An OWA-Based Distance Measure for Ordered Frames of Discernment

Xiong Zhao, Liyao Ma[✉], Yiyang Wang, and Shuhui Bi

School of Electrical Engineering, University of Jinan, Jinan 250022, China
cse_maly@ujn.edu.cn

**Abstract.** As ordinal variables are involved in many applications, in this paper, a distance measure is proposed within the ordered frame of discernment. The relation between elements of the frame of discernment is quantified considering their order information. A relation matrix for the ordered power set is then generated using the OWA operator. The Euclidean-type distance based on the relation matrix is defined and verified with numerical experiments. Results show that this measure exhibits greater flexibility in describing the ordering relationships of elements.

**Keywords:** Ordered frame of discernment · Ordered Weighted Averaging · Distance measure · Ordinal variable

## 1 Introduction

As a flexible framework for modelling and reasoning with epistemic uncertainty, the theory of belief functions [1,2] is widely used in applications such as data fusion [3], machine learning [4], etc. Many works have emerged on one of the key topics of the theory, measuring the distance between belief functions. Jousselme et al. [5] made a comprehensive survey on dissimilarity measures considering their formal properties and proposed new measures. Concerning the internal discrepancy of subsets, Huang et al. [6] designed the enhanced belief logarithmic similarity measure. Xiao [7] proposed a reinforced belief divergence measure to quantify the discrepancy between basic belief assignments and then applied it in multi-sensor data fusion.

However, in applications such as fruit grading and questionnaire studying, the labels are naturally ordinal variables the order relations among elements in such situations lead to the study of ordered frames of discernment. Cheng et al. [8] defined the similarity coefficient to quantify the distance between orderable focal elements. Martin [9] explored multiple aspects of the ordered frame of discernment, such as the combination, distances, and conflicts on ordered power set. Within the ordered frame of discernment, there have been researches on evidential software risk assessment, evidence aggregation and ranking, etc. [10,11], yet few studies have focused on the distance measure in the new frame. In this paper, considering both the ordering relation and uncertainty, we propose a distance measure taking advantage of the OWA operator. With Sect. 2 providing basic knowledge, we design the relation matrix in Sect. 3. The OWA-based distance measure is proposed in Sect. 4 and verified by numerical experiments in Sect. 5, followed by the conclusion

## 2 Preliminaries

### 2.1 Ordered Frames of Discernment

Let $\Theta = \{1, 2, \cdots, K\}$ be the ordered frames of discernment, where the $K$ exclusive and exhaustive values that an ordinal variable can take have the ordered relation $1 < 2 < \cdots < K$. Take exam score as an example, the five grades A, B, C, D and F, respectively, will be marked as 5, 4, 3, 2 and 1. The ordered power set $oPS^\Theta$ consists of the empty set and all the disjunctions of consecutive elements in $\Theta$, denoted as $oPS^\Theta = \{\emptyset, [\![i,j]\!]_{i,j=1,\cdots,K}\}$, where $[\![i,j]\!] = \{i, i+1, \cdots, j-1, j\}, 1 \leq i \leq j \leq K$. The uncertainty is modelled by the mass function that mapping from the ordered power set to [0,1], such that $m(\emptyset) = 0$ and

$$\sum_{A \in oPS^\Theta} m(A) = 1. \tag{1}$$

A is called a focal element with $m(A) > 0$. Readers can refer to [9] for more details within ordered frames of discernment such as credibility function, plausibility function and pignistic probability.

### 2.2 The OWA Operator

Proposed by Yager [12] as a parametrized class of mean type aggregation operators, the OWA operators have been widely used [13,14]. Given elements $a_1, \cdots, a_n$, an OWA operator of dimension $n$ aggregates them with a collection of positive weights $\boldsymbol{w} = (w_1, \cdots, w_n)$ such that

$$F_{\boldsymbol{w}}(a_1, \cdots, a_n) = \sum_{i=1}^{n} w_i a_{(i)}, \tag{2}$$

where $a_{(i)}$ is the $i$-th largest element of $a_1, \cdots, a_n$. The varying weights lead to different aggregating operators, including commonly-used operators such as the maximum/minimum and the arithmetic average. The weights are determined by searching for the weight vector $\boldsymbol{w}^*$ maximizing the entropy

$$H(\boldsymbol{w}) = -\sum_{i=1}^{n} w_i \log w_i, \tag{3}$$

under the constraints of weights summing up to one $\sum_{i=1}^{n} w_i = 1$ and $\mathsf{orness}(\boldsymbol{w}) = \gamma$, a given value $\gamma \in [0, 1]$. Fixing $\gamma$ at 1, 0.5 and 0, respectively, the maximum, the arithmetic average and the minimum operators can be retrieved. The measure of *orness* describes the attitude of operator, being defined as

$$\mathsf{orness}(\boldsymbol{w}) = \frac{1}{n-1} \sum_{i=1}^{n} (n-i) w_i. \tag{4}$$

## 3 Relation Matrix Design for the Ordered Power Set

In traditional distance measures, the Jaccard index is often adopted to describe the interaction between focal elements. In this paper, taking both the order and epistemic uncertainty into consideration, we propose the relation matrix to depict the mutual relations among all focal elements on the ordered power set. Based on the quantified relations between singleton focal elements, relations involving set-valued focal elements are calculated with the OWA operator.

### 3.1 Relations Between Singleton Focal Elements

Let the ordered frame of discernment be $\Theta = \{1, 2, \cdots, K\}$. Inspired by the cost of mislabelling in ordinal classification, we first define the relation of singleton focal elements. Since all possible values in the frame of discernment are ordered, we regard the dissimilarity between two singletons as a function of their disparity. The dissimilarity between a singleton focal element and itself should be 0. For two singleton focal elements, when one is fixed, say at 1, as the other one moves farther away from 2 to $K$, their dissimilarity increases (such as compared to score B, F should be much farther from A). Given singletons $i, j \in \Theta$, their relation is quantified as

$$r(i,j) = e^{-|\mu|}, \qquad (5)$$

where $\mu = i - j$. The changing trend of dissimilarity between singleton focal elements is shown in Fig. 1. The farther away these two elements are from each other, the greater the relation value. The value ranges in $[0, 1]$ non-linearly. When two singletons are far enough (such as $|\mu|>5$), the dissimilarity approaches 1. The increase of their disparity has little effect on the relation value. The relation is designed to be unaffected by the cardinality of $\Theta$, the relation value of two adjacent elements remains unchanged in different frames of discernment.

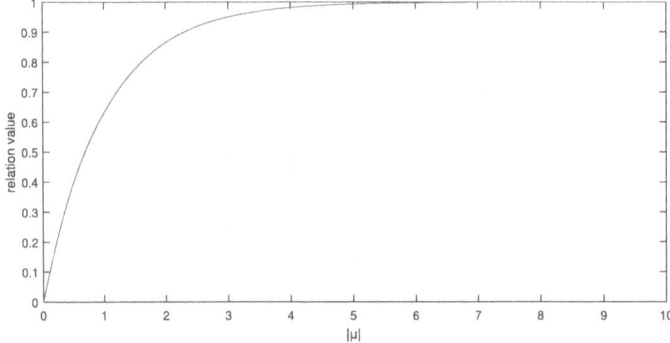

**Fig. 1.** Relation value between singleton focal elements.

Take $K = 4$ for instance ($\Theta = \{1, 2, 3, 4\}$), the singleton part of the relation matrix achieved by Eq. (5) is calculated as

$$R = \begin{bmatrix} 0 & 0.6321 & 0.8647 & 0.9502 \\ 0.6321 & 0 & 0.6321 & 0.8647 \\ 0.8647 & 0.6321 & 0 & 0.6321 \\ 0.9502 & 0.8647 & 0.6321 & 0 \end{bmatrix}, \quad (6)$$

where $R_{ij} = r(i,j)$ quantifies the relation between singleton focal elements $\{i\}$ and $\{j\}$, $i, j = 1, \cdots, 4$.

## 3.2 Relations Between Singleton and Set Focal Elements

In our view, the relation between a singleton focal element $\{i\}$ and a set-valued one $S$ should be a function of previously-designed relations between $\{i\}$ and all singletons in $S$. To aggregate these relation values, the OWA operator, which provides a class of mean type aggregation operators, is adopted in this paper to quantify the relations between singleton and set focal elements. Given relation matrix of singletons $R_{K \times K}$, the quantified relation between a singleton focal element $\{i\}$ and a set focal element $S$ is computed as

$$r(i, S) = \sum_{j=1}^{|S|} r_{(j)}(i, p) \omega_j, \ p \in S, i = 1, \cdots, K \quad (7)$$

in which $\omega_j$ are weights, $r_{(j)}(i,p)$ represents the $j$-th largest element among all relation values calculated between singleton $i$ and each singleton $p \in S$. Weights $\boldsymbol{\omega} = \{\omega_1, \cdots, \omega_{|S|}\}$ are achieved by maximizes the entropy

$$H(\boldsymbol{\omega}) = -\sum_{i=1}^{|S|} \omega_i \log \omega_i \quad (8)$$

under constraints $\sum_{i=1}^{|S|} \omega_i = 1$ and $\sum_{i=1}^{|S|} \frac{|S|-i}{|S|-1} \omega_i = \gamma$.

The parameter $\gamma \in [0,1]$ measures one's attitude towards imprecision. Take singleton 2 and set $\{2,3\}$ as an example, $r(2, \{2,3\})$ is a combination of two values $r(2,2) = 0$ and $r(2,3) = 0.6321$. As $\gamma$ increases, the largest element obtains larger weight, making the calculated $r(2, \{2,3\})$ more close to $r(2,3)$. Back to the meaning of $m(\{2,3\})$, its value can transfer to 2 or 3 precisely with more information. So a larger $\gamma$ infers that one prefers to believe the true value within the set being the most far away element, which is a more pessimistic view. $\gamma = 1$ makes $r(i, S)$ the maximum operator, which regards the set as definitely the most far away singleton from $i$ in the set $S$. $\gamma = 0$ gives oppositely the minimum operator and the nearest singleton in the set. $\gamma = 0.5$ leads to the arithmetic average of relation values in the set. Being suspicious of set focal elements, we set $\gamma = 0.8$ in this paper, one can also adopt another value according to different attitude toward imprecision.

## 3.3 Relations Between Set Focal Elements

The relation between two set-valued focal elements can also be aggregated from previous results by the OWA operator. Similar to the procedure in Sect. 3.2, relations between sets $T$ and $S$ is defined as

$$r(T, S) = \sum_{j=1}^{|T|} r^{[j]}(p, S)\omega_j, \ p \in T, \tag{9}$$

where weights $\boldsymbol{\omega} = \{\omega_1, \cdots, \omega_{|T|}\}$ are computed in the same optimization way of Sect. 3.2.

Let $\Theta = \{1, 2, 3, 4\}$ and set $\gamma = 0.8$, the finally obtained relation matrix $\tilde{R}_o$ on the ordered power set is depicted in Table 1 ($\emptyset$ is not included as $m(\emptyset) = 0$). According to the definitions, the matrix is symmetric. The upper-left, lower-left (upper-right), and lower-right parts of the matrix, respectively, are calculated by methods detailed in Sect. 3.1, Sect. 3.2 and Sect. 3.3.

For the calculation of $r(i, S)$ in Eq. (7), when $i \in S$ such as $r(1, [\![1, 2]\!])$, since $r(1, 1) = 0$ is included in the OWA aggregation, $r(1, [\![1, 2]\!])$ obtains a relatively small value 0.5057. When $i \notin S$, $r(1, [\![2, 4]\!]) = 0.9094$ with $r(1, 1)$ excluded in the calculation. Similarly for Eq. (9), considering all sets $S$ of the same cardinality, when $T \subset S$, $r(T, S)$ obtains the smallest value (such as $r([\![1, 2]\!], [\![1, 2]\!]) = 0.5057$ compared with $r([\![1, 2]\!], [\![2, 3]\!]) = 0.7557$ and $r([\![1, 2]\!], [\![3, 4]\!]) = 0.9101$). When $T \not\subset S$, if $T \cap S \neq \emptyset$, $r(T, S)$ can achieve a relatively small value (such as $r([\![1, 2]\!], [\![2, 3]\!]) = 0.7557$ compared with $r([\![1, 2]\!], [\![3, 4]\!]) = 0.9101$).

Table 1. Relation matrix $\tilde{R}_o$ obtained by the OWA operator.

| Focal element | 1 | 2 | 3 | 4 | $[\![1,2]\!]$ | $[\![2,3]\!]$ | $[\![3,4]\!]$ | $[\![1,3]\!]$ | $[\![2,4]\!]$ | $[\![1,4]\!]$ |
|---|---|---|---|---|---|---|---|---|---|---|
| 1 | 0 | 0.6321 | 0.8647 | 0.9502 | 0.5057 | 0.8182 | 0.9331 | 0.7390 | 0.9040 | 0.8520 |
| 2 | 0.6321 | 0 | 0.6321 | 0.8647 | 0.5057 | 0.5057 | 0.8182 | 0.5804 | 0.7390 | 0.7424 |
| 3 | 0.8647 | 0.6321 | 0 | 0.6321 | 0.8182 | 0.5057 | 0.5057 | 0.7390 | 0.5804 | 0.7424 |
| 4 | 0.9502 | 0.8647 | 0.6321 | 0 | 0.9331 | 0.8182 | 0.5057 | 0.9040 | 0.7390 | 0.8520 |
| $[\![1,2]\!]$ | 0.5057 | 0.5057 | 0.8182 | 0.9331 | 0.5057 | 0.7557 | 0.9101 | 0.7073 | 0.8710 | 0.8301 |
| $[\![2,3]\!]$ | 0.8182 | 0.5057 | 0.5057 | 0.8182 | 0.7557 | 0.5057 | 0.7557 | 0.7073 | 0.7073 | 0.7424 |
| $[\![3,4]\!]$ | 0.9331 | 0.8182 | 0.5057 | 0.5057 | 0.9101 | 0.7557 | 0.5057 | 0.8710 | 0.7073 | 0.8301 |
| $[\![1,3]\!]$ | 0.7390 | 0.5804 | 0.7390 | 0.9040 | 0.7073 | 0.7073 | 0.8710 | 0.7261 | 0.8386 | 0.8172 |
| $[\![2,4]\!]$ | 0.9040 | 0.7390 | 0.5804 | 0.7390 | 0.8710 | 0.7073 | 0.7073 | 0.8386 | 0.7261 | 0.8172 |
| $[\![1,4]\!]$ | 0.8520 | 0.7424 | 0.7424 | 0.8520 | 0.8301 | 0.7424 | 0.8301 | 0.8172 | 0.8172 | 0.8354 |

## 4 OWA-Based Distance on Ordered Power Set

Based on the relation matrix designed in Sect. 3, considering the ordered frame of discernment, the distance between mass functions $m_1$ and $m_2$ is defined as

$$d_{OWA}(m_1, m_2) = \sqrt{\frac{1}{2}(m_1 - m_2)^T(\mathbf{1} - \tilde{R}_o)(m_1 - m_2)}, \qquad (10)$$

where $\tilde{R}_o$ is the $\frac{K(K+1)}{2} \times \frac{K(K+1)}{2}$ relation matrix, which quantifies the relation between any two focal elements on the ordered power set. The proposed distance comes from the Euclidean distance. Here is a discussion about its property.

- Non-negativity

According to the property of defined focal element relation matrix, matrix $(\mathbf{1} - \tilde{R}_o)$ is positive definite. This leads to the matrix decomposition $(\mathbf{1} - \tilde{R}_o) = C^T C$, where $C$ is an $\frac{K(K+1)}{2} \times \frac{K(K+1)}{2}$ invertible matrix. Then we have

$$\begin{aligned} d_{OWA}(m_1, m_2) &= \sqrt{\frac{1}{2}(m_1 - m_2)^T C^T C(m_1 - m_2)} \\ &= \frac{\sqrt{2}}{2}||C(m_1 - m_2)|| \geq 0. \end{aligned} \qquad (11)$$

$d_{OWA}(m_1, m_2) = 0$ if and only if $C(m_1 - m_2) = 0$. Since $C$ is non-singular, we have $d_{OWA}(m_1, m_2) = 0 \Rightarrow m_1 = m_2$.

- Symmetry

Since $(m_2 - m_1)$ is equal to $-(m_1 - m_2)$ and $\tilde{R}_o$ is symmetric, we have

$$d_{OWA}(m_2, m_1) = \sqrt{\frac{1}{2}[-(m_1 - m_2)]^T(\mathbf{1} - \tilde{R}_o)[-(m_1 - m_2)]}. \qquad (12)$$

Therefore, $d_{OWA}(m_1, m_2) = d_{OWA}(m_2, m_1)$, satisfying the symmetry property.

- Triangle inequality

Consider three mass functions $m_1$, $m_2$ and $m_3$. As they have the same starting point in the vector space, when they are non-collinear, a triangle with sides $(m_1 - m_2)$, $(m_1 - m_3)$ and $(m_2 - m_3)$ is formed. When calculating distances, the matrix $\mathbf{1} - \tilde{R}_o$ only changes the length of the vectors but does not alter their directions. Therefore, $d_{OWA}(m_1, m_2)$, $d_{OWA}(m_1, m_3)$ and $d_{OWA}(m_2, m_3)$ still satisfy the triangle inequality. For new vectors obtained through left multiplying the mass vectors by invertible matrices $C$, we have

$$||C(m_1 - m_2)|| \leq ||C(m_1 - m_3)|| + ||C(m_2 - m_3)||. \qquad (13)$$

Therefore, the proposed distance satisfies the triangle inequality.

## 5  Numerical Experiments and Discussion

The performance of proposed OWA-based distance measure is demonstrated by numerical experiments. We compared the distance calculated by our method

$d_{OWA}$ with other two methods: the evidence distance $d_J$ based on the Jaccard coefficient within a traditional frames of discernment [15], and the evidence distance $d_{Jo}$ based on improved Jaccard coefficient within an ordered frames of discernment [9]. Let $\Theta = \{1,2,3,4\}$ and $\gamma = 0.8$, the mass functions for distance calculating are shown in Table 2. Mass function $m_1^*$ is fixed, with a relatively strong support for singleton focal element $\{2\}$. To calculate the distance between two mass functions, the other mass function changes from $m_1(\cdot)$ to $m_9(\cdot)$, each mainly supporting a focal element (except $\Theta$) on the ordered power set. Adopting the relation matrix in Table 1, the distance between $m_1^*(\cdot)$ and $m_i(\cdot), i = 1, ..., 9$ are calculated and shown in Fig. 2.

**Table 2.** Details of mass functions $m_1^*(\cdot)$ and $m_i(\cdot), i = 1, ..., 9$.

| Focal element | $m_1^*(\cdot)$ | mass function | | | | | | | | |
|---|---|---|---|---|---|---|---|---|---|---|
| | | $m_1(\cdot)$ | $m_2(\cdot)$ | $m_3(\cdot)$ | $m_4(\cdot)$ | $m_5(\cdot)$ | $m_6(\cdot)$ | $m_7(\cdot)$ | $m_8(\cdot)$ | $m_9(\cdot)$ |
| 1 | 0.1 | 0.8 | 0 | 0 | 0 | 0 | 0 | 0 | 0 | 0 |
| 2 | 0.8 | 0 | 0.8 | 0 | 0 | 0 | 0 | 0 | 0 | 0 |
| 3 | 0 | 0 | 0 | 0.8 | 0 | 0 | 0 | 0 | 0 | 0 |
| 4 | 0.1 | 0 | 0 | 0 | 0.8 | 0 | 0 | 0 | 0 | 0 |
| $[1,2]$ | 0 | 0 | 0 | 0 | 0 | 0.8 | 0 | 0 | 0 | 0 |
| $[2,3]$ | 0 | 0 | 0 | 0 | 0 | 0 | 0.8 | 0 | 0 | 0 |
| $[3,4]$ | 0 | 0 | 0 | 0 | 0 | 0 | 0 | 0.8 | 0 | 0 |
| $[1,3]$ | 0 | 0 | 0 | 0 | 0 | 0 | 0 | 0 | 0.8 | 0 |
| $[2,4]$ | 0 | 0 | 0 | 0 | 0 | 0 | 0 | 0 | 0 | 0.8 |
| $[1,4]$ | 0 | 0.2 | 0.2 | 0.2 | 0.2 | 0.2 | 0.2 | 0.2 | 0.2 | 0.2 |

The horizontal axis of Fig. 2 marks the mostly supported focal elements of mass functions $m_i(\cdot), i = 1, ..., 9$. Since $m_2$ and $m_1^*$ both strongly support singleton $\{2\}$, they obtain the smallest distance. $m_1$, $m_3$ and $m_4$ support conflicting singletons, achieving greater distance. With the supported singleton being further from $\{2\}$, $m_4$ achieves a larger distance than $m_3$. Comparing the distances $d_{OWA}(m_1^*, m_5)$, $d_{OWA}(m_1^*, m_6)$ and $d_{OWA}(m_1^*, m_7)$, the last one has the largest value because the supported set $\{3,4\}$ has no intersection with $\{2\}$.

In our method, $d_{OWA}(m_1^*, m_8) < d_{OWA}(m_1^*, m_9)$, being quite different from other two methods. Although $m_8$ and $m_9$ both have the supported set of 3 elements, the mass of $\{2,3,4\}$ possibly goes to singleton $\{4\}$, which is further from $\{2\}$, resulting in larger distance. Besides, consider the distances $d_{OWA}(m_1^*, m_1)$, $d_{OWA}(m_1^*, m_5)$ and $d_{OWA}(m_1^*, m_8)$. Using the OWA-based distance, we has $d_{OWA}(m_1^*, m_1) > d_{OWA}(m_1^*, m_5)$, as the main support transfer from a conflict singleton $\{1\}$ to a set containing the same singleton $\{2\}$. $d_{OWA}(m_1^*, m_5) > d_{OWA}(m_1^*, m_8)$ comes from the fact that when the cardinality of set increases, OWA operator gives lower weight to the largest element of the values to be aggregated. To some extent, imprecise focal elements lead to smaller distance in our method.

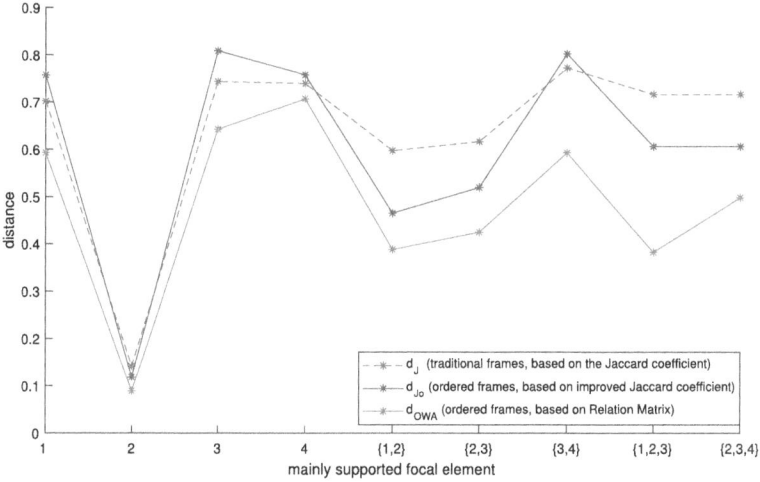

**Fig. 2.** Comparison of distances between $m_1^*(.)$ and $m_i(.), i = 1, ..., 9$.

**Table 3.** Some special cases of distance between logical mass functions

| # | $m_1$ | $m_2$ | $d_J$ | $d_{J_o}$ | $d_{OWA}$ |
|---|---|---|---|---|---|
| 1 | $m_1(1) = 1$ | $m_2(2) = 1$ | 1 | 0.9129 | 0.7950 |
| 2 | $m_1(1) = 1$ | $m_2(3) = 1$ | 1 | 0.9574 | 0.9299 |
| 3 | $m_1(1) = 1$ | $m_2(4) = 1$ | 1 | 1 | 0.9748 |
| 4 | $m_1(1) = 1$ | $m_2(1,2) = 1$ | 0.5774 | 0.7071 | 0.5028 |
| 5 | $m_1(1) = 1$ | $m_2(2,3) = 1$ | 1 | 0.9354 | 0.7519 |
| 6 | $m_1(1) = 1$ | $m_2(3,4) = 1$ | 1 | 0.9789 | 0.8248 |
| 7 | $m_1(1) = 1$ | $m_2(1,2,3) = 1$ | 0.7071 | 0.8165 | 0.6131 |
| 8 | $m_1(1) = 1$ | $m_2(2,3,4) = 1$ | 1 | 0.9574 | 0.7355 |
| 9 | $m_1(1) = 1$ | $m_2(\Theta) = 1$ | 0.7746 | 0.8660 | 0.6590 |
| 10 | $m_1(1,2) = 1$ | $m_2(\Theta) = 1$ | 0.5774 | 0.7071 | 0.3994 |
| 11 | $m_1(1,2,3) = 1$ | $m_2(\Theta) = 1$ | 0.3780 | 0.5000 | 0.1909 |
| 12 | $m_1(\Theta) = 1$ | $m_2(\Theta) = 1$ | 0 | 0 | 0 |

Distances calculated in some special cases of logical mass functions ($\Theta = \{1, 2, 3, 4\}$ and $\gamma = 0.8$) are demonstrated in Table 3. Both $d_{OWA}$ and $d_{J_o}$ can effectively distinguish the difference among ordered labels. For cases 1–3, as the difference between supported singletons becomes larger, distance grows larger. Once set focal elements are involved, when possible singleton in the set moves further, the corresponding distance becomes larger. Comparing case 9–12, as the focal element of $m_1$ becomes gradually imprecise, the distance decreases. Overall, the distances calculated by our method are relatively smaller in the presence of set focal elements. And our distance varies in a larger range compared to $d_{Jo}$.

Distance becomes 0 when two mass functions are exactly the same. In this case of $K = 4$, the maximum distance value only become close to 1 but not 1. It results from the relation between singletons defined in Eq. (5). For our method, distance of 1 can be obtained only when the framework of discernment contains more elements and the mass functions explicitly support two singletons with large differences.

## 6 Conclusion

In this paper, a new distance measure based on the OWA operator is defined within the ordered frames of discernment. Considering the relation between ordered elements, a relation matrix is designed on the ordered power set and further used to calculate evidence distance. The proposed Euclidean-type distance has several good properties and shows flexibility and rationality in dealing with ordinal variables. The method has a parameter $\gamma$ to adjust, making it possible to take one's attitude towards imprecision into consideration. In future work, the OWA-based distance will be applied in real-world applications such as fruit grading. Other distance measures will also be studied within the ordered frames of discernment.

**Acknowledgments.** This study was funded by the Natural Science Foundation of Shandong Province (ZR2021MF074, ZR2023MF094) and the Shandong Province Higher Education Institution Youth Innovation and Entrepreneurship Talents Cultivation Program (TJY2114).

**Disclosure of Interests.** The authors have no competing interests to declare that are relevant to the content of this article.

## References

1. Dempster, A.P.: A generalization of Bayesian inference. J. Roy. Stat. Soc.: Ser. B (Methodol.) **30**(2), 205–232 (1968)
2. Shafer, G.: A Mathematical Theory of Evidence. Princeton University, Princeton (1976)
3. Li, Z.W., Zhang, Q.L., Liu, S.P., Peng, Y.C., Li, L.L.: Information fusion and attribute reduction for multi-source incomplete mixed data via conditional information entropy and DS evidence theory. Appl. Soft Comput. **151**, 111149 (2024)
4. Cai, M.J., Wu, Z.S., Li, Q.G., Xu, F., Zhou, J.: GFDC: A granule fusion density-based clustering with evidential reasoning. Int. J. Approximate Reasoning **164**, 109075 (2024)
5. Jousselme, A.-L., Maupin, P.: Distances in evidence theory: comprehensive survey and generalizations. Int. J. Approximate Reasoning **53**(2), 118–145 (2012)
6. Huang, H.J., Liu, Z., Han, X., Yang, X.Y., Liu, L.S.: A belief logarithmic similarity measure based on dempster-shafer theory and its application in multi-source data fusion. J. Intell. Fuzzy Syst. **45**, 1–13 (2023)

7. Xiao, F.: A new divergence measure for belief functions in D-S evidence theory for multisensor data fusion. Inf. Sci. **514**, 462–483 (2020)
8. Cheng, C., Xiao, F.: A distance for belief functions of orderable set. Pattern Recogn. Lett. **145**, 165–170 (2021)
9. Martin, A.: Belief functions on ordered frames of discernment. In: Le Hégarat-Mascle, S., Bloch, I., Aldea, E. (eds.) Belief Functions: Theory and Applications, BELIEF 2022, LNCS, vol. 13506, pp. 129–138. Springer, Cham (2022). https://doi.org/10.1007/978-3-031-17801-6_13
10. Chen, X.Y., Deng, Y.: Evidential software risk assessment model on ordered frame of discernment. Expert Syst. Appl. **250**, 123786 (2024)
11. Lepskiy, A.: Evidence-based aggregation and ranking in an ordinal scale. Procedia Comput. Sci. **221**, 1066–1073 (2023)
12. Yager, R.R.: On ordered weighted averaging aggregation operators in multicriteria decisionmaking. IEEE Trans. Syst. Man Cybern. **18**(1), 183–190 (1988)
13. Ma, L.Y., Denoeux, T.: Partial classification in the belief function framework. Knowl.-Based Syst. **214**, 106742 (2021)
14. Yu, D.J., Pan, T.X., Xu, Z.S., Yager, R.R.: Exploring the knowledge diffusion and research front of OWA operator: a main path analysis. Artif. Intell. Rev. **56**(10), 12233–12255 (2023)
15. Jousselme, A.-L., Grenier, D., Bossé, É.: A new distance between two bodies of evidence. Inf. fusion. **2**(2), 91–101 (2001)

# Automated Hierarchical Conflict Reduction for Crowdsourced Annotation Tasks Using Belief Functions

Constance Thierry[✉], David Gross-Amblard, Yolande Le Gall, and Jean-Christophe Dubois

Univ Rennes, IRISA, CNRS, Rennes, France
{constance.thierry,david.gross-amblard,yolande.le-gall,
jean-christophe.dubois}@irisa.fr

**Abstract.** A typical crowdsourcing task is concept labeling, where participants annotate e.g. images using a list of predefined concepts. Recent popular campaigns for environmental bird monitoring even use hierarchies of concepts (taxonomies of species) to obtain the most precise labeling of bird images. But in most applications, volunteer opinions are isolated from each other, and decision is taken upon majority voting. In this work we propose a new iterative labeling process where participants express their opinions together, on ascending levels of the taxonomy. Level changes are performed to minimize opinion conflict, according to the belief function theory. This complex task is orchestrated by a finite-state automaton driven by conflict measures.

**Keywords:** Belief functions · Crowdsourcing · Taxonomy · Automaton

## 1 Introduction

The crowds on crowdsourcing platforms are very diversified [11,16]. In order to ensure the quality of the data received and to limit any noise, the employer is obliged to recruit a large number of contributors. In a classic industrial platforms (Amazon mechanical turk[1], Wirk[2], etc.), the employer proposes a finite number of tasks to a finite number of contributors. Some elements of the literature [1,3,7,17,22] choose to authorize or assign a task to contributors according to their level of knowledge in the field, which can limit the size of the crowd. However, this remains problematic in the event of there not being enough expert on the platform. In addition, crowdsourcing campaigns tend to be linear, and current industry platforms don't allow for a more advanced workflow. The same questions are asked of all contributors, with no dynamic processing and consequently no modulation of questions according to the difficulties encountered.

In the approach proposed here, the crowdsourcing campaign is more flexible: there's no size imposed on the crowd, and there's no finite number of tasks either.

---

[1] https://www.mturk.com/.
[2] https://www.wirk.io/.

We hypothesize that the campaign enables the use of a taxonomy, such as the annotation of bird photos. The idea is as follows: we allow contributors to participate in a task (with no restrictions on their expertise) as long as the aggregated responses present too significant conflict. If no agreement can be reached, the task is reformulated thanks to the taxonomy and sent out to the crowd again. The aim here is to reduce the size of the crowd required for a crowdsourcing campaign, while increasing the degree of trust placed in contributions.

Headwork[3], defined by Gross-Amblard et al. [8], enables crowdsourcing of complex tasks by addressing a workflow with a state automaton. In this way, crowdsourcing tasks are states of the automaton, and the task workflow will depend on the automaton's rules. To achieve our goal, we are using an automaton as Headwork and a taxonomy to manage the crowdsourcing campaign workflow. A state in the automaton corresponds to a task to be carried out by the contributor where the answers given are related to a taxonomy level. The change of state is decided thanks to a conflict measure estimated by the theory of belief functions [5,18]. As shown in previous work [19–21], this theory can be used to model contributor responses on crowdsourcing platforms. In particular, it enables contribution aggregation and decision-making. We propose to go a step further by using this theory to define when to stop asking contributors, or how to refine a question to improve the quality of the results obtained.

The contributions in this article are as follows:

- New dynamic crowdsourcing campaign model
- Workflow management of a crowdsourcing campaign by an automaton using a taxonomy
- Change of state of the automaton by computing conflict within the crowd using belief function theory

The article is structured as follows. Section 2 reviews the key elements of the belief functions required for the proposed model, while Sect. 3 presents the state of the art. Section 4 introduces the model and Sect. 5 concludes.

## 2 Belief Functions

The section begins with a review of the basics of mass functions (Sect. 2.1), followed by a discussion of changes in the frame of discernment (Sect. 2.2).

### 2.1 Mass Function Basics

The set of classes or hypotheses $r_i$ that are exclusive and exhaustive is the frame of discernment $\Omega = \{r_0, ..., r_n\}$. The mass functions $m^\Omega : 2^\Omega \to [0, 1]$ model the elementary degree of belief of the source and respect the normalization condition: $\sum_{X \in 2^\Omega} m^\Omega(X) = 1$. An element $X \in 2^\Omega$ such that $m^\Omega(X) > 0$ is called focal element. If only the singletons of $\Omega$ are focal elements then $m^\Omega$ is

---

[3] https://headwork.irisa.fr/headwork/.

a probability, the function is then called Bayesian mass function, it is a specific mass function but there are others such as the simple support mass functions ($X^\omega$). This mass function reflects an uncertain and imprecise response from the information source.

$$\begin{cases} m^\Omega(X) = \alpha \text{ with } X \in 2^\Omega \setminus \Omega, \omega \in [0,1] \\ m^\Omega(\Omega) = 1 - \alpha \\ m^\Omega(Y) = 0, \forall Y \in 2^\Omega \setminus \{X, \Omega\} \end{cases} \quad (1)$$

For information fusion, the sources all report on the same frame of discernment $\Omega$. Numerous combination operators are available [14], from the average of mass functions to disjunctive and conjunctive combinations. As an example, the conjunctive rule (Eq. 2) reduces the imprecision on the focal elements and increases the belief on the concordant ones.

$$m^\Omega_{Conj}(X) = \left(\bigcap_{c=1}^{K} m_c^\Omega\right)(X) = \sum_{Y_1 \cap \ldots \cap Y_K = X} \prod_{c=1}^{K} m_c^\Omega(Y_c) \quad (2)$$

The mass $m^\Omega_{Conj}(\emptyset)$ represents the global conflict of the combination.

## 2.2 Change of Discernment Frame

If two sources express themselves on two different, but compatible, frames of discernment ($\Omega$ and $\Theta$), it is possible to combine information by refining or expending one of the frames of discernment. Thanks to a refinement function $R: 2^\Omega \to 2^\Theta$:

$$m^\Theta(R(X)) = m^\Omega(X), \forall X \in 2^\Omega \quad (3)$$

For an expending of the mass function, use the reciprocal of R. However, these functions only apply to compatible discernment frames. Otherwise, it is necessary to turn to other methods. Thanks to vacuous extension and marginalization present by Delmotte et al. [4], it is possible to perform projections of $\Omega$ onto $\Omega \times \Theta$ and conversely. These operations are extremely useful for combination of belief functions with different discernment frameworks.

## 3 State of the Art

Karampinas and Triantafillou [10] propose an algorithm for crowdsourcing taxonomy creation. The authors believe that humans can be useful in creating taxonomies of their knowledge domain. The objective is to aggregate a set of tags to create the taxonomy.

Farrell and Knapp [6] propose a multi-source classification based on a taxonomy using belief functions. To achieve this, the approach considers the fusion of heterogeneous input classifications that accommodate a taxonomy of one million leaf nodes with an output that can be used as a basis for decision-making. Belief functions are used to move from one level of taxonomy to another, while taking into account the possibility of opening up to the world ($\emptyset$). Each level of the taxonomy $T^i = t_1^i, ..., t_j^i$ can be interpreted as a frame of discernment ($T^i = \Omega$). $T^0$ is the root level and $T^n$ the most specific level. The approach considers a taxonomy node to be the union of its child nodes. To choose the appropriate taxonomic framework for decision-making, the authors compute the information content value. We note that the authors do not use Dempster's conjunctive combination but a Bayesian combination, and the focal elements of the mass functions are exclusively singletons.

In our approach, we propose to use the levels of a taxonomy as a frame of discernment, as Farrell and Knapp do for classification. While the authors propose a static generalist approach, we propose a dynamic model in a crowdsourcing context. Indeed, whereas the authors have a constant number of sources, our objective is to limit the number of sources and integrate more contributors only if necessary. Furthermore, we only consider contributors' answers to the most detailed $\Omega = T^n$ discernment framework, in order to facilitate the choice of answers for non-expert contributors.

In the work of Thierry et al. [21] a crowdsourcing campaign was carried out to annotate bird photos. The identification process takes place in two stages. First, the contributor is asked to choose one or more species from the 10 proposed, and to provide a degree of certainty in his answer. Two alternatives are possible: if the contributor has chosen only one species but is not totally certain of his answer, he is then asked to complete his choice by adding other species to increase his certainty. On the other hand, if the contributor has chosen several species, he is asked whether or not he is able to narrow down his choice. A limitation of this method is that no structure is used to organize the bird species proposed. So, if the contributor hesitates during the first identification stage, the system has no information to help him make his choice.

In the next section of this article, we propose not to use the scale of the contributor but the scale of the crowd to reformulate the question we wish to ask (i.e. the contributor may not find his previous selection in the new set proposed).

## 4 Proposed Approach

In this section, we first introduce (Sect. 4.1) the proposed theoretical model, then illustrate our ideas with a case study (Sect. 4.2).

### 4.1 Proposed Model

Following on from previous studies [19–21], we model the responses $X$ of contributors $c$ to crowdsourcing Multiple Choice Questionnaire (MCQ) $q$ using the theory of belief functions. To do this, the set of MCQ choices proposed to the contributor composes the frame of discernment $\Omega$. The contributor indicates his certainty $\alpha$ in his answer $X$, and his contribution is modeled by a simple support mass function $m_{cq}^{\Omega}$ (Eq. 1). The responses of all contributors are then aggregated on their frame of discernment by a combination operator to obtain the mass function $m_q^\Omega$. Then, a decision on the answer is made in a probabilistic framework.

In the context of crowdsourcing, the modeling of contributions using belief function theory is usually carried out with a fixed frame of discernment. We consider a crowdsourcing campaign whose response proposals can be extracted from the taxonomy $T$, assumed to be known at the time of campaign design. This taxonomy associated with the proposed choices is used to refine or extend the frame of discernment using the Eq. 3. The campaign is directed by a deterministic state automaton. A state of the automaton corresponds to a question, address to the crowd, whose answers depend on the taxonomy $T$. Since the crowd has varying levels of expertise on crowdsourcing platforms, we choose as our answer set $\Omega_n = T^n$ where $T^n$ is the taxonomy level furthest from the root and therefore the most detailed. When a minimal number of contributors, defined by the employer, have realized the question of a state, we calculate the conflict on the answers $conf(\Omega_n)$ and compared it to a threshold $\epsilon$, then:

- If $conf(\Omega_n) < \epsilon$ then we can aggregate the crowd answers and take a decision.
- Else $(conf(\Omega_n) > \epsilon)$, we compute the conflict again by changing the frame of discernment to use the values of the $l < n$ level of the taxonomy ($\Omega_l = T^l$). We calculate this new conflict until we find $\Omega_l$ such that $conf(\Omega_l) < \epsilon$. This calculation must be based on a reasonable number of levels.

- If there is no $\Omega_l$ such as $conf(\Omega_l) < \epsilon$ then we allow new participants join the crowd and answer the question $q$.
- Else $(conf(\Omega_l) < \epsilon)$, we take the decision on $\Omega_l$. Then ask the crowd the question $q'$ which is the same as $q$ but change the answer set to $\Omega'_n \subset \Omega_n$ such that $\Omega'_n$ is the set of leaves of the $T^l$ node chosen. To do so, we ask contributors for whom their initial response $X$ to $q$ was not in the proposed answers of $q'$ (i.e. $X \not\subseteq \Omega'_n$).

And so we continue until there is no longer any conflict over decision-making.

There are many measures of conflict in the theory of belief functions [9,13,15]. Among them, the value of the aggregated mass $m_q^\Omega(\emptyset)$, resulting from conjunctive rule (Eq. 2), can symbolizes the conflict between the sources of information (in this case the contributors), the higher this value, the greater the conflict. So we proposed that in the following case study the transition from one state to another in the automaton depends on the value of $m_q^\Omega(\emptyset)$, which must be below a threshold value $\epsilon$.

We assume that only responses from serious contributors are processed by the model. There are various methods for estimating the contributor's profile and seriousness [2,7,12,20] that can be used to complement the model proposed in this paper.

### 4.2 Case Study

A taxonomy of birds creates different categories to determine their relationship. Several similar bird species belong to the same genus; several similar genus belong

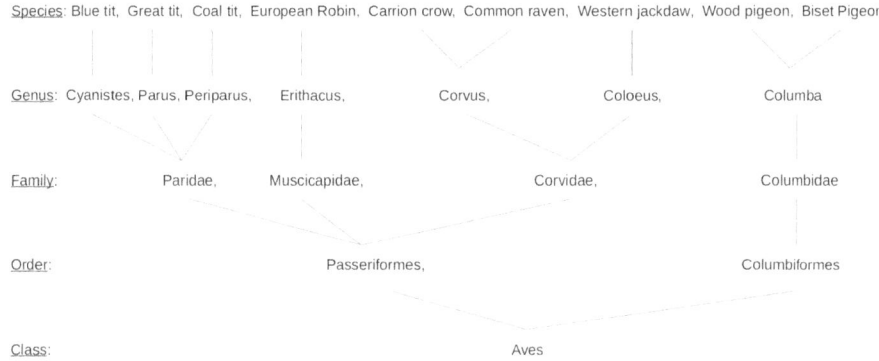

**Fig. 1.** Example of bird taxonomy for crowdsourcing campaign

to the same family; and several families make up an order. Taxonomists use other categories that allow more sophisticated classification, but in this case study we'll keep this simplified version. All birds together make up what is known as the bird class (Aves). In our case study, we consider a crowdsourcing campaign for bird photo annotation using the taxonomy shown in Fig. 1. We therefore have the following associated frame of discernment:

- $\Omega_S$ for the **Species**: Blue tit, Great tit, Coal tit, European Robin, Carrion crow, Common raven, Western jackdaw, Wood pigeon, Biset pigeon.
- $\Omega_G$ for the **Genius**: Cyanistes, Parus, Periparus, Erithacus, Corvus, Coloeus, Columba.
- $\Omega_F$ for the **Family**: Paridae, Muscicapidae, Corvidae, Columbidae
- $\Omega_O$ for the **Order**: Passeriformes, Columbiformes

Consider a photo of a bird presented to the contributor with the question 1 ($Q_1$) : "What species does this bird belong to ?". The proposed set of answers is composed of the elements of $\Omega_S$. We have chosen to use the common names of the proposed species rather than the scientific names, which are more difficult for non-experts to know.

Now let's consider a crowdsourcing campaign run by a deterministic automaton whose questions will depend on the contributors' answers (Fig. 2). The first question ($Q_1$) is open to a set of contributors with the set of answers from the frame of discernment $\Omega_S$. There are then 3 possible scenarios.

**Case 1:** For the $\Omega_S$ discernment framework, the conflict on all responses is lower than the threshold value ($m^{\Omega_S}(\emptyset) < \epsilon$), so a decision can be made.

If there is too much conflict on the frame of discernment $\Omega_S$ ($m^{\Omega_S}(\emptyset) > \epsilon$), then we perform a discernment frame change for $\Omega_l$ with $l \in \{G, F, O\}$ (possible thanks to marginalization).

**Case 2:** If there is still too much conflict after enlarging $\Omega_S$ to $\Omega_l$ ($m^{\Omega_l}(\emptyset) > \epsilon \forall l$) then we invite new contributors to answer question $Q_1$.

**Case 3:** After refinement, the conflict on $\Omega_l$ is acceptable. A decision can be made on $\Omega_l$, which allows the question to be rephrased to the crowd. For example,

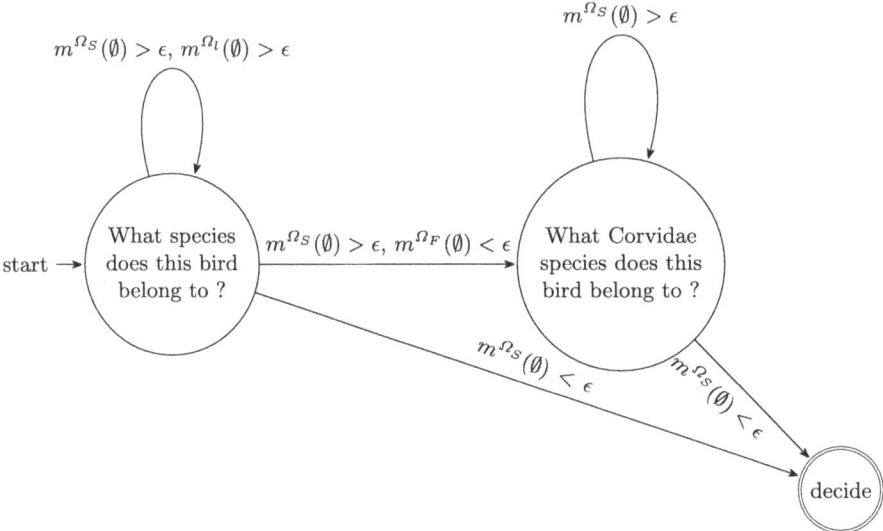

**Fig. 2.** Workflow with BF Example

if there is too much conflict on $\Omega_S$ and $\Omega_G$ but not on $\Omega_F$. The decision on $\Omega_F$ indicates that the bird is a *Corvidae*. Then it's possible to ask the crowd with the same photo question 2 ($Q_2$): What species of Corvidae does this bird belong to? The discernment framework $\Omega'_S \subset \Omega_S$ used will be composed exclusively of Corvidae specie.

$$\Omega'_S = \{Carrion\ crow, Common\ raven, Western\ jackdaw\}$$

When in this 3rd case, we can calculate the $\Omega_S$ frame again using an expending. We check if ($m^{\Omega_S}(\emptyset) < \epsilon$), then we proceed with the decision; otherwise, we continue questioning the crowd.

## 5 Conclusion and Perspectives

In this paper we propose a model for dynamically managing the workflow of a crowdsourcing campaign using a taxonomy and an automaton. The automaton's state changes are decided on the basis of conflict within a crowd of contributors, calculated using belief function theory. This model will soon be integrated into the Headwork platform to be tested during real crowdsourcing campaigns and to evaluate its performance. We particularly plan to conduct a study to determine the most relevant conflict calculation in this context. Our future work also focuses on adapting the level of taxonomy to be adopted according to the profile and expertise of the contributor: more specific for experts and more general for non-experts.

# References

1. Amsterdamer, Y., Davidson, S.B., Milo, T., Novgorodov, S., Somech, A.: Oasis: query driven crowd mining. In: Proceedings of the 2014 ACM SIGMOD International Conference on Management of Data, pp. 589–600. ACM (2014)
2. Blanco, H.H.R.: Machine-learning for spammer detection in crowd-sourcing. Human Computation AAAI Technical Report (2012)
3. Boim, R., Greenshpan, O., Milo, T., Novgorodov, S., Polyzotis, N., Tan, W.C.: Asking the right questions in crowd data sourcing. In: 2012 IEEE 28th International Conference on Data Engineering, pp. 1261–1264. IEEE (2012)
4. Delmotte, F., Smets, P.: Target identification based on the transferable belief model interpretation of dempster-shafer model. IEEE Trans. Syst. Man Cybern.-Part A: Syst. Hum. **34**(4), 457–471 (2004)
5. Dempster, A.P.: Upper and lower probabilities induced by a multivalued mapping. Ann. Math. Stat. **38**, 325–339 (1967)
6. Farrell III, W.J., Knapp, A.M.: Multisource taxonomy-based classication using the transferable belief model. In: Multisensor, Multisource Information Fusion: Architectures, Algorithms, and Applications 2012. vol. 8407, pp. 35–41. SPIE (2012)
7. Folorunso, O., Mustapha, O.A.: A fuzzy expert system to trust-based access control in crowdsourcing environments. Appl. Comput. Inf. **11**(2), 116–129 (2015)
8. Gross-Amblard, D., Tommasi, M., Rakotoniaina, I., Thierry, C., Singh, R., Jacoboni, L.: Headwork: a data-centric crowdsourcing platform for complex tasks and participants. In: EDBT 2024 (2024)
9. Jousselme, A.L., Grenier, D., Bossé, É.: A new distance between two bodies of evidence. Inf. Fusion **2**(2), 91–101 (2001)
10. Karampinas, D., Triantafillou, P.: Crowdsourcing taxonomies. In: Simperl, E., Cimiano, P., Polleres, A., Corcho, O., Presutti, V. (eds.) The Semantic Web: Research and Applications, ESWC 2012, LNCS, vol. 7295, pp. 545–559. Springer, Berlin (2012). https://doi.org/10.1007/978-3-642-30284-8_43
11. Kazai, G., Kamps, J., Milic-Frayling, N.: The face of quality in crowdsourcing relevance labels: Demographics, personality and labeling accuracy (2012)
12. Khattak, F.K., Salleb-Aouissi, A.: Quality control of crowd labeling through expert evaluation. In: Proceedings of the NIPS 2nd Workshop on Computational Social Science and the Wisdom of Crowds, vol. 2, p. 5 (2011)
13. Martin, A.: About conflict in the theory of belief functions. In: Denoeux, T., Masson, MH. (eds.) Belief Functions: Theory and Applications. Advances in Intelligent and Soft Computing, vol. 164, pp. 161–168. Springer, Berlin, Heidelberg (2012). https://doi.org/10.1007/978-3-642-29461-7_19
14. Martin, A.: Conflict Management in Information Fusion with Belief Functions. In: Bossé, É., Rogova, G. (eds.) Information Quality in Information Fusion and Decision Making. Information Fusion and Data Science, pp. 79–97. Springer, Cham (2019). https://doi.org/10.1007/978-3-030-03643-0_4
15. Martin, A., Jousselme, A.L., Osswald, C.: Conflict measure for the discounting operation on belief functions. In: Information Fusion, 2008 11th International Conference on, pp. 1–8. IEEE (2008)
16. Ross, J., Zaldivar, A., Irani, L., Tomlinson, B.: Who are the turkers? worker demographics in amazon mechanical turk. Department of Informatics, University of California, Irvine, USA, Technical Report (2009)

17. Roy, S.B., Lykourentzou, I., Thirumuruganathan, S., Amer-Yahia, S., Das, G.: Crowds, not drones: modeling human factors in interactive crowdsourcing. In: DBCrowd 2013-VLDB Workshop on Databases and Crowdsourcing, pp. 39–42. CEUR-WS (2013)
18. Shafer, G.: A mathematical theory of evidence, vol. 42. Princeton University Press, Princeton (1976)
19. Thierry, C., Dubois, J.C., Le Gall, Y., Martin, A.: Modeling uncertainty and inaccuracy on data from crowdsourcing plateforms: Monitor. In: Proceedings of the 31st International Conference on Tools with Artificial Intelligence (2019)
20. Thierry, C., Martin, A., Dubois, J.C., Le Gall, Y.: Estimation of the qualification and behavior of a contributor and aggregation of his answers in a crowdsourcing context. Expert Syst. Appl. **216**, 119496 (2023)
21. Thierry, C., Martin, A., Le Gall, Y., Dubois, J.C.: Modeling evolutionary responses in crowdsourcing MCQ using belief function theory. Procedia Comput. Sci. **225**, 2575–2584 (2023)
22. Yongxin, T., Caleb, C.C., Chen, J.Z., Yatao, L., Lei, C.: Crowdcleaner: data cleaning for multi-version data on the web via crowdsourcing. In: 2014 IEEE 30th International Conference Data Engineering (ICDE) (2014)

# Continuous Belief Functions, Logics, Computation

# Gamma Belief Functions

Liping Liu(✉)

The University of Akron, Akron, USA
`liu@acm.org`

**Abstract.** This paper proves that the combination of gamma distributions via Dempster's rule is also a gamma and therefore establishes the notion of gamma belief functions or gamma evidence, extending the scope of continuous belief functions from Gaussian belief functions to three additional members of the exponential family of distributions: gamma, exponential, and Erlang. It applies the result to the combination of generalized gamma regressions for evidence-based ensemble learning. Using simulated data and thousands of replications, the paper shows that the predictions by the combined model is very close to that made by regression models built by merging partial data sets but surprisingly outperforms in predicting actual responses.

## 1 Introduction

In theory, any probability distribution is a belief function with singletons or hyperplanes as focal elements. Yet, so far only Gaussian distributions are known to be closed under combination, i.e., the combination of Gaussian distribution via Dempster's rule is still Gaussian, and thus we have the notion of Gaussian belief functions [4,5]. There has been no proof that any other family of distributions has the same property. This paper attempts to push the boundary further by bringing in gamma distributions.

A gamma distribution is flexible for capturing a variety of distributions and uniquely capable of handling positive and skewed data, which is a common scenario in practical applications involving time and money. A gamma distribution has been used to model waiting time until a repair is necessary for a hydraulic system [11], amount of insurance claims [1], the amount of rainfall accumulated in a reservoir [7,12], etc. It also acts as a conjugate prior for the inverse variance of a Gaussian distribution and a conjugate prior for an exponential distribution. Additionally, when the shape parameter is held constant, varying the scale or rate parameter can capture different levels of variance, allowing for a more nuanced understanding of data dispersion. Thanks to the central limit theorem, a Gaussian distribution is the foundation of modern data analytic models and tools, but gamma distributions are not a less important member in the exponential family of distributions in practice.

A study on whether the combination of gamma distributions is gamma or not is not only interesting in theory but also has practical implications. In a sense, gamma regression, a special type of generalized linear models [8], is to build a continuum of conditional gamma distributions of a response variable given predictors. If two gamma distributions can be combined into another gamma, it will empower ensemble learning with a new evidence-based bootstrapping tool to combine gamma regressions for better predictions without data merging. Two possible applications may be imagined. First, when the volume of data outgrows computing power, data analysts may conduct gamma regressions based on partial data sets and then combine the models instead of merging data sets. If the velocity of data is high, one can even perform dynamic gamma regression using new data and use it to update the regression from the past data. There will be no need to merge data for a better model. Second, it is often the case that different scientific studies may conduct gamma regressions on the same variable of interest using different data sets. Combining regressions will allow a quantitative review of these studies by combining their regression results as what classic meta analysis [10] does with linear regressions.

The rest of the paper is organized as follows. Section 2 will briefly introduce Dempster's rule and apply it to the combination of gamma distributions. Section 3 will explore the combination of generalized gamma regression on simulated data. Finally, I will make a conclusion remark in Sect. 4.

## 2 Gamma Belief Functions

The key to the concept of Dempster-Shafer theory is limited divisibility of beliefs [6]: a belief function is made of indivisible atomic subsets, called *focal elements*, and indivisible *probability mass assignments*. Dempster-Shafer theory assumes a finite frame of discernment. Recent research has led to two extensions to the continuous variables of interest: Gaussian belief functions conceptualized as Gaussian distributions over hyperplanes as focal elements [4,5] and belief functions on real numbers with closed intervals in $\Re = [-\infty, \infty]$ as focal elements [9]. In any extensions to continuous variables of interest, mass assignments are replaced by belief densities. Assume $f(x)$ and $g(x)$ are the density functions for two pieces of independent evidence. Since focal elements are singletons, their three-step combination via Dempster's rule, namely intersection of focal elements, multiplication of mass numbers, and normalization of the products of the mass numbers, are expressed as follows [5]:

$$f(x) \oplus g(x) = \frac{f(x)g(x)}{\int f(x)g(x)dx} \tag{1}$$

A random variable $X$ is gamma with shape parameter $\alpha$ and rate parameter $\beta$, denoted by $X \sim \Gamma(\alpha, \beta)$, if it has a probability density function

$$f(x|\alpha, \beta) = \frac{\beta^\alpha x^{\alpha-1} e^{-\beta x}}{\Gamma(\alpha)}$$

for any $x \geq 0$, where $\alpha, \beta > 0$, and $\Gamma(\alpha)$ is the gamma function. In econometrics, alternative parameterization $(k, \theta)$ is often preferred, where $k = \alpha$ and $\theta = 1/\beta$, and a gamma distribution models the waiting time until the $k$th arrival in a one-dimensional Poisson process with average arrival time $\theta$.

Assume there are two independent pieces of evidence justifying that a random variable $X$ is respectively modeled by two gamma distributions with respectively parameters $(\alpha_1, \beta_1)$ and $(\alpha_2, \beta_2)$. Then their combination via Dempster's rule is

$$\frac{f(x|\alpha_1,\beta_1)f(x|\alpha_2,\beta_2)}{\int_0^\infty f(x|\alpha_1,\beta_1)f(x|\alpha_2,\beta_2)dx} = \frac{x^{\alpha_1+\alpha_2-2}e^{-(\beta_1+\beta_2)x}}{\int_0^\infty x^{\alpha_1+\alpha_2-2}e^{-(\beta_1+\beta_2)x}dx}$$

because

$$f(x|\alpha_1,\beta_1)f(x|\alpha_2,\beta_2) = \frac{\beta_1^{\alpha_1}x^{\alpha_1-1}e^{-\beta_1 x}}{\Gamma(\alpha_1)}\frac{\beta_2^{\alpha_2}x^{\alpha_2-1}e^{-\beta_2 x}}{\Gamma(\alpha_2)}$$

$$= \frac{\beta_1^{\alpha_1}\beta_2^{\alpha_2}x^{\alpha_1+\alpha_2-2}e^{-(\beta_1+\beta_2)x}}{\Gamma(\alpha_1)\Gamma(\alpha_2)}$$

Let $y = (\beta_1 + \beta_2)x$.

$$\int_0^\infty x^{\alpha_1+\alpha_2-2}e^{-(\beta_1+\beta_2)x}dx = (\frac{1}{\beta_1+\beta_2})^{\alpha_1+\alpha_2-1}\int_0^\infty y^{\alpha_1+\alpha_2-2}e^{-y}dy$$

$$= (\frac{1}{\beta_1+\beta_2})^{\alpha_1+\alpha_2-1}\Gamma(\alpha_1+\alpha_2-1)$$

Thus, the combined distribution

$$f(x|\alpha,\beta) = \frac{(\beta_1+\beta_2)^{\alpha_1+\alpha_2-1}x^{\alpha_1+\alpha_2-2}e^{-(\beta_1+\beta_2)x}}{\Gamma(\alpha_1+\alpha_2-1)}$$

is also a gamma distribution with shape parameters $\alpha = \alpha_1 + \alpha_2 - 1$ and rate parameter $\beta = \beta_1 + \beta_2$.

This result states that gamma distributions are closed under combination operations. Also, for any gamma distribution $\Gamma(\alpha, \beta)$, its combination with $\Gamma(1,0)$ will not change the distribution: $\Gamma(\alpha,\beta) \oplus \Gamma(1,0) = \Gamma(\alpha,\beta)$. In words, $\Gamma(1,0)$ does not add new knowledge to any gamma distribution when combined. Thus, in the parlance of Dempster-Shafer theory of belief functions, $\Gamma(1,0)$ is a vacuous belief function, representing the state of full ignorance on the variable of interest, and so we call $\Gamma(1,0)$ the *vacuous gamma belief function*. Note that $\Gamma(1,0)$ is not a gamma distribution, and so its density function does not have a finite expression. One may imagine it as a gamma distribution with an infinite scale or the limit case of the gamma distribution when the scale parameter $\theta$ approaches $\infty$. A probability distribution with a larger scale parameter models a random variable with a larger uncertainty and a distribution with an infinite scale models a variable on which we have complete uncertainty or full ignorance, which is identical to assigning the whole mass to the entire frame of discernment. Therefore, the vacuous gamma belief function is well justified semantically.

Therefore, I propose the concept of gamma belief functions as a new distinct family of continuous belief functions for representing knowledge on variables with positive and skewed distributions. The family includes all gamma probability distributions and $\Gamma(1,0)$ as the vacuous belief function. And the combination of any two gamma belief functions is also gamma.

**Theorem 1.** *The combination of two gamma belief functions via Dempster's rule is also gamma with the combined rate parameter as the sum of the component rate parameters and the combined shape parameter as the sum of the component shape parameters minus one. In mathematical notations,*

$$\Gamma(\alpha_1, \beta_1) \oplus \Gamma(\alpha_2, \beta_2) = \Gamma(\alpha_1 + \alpha_2 - 1, \beta_1 + \beta_2)$$

As rate parameter $\beta$ is the inverse of a scale parameter $\theta$, $\beta = \frac{1}{\theta}$. Let $\theta_1 = \frac{1}{\beta_1}$ and $\theta_2 = \frac{1}{\beta_2}$ be respectively the scale parameters of the two component distributions. Then, a combined gamma distribution has scale parameter

$$\theta = \frac{1}{\beta_1 + \beta_2} = \frac{1}{\frac{1}{\theta_1} + \frac{1}{\theta_2}} = \frac{\theta_1 \theta_2}{\theta_1 + \theta_2}$$

In words, the inverse of the scale of the combined gamma distribution is the sum of the inverse component scales: $\frac{1}{\theta} = \frac{1}{\theta_1} + \frac{1}{\theta_2}$. Since component scales are all positive, thus we have $\frac{1}{\theta} > max(\frac{1}{\theta_1}, \frac{1}{\theta_2})$ or alternatively, $\theta < min(\theta_1, \theta_2)$, which means that the scale of the combined gamma distribution is smaller than the scale of any component distribution. Since scale parameter determines the spread of a distribution, it makes an intuitive sense that the combination reduces the distributional spread; or the combined knowledge reduces uncertainty, which is the very purpose of combination.

In terms of shape and scale parameters, the mean of a gamma distribution is $\mu = E(X) = k\theta$ and the variance is $\sigma^2 = var(X) = k\theta^2$. Then, the combined mean and variance are

$$\mu = \frac{\alpha_1 + \alpha_2 - 1}{\beta_1 + \beta_2}, \sigma^2 = \frac{\alpha_1 + \alpha_2 - 1}{(\beta_1 + \beta_2)^2} \tag{2}$$

Gamma distributions generalize exponential, Erlang, and $\chi^2$ distributions: If $X \sim \Gamma(n, \frac{1}{\lambda})$ with shape parameter $\alpha = n$ being an integer, then the gamma distribution is Erlang distribution $E_n(\lambda)$ with shape parameter $n$ and rate parameter $\lambda$; If $X \sim \Gamma(1, \frac{1}{\lambda})$, then $X$ has exponential distribution $E(\lambda)$ with rate parameter $\lambda$; If $X \sim \Gamma(n/2, 2)$, then $X$ has $\chi^2$ distribution with degree of freedom $n$.

Exponential and Erlang distributions are both concerned with a Poisson process in which independent events occur randomly with average rate $\lambda$ (either in time or space). Random variable $X$ is exponential, denoted by $X \sim E(\lambda)$ if its density function is $f(x) = \lambda e^{-\lambda x}$ for $x \geq 0$. An exponential variable measures the time (or some metric measures of space such as distance, area, or volume) between occurrences of events. Thus, exponential distributions are often used to model the time between machine failures, the distance between breaks in pipelines, or the longitivity of a person who is diagnosed with a certain disease.

If we apply Theorem 1 to these special case, it is easy to verify that combination via Dempster's rule is a closed operation for both Erlang and exponential distributions. Thus, for the same exponential or Erlang variable of interest, if there are multiple pieces of evidence justifying different distributional knowledge, the knowledge may be combined via Dempster's rule.

**Corollary 1.** *The combination of independent exponential distributions $E(\lambda_1)$ and $E(\lambda_2)$ is also exponential with the combined rate parameter as the sum of the rate parameters of the components. I.e., $E(\lambda_1) \oplus E(\lambda_2) = E(\lambda_1 + \lambda_2)$.*

*The combination of independent Erlang distributions $E_n(\lambda_1)$ and $E_m(\lambda_2)$ is also Erlang with the combined shape parameter $n + m - 1$ and the combined rate parameter as $\lambda_1 + \lambda_2$. I.e., $E_n(\lambda_1) \oplus E_m(\lambda_2) = E_{n+m-1}(\lambda_1 + \lambda_2)$.*

Exponential variable $X$ with rate parameter $\lambda$ has both mean and standard deviation as the inverse of rate parameter: $\mu = \sigma = \frac{1}{\lambda}$. Thus, the mean and the standard deviation of the combined exponential distribution are geometric mean of the component means and component standard deviations:

$$\mu = \frac{\mu_1 \mu_2}{\mu_1 + \mu_2}, \sigma = \frac{\sigma_1 \sigma_2}{\sigma_1 + \sigma_2}$$

Note that the combination of two $\chi^2$ distributions will produce a general gamma distribution with combined rate parameter 4, which is not $\chi^2$ anymore. Thus, $\chi^2$ distributions do not constitute a closed family of belief functions.

**Corollary 2.** *The combination of $\chi^2$ distributions $\chi_n^2$ and $\chi_m^2$ is gamma distribution $\Gamma(\frac{n}{2} + \frac{m}{2} - 1, 4)$, which is not a $\chi^2$ distribution.*

## 3 Combining Gamma Regressions

Gamma regressions extend usual linear ones to the case when a response variable follows a gamma distribution: the response variable is predicted by a linear combination of regressors such that the mean $\mu$ of the response variable is related to the linear combination of regressors through a link function as $g(\mu) = b_0 + \Sigma_{i=1}^n b_i X_i$, where $X_1, X_2, ..., X_n$ are regressors, and $b_0, b_1, ..., b_n$ are regression coefficients. A gamma regression model is a continuum of conditional gamma distributions of the response variable given the regressors, and so two or more regression models may be combined as gamma belief functions per Theorem 1. This leads to a new evidence-based bootstrapping method for ensemble learning of gamma regression models. Note that in big data and meta analytics, we may not load the whole data due to its volume, velocity, and variety or may not access all data due to data privacy, but we can have gamma regressions from partial data sets. We can combine the regressions from the partial data sets, and it is desirable that the combination of these regression models based on the partial data sets approximates the regression model based on the whole data.

The classic gamma regression technique and glm() implementation in R core assume invariant dispersion or inverse shape parameter. The doctoral thesis [3] attempts to remove the assumption and fit gamma regression models with both mean and shape parameters following regression structures. A recent package "Gammareg" implements the algorithm for R [2]. Since my formula allows any shapes and scales of gamma distributions, it can be applied to combine gamma regressions with or without varying dispersion. Here, due to the space limitation, I will examine how well the combined model performs with generalized gamma regressions.

Following the examples in [2], I generated two pairs of simulated data sets using respectively log-link and identity-link for mean parameters: $\mu = 5 + 2X_2 + 3X_3$ and $log(\mu) = -5 + 0.2X_2 - 0.03X_3$ and log-link for the shape parameters: $log(\alpha) = 0.2 + 0.1X_2 + 0.3X_4$. The data on regressors were generated using uniform distributions: $X_2 \sim U(0,30)$, $X_3 \sim U(0,15)$, and $X_4 \sim U(10,20)$, and the data on the response variable is generated by gamma distribution $Y \sim \Gamma(\alpha, \beta)$ with parameters determined by the link function: $\alpha = e^{0.2+0.1X_2+0.3X_4}$, and $\beta = \frac{\alpha}{\mu} = \frac{e^{0.2+0.1X_2+0.3X_4}}{5+2X_2+3X_3}$ when using identity-link for mean, or $\beta = \frac{\alpha}{\mu} = \frac{e^{0.2+0.1X_2+0.3X_4}}{e^{-5+0.2X_2-0.03X_3}}$ when using log-link for mean. Each pair consists of two data sets of 500 and 800 samples each, and I built four regression models: two component models $m_1$ and $m_2$ using two data sets separately, model $m_c$ by combining $m_1$ and $m_2$ using Theorem 1, and model $m$ by doing regression after merging both data sets together.

All models have varying shape and rate parameters, and so I computed the parameters for each observation individually. Using the log-link for mean, I obtained two component models as $m_1$: $log(\alpha) = -0.349 + 0.1081X_2 + 0.3303X_4$ and $log(\mu) = -4.999 + 0.1.998X_2 - 0.02985X_3$, and $m_2$: $log(\alpha) = 0.3924 + 0.1009X_2 + 0.2845X_4$ and $log(\mu) = -4.9998 + 0.2001X_2 - 0.02845X_3$. By merging two data sets, I obtained regression $m$: $log(\alpha) = 0.1549 + 0.1028X_2 + 0.2995X_4$ and $log(\mu) = -4.998 + 0.2000X_2 - 0.03025X_3$. Using sample regressor values $(X_2, X_3, X_4) = (20.37, 1.0447, 15.07)$, for example, we can compute $\alpha_1 = 926.62$, $\mu_1 = 0.382302$, and $\beta_1 = \alpha_1/\mu_1 = 2423.8$ from model $m_1$ and $\alpha_2 = 843.10$, $\mu_2 = 0.384749$, and $\beta_2 = \alpha_2/\mu_2 = 2191.3$ from model $m_2$. Per Theorem 1, the combined model will have $\alpha_C = \alpha_1 + \alpha_2 - 1 = 1768.72$ and $\beta_C = \beta_1 + \beta_2 = 4615.1$. Then the predicted mean value per the combined model is $\mu_C = \alpha_C/\beta_C = 0.383247$. From model $m$, we get the estimate as $\mu = 0.383924$. So the difference between the predictions by merged-data model $m$ and combined model $m_c$ is $|\mu - \mu_C|/\mu = 0.00176$ or $0.176\%$. By comparing with the actual response value $Y = 0.356352$, we can calculate the prediction error by the combined model $m_c$ as $|Y - \mu_C|/Y = 0.0754736$ and by the merged-data model $m$ as $|Y - \mu|/Y = 0.07737$.

Doing the same calculation for all sample observations, we can compare the overall performance of the models. For the pair of data sets using log-link for the mean parameter, I found that the predictions by the combined model $m_c$ and the merged-data regression $m$ has average difference rate of $0.18052\%$ with $95\%$ quantile range of $(0.000375\%, 0.89455\%)$. The two models made almost identical predictions. The prediction error against actual response value by the combined model $m_c$ has average error rate of $4.275\%$ with $95\%$ quantile range

(0.1668%, 16.7531%). The prediction error by the merged-data regression $m$ has average error rate of 4.2984% and 95% quantile range (0.1637%, 17.0212%). The left chart in Fig. 1 shows the prediction error distributions of both models, which are almost identical. In 647 of 1300 cases, the combined model has smaller prediction errors than the merged-date regression.

When using identity-link for the mean parameter, I found that the predictions by the combined model $m_c$ and the merged-data regression $m$ has average difference rate of 0.19585% with 95% quantile range of (0.0036%, 0.9650%). Again both models made almost identical predictions. The prediction error by the combined model $m_c$ has average error rate of 4.5127% with 95% quantile range (0.1522%, 17.82%). The prediction error by the merged-data regression m has average error rate of 4.516% with 95% quantile range (0.1522%, 18.068%). The prediction error distribution is shown by the left chart in Fig. 1. In 600 of 1300 cases, the combined model performed better than the merged-data regression model.

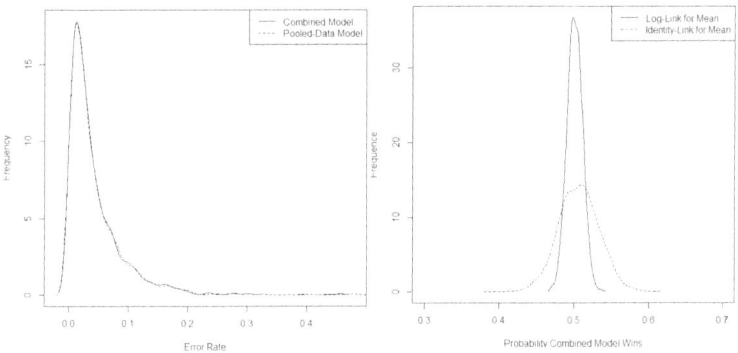

**Fig. 1.** Distributions of Prediction Error Rates and Winning Ratios

The above result was based on one experiment. By replicating the experiment many times, each time I generated four samples of random sizes respectively determined by Poisson(500) and Poisson(800) and counted the number of observations in which the combined model predicts better than the merged-data regression. Using log-link for the mean parameter, there were 530 of 1000 times in the first and 1043 of 2000 in the second simulation in which the combined model performed better. Using identity-link for the mean parameter, there were 580 of 1000 times in the first and 1200 of 2000 in the second simulation in which the combined model did better. Thus, the combined model outperformed the merged-data regression statistically. The right chart in Fig. 1 shows the frequencies of the winning ratios by the combined model in 2000 replications. t tests on the hypotheses: $H_0$: winning ratio = 0.5 and $H_1$: winning ratio > 0.5 have p-values $p = 2 \times 10^{-16}$ for both sets of 2000 replications, meaning that the combined model significantly outperformed the merged-data regression using both log-link and identity-link functions.

## 4 Conclusion

This paper made two contributions to the theory and applications of belief functions. First, it proves that gamma distributions are closed under combination via Dempster's rule and thus validates the notion of gamma belief functions with $\Gamma(1, 0)$ as the representation of ignorance. Within the family gamma belief functions, exponential and Erlang belief functions form two closed sub families. However, $\chi^2$ distributions are not closed under combination and will not from a distinct sub family. Second, it shows that the combination of gamma distributions can be applied to the combination of gamma regressions. Combining generalized gamma regressions with varying dispersion from simulated data, I found the combined model statistically outperforms the merged-data regression model. This result has an important implication to big data analytics where data volume, velocity, and variety often prohibit doing a gamma regression on the whole data and to meta analysis or quantitative integration of research results when we may not have access to all the data sets due to data privacy.

## References

1. Boland, P.J.: Statistical and Probabilistic Methods in Actuarial Science. Chapman & Hall CRC, Boca Raton (2007)
2. Bossio, M.C., Cuervob, E.C.: Gamma regression models with the gammareg rpackage. Comunicaciones en Estadstica **8**(2), 211–223 (2015)
3. Cepeda-Cuervo, E.: Modelagem de variabilidade em modelos lineares generalizados. Ph.D. thesis, Mathematics Institute, Universida d eFederal Rio de Janeiro (2001)
4. Dempster, A.P.: Normal belief functions and the Kalman filter. In: Saleh, A.K.M.E. (ed.) Data Analysis from Statistical Foundations, pp. 65–84. Nova Science Publishers, Hauppauge, New York (2001)
5. Liu, L.: A theory of Gaussian belief functions. Int. J. Approximate Reasoning **14**, 95–126 (1996)
6. Liu, L., Yager, R.: Classic works on the Dempster-Shafer theory of belief function: An introduction. In: Yager, R., Liu, L. (eds.) Classic Works of the Dempster-Shafer Theory of Belief Functions, pp. 1–34. Springer-Verlag, New York, NY (2008)
7. Mathai, A.: Storage capacity of a dam with gamma type inputs. Ann. Inst. Stat. Math. **34**(3), 591–597 (1982)
8. McCullagh, J., Nelder, J.: Generalized Linear Models, 2nd edn. Chapman and Hall, London (1989)
9. Smets, P.: Belief functions on real numbers. Int. J. Approximate Reasoning **40**(3), 181–223 (2005)
10. Smith, M.L., Glass, G.V.: Meta-analysis of psychotherapy outcome studies. Am. Psychol. **32**, 752–760 (1977)

11. Vineyard, M., Amoako-Gyampah, K., Meredith, J.R.: Failure rate distributions for flexible manufacturing systems: an empirical study. Eur. J. Oper. Res. **116**(1), 139–155 (1999)
12. Wilks, D.S.: Maximum likelihood estimation for the gamma distribution using data containing zeros. J. Clim. **3**(12), 1495–1501 (1990)

# Combination of Dependent Gaussian Random Fuzzy Numbers

Thierry Denœux[1,2]

[1] Université de Technologie de Compiègne, CNRS, Heudiasyc, Compiègne, France
tdenoeux@utc.fr
[2] Institut Universitaire de France, Paris, France

**Abstract.** Gaussian random fuzzy numbers are random fuzzy sets generalizing Gaussian random variables and possibility distributions. They define belief functions on the real line that can be conveniently combined by the product-intersection rule under the independence assumption. In this paper, we provide formulas for the combination of an arbitrary number of Gaussian random fuzzy numbers whose dependence is described by a correlation matrix.

**Keywords:** Evidence theory · Dempster-Shafer theory · belief functions · random fuzzy sets · information fusion

## 1 Introduction

In recent papers [3,5,7], I introduced a theory of epistemic random fuzzy sets as an extension of both the Dempster-Shafer theory of evidence [9] and possibility theory [10]. In this new theoretical framework, uncertain and/or fuzzy pieces of evidence are represented by random fuzzy sets inducing belief functions, and independent items of evidence are combined by the product-intersection rule generalizing both Dempster's rule of combination and the normalized product-intersection operator of possibility theory.

Gaussian random fuzzy numbers (GRFNs) introduced in [5] are an important practical model making it possible to represent evidence about continuous real variables. A GRFN is characterized by its mean, its variance and its imprecision. Gaussian random variables and Gaussian fuzzy numbers are recovered, respectively, in the special cases of infinite precision and zero variance. As shown in [5], GRFNs define a parametric family of belief functions on the real line, closed under the product-intersection rule.

As Dempster's rule, the product-intersection rule introduced in [3,5] assumes independence of the combined pieces of evidence. Formulas for the combination of independent GRFNs are given in [5]. However, the independence assumption is sometimes too restrictive in applications. For instance, opinions from different experts, or predictions based on correlated features or overlapping datasets (using, e.g., the ENNreg model [4]) often cannot be treated as independent. The combination of dependent GRFNs is, thus, an important problem; it is addressed in this paper.

The rest of this paper is organized as follows. Necessary notions about epistemic random fuzzy sets are first recalled in Sect. 2. Formulas for the combination of $n$ GRFNs with an arbitrary covariance matrix are then derived in Sect. 3. Finally, examples are presented in Sect. 4 and Sect. 5 concludes the paper.

## 2 Background

General definitions and results about random fuzzy sets will first be recalled in Sect. 2.1. GRFNs will then be addressed in Sect. 2.2.

### 2.1 Epistemic Random Fuzzy Sets

Epistemic Random Fuzzy Set (ERFS) theory is based on two main components: the representation of evidence by random fuzzy sets (inducing belief and plausibility functions), and a combination mechanism: the product-intersection rule for pooling independent evidence.

Let $(\Omega, \Sigma_\Omega, P)$ denote a probability space, $(\Theta, \Sigma_\Theta)$ a measurable space, and $\widetilde{X}$ a mapping from $\Omega$ to the set $[0,1]^\Theta$ of fuzzy subsets of $\Theta$. For any $\alpha \in [0,1]$, let $^\alpha\widetilde{X}$ be the mapping from $\Omega$ to $2^\Theta$ that maps each $\omega \in \Omega$ to the (weak) $\alpha$-cut of $\widetilde{X}(\omega)$. If, for any $\alpha \in [0,1]$, $^\alpha\widetilde{X}$ is $\Sigma_\Omega - \Sigma_\Theta$ strongly measurable [8], the tuple $(\Omega, \Sigma_\Omega, P, \Theta, \Sigma_\Theta, \widetilde{X})$ is said to be a *random fuzzy set* (RFS) [2].

In ERFS theory, a RFS represents a piece of evidence, which may be unreliable, vague (fuzzy), or both. The set $\Omega$ is seen as a *set of interpretations* of a piece of evidence about a variable $\boldsymbol{\theta}$ taking values in $\Theta$. If interpretation $\omega \in \Omega$ holds, we only know that $\boldsymbol{\theta}$ is constrained by the possibility distribution defined by fuzzy set $\widetilde{X}(\omega)$. Standard Dempster-Shafer theory only considers the case of unambiguous evidence, in which every image $\widetilde{X}(\omega)$ is crisp; mapping $\widetilde{X}$ is then a random set. In contrast, possibility theory only imposes a flexible constraint on the unknown quantity, without considering that this constraint may be itself uncertain. By considering both vagueness and uncertainty, ERFS is, thus, more flexible, allowing for faithful representation of many different kinds of evidence.

To any RFS, we can be associate a belief function representing one's beliefs based on the available evidence. For technical reasons, we assume hereafter any RFS $\widetilde{X}$ to verify the following normalization conditions: (1) For all $\omega \in \Omega$, $\widetilde{X}(\omega)$ is either the empty set, or a normal fuzzy set, and (2) the image $\widetilde{X}(\omega)$ is almost surely nonempty, i.e., $P(\{\omega \in \Omega : \widetilde{X}(\omega) = \emptyset\}) = 0$. For any $\omega \in \Omega$, a conditional possibility measure $\Pi_{\widetilde{X}(\omega)}$ and a dual conditional necessity measure $N_{\widetilde{X}(\omega)}$ on $\Theta$ can be defined as follows: for any $B \subseteq \Theta$, $\Pi_{\widetilde{X}(\omega)}(B) = \sup_{\theta \in B} \widetilde{X}(\omega)(\theta)$, and $N_{\widetilde{X}(\omega)}(B) = 1 - \Pi_{\widetilde{X}(\omega)}(B^c)$ if $\widetilde{X}(\omega) \neq \emptyset$ and $N_{\widetilde{X}(\omega)}(B) = 0$ otherwise. For any $B \in \Sigma_\Theta$, let $Bel_{\widetilde{X}}(B)$ and $Pl_{\widetilde{X}}(B)$ denote, respectively, the *expected necessity* and the *expected possibility* of $B$ wrt $P$. The corresponding mappings $Bel_{\widetilde{X}} : \Sigma_\Theta \to [0,1]$ and $Pl_{\widetilde{X}} : \Sigma_\Theta \to [0,1]$, are, respectively, belief and plausibility functions [2,11].

*Product-intersection Rule.* Given two normal fuzzy subsets $\widetilde{F}$ and $\widetilde{G}$ of $\Theta$, their normalized product intersection is defined as

$$(\widetilde{F} \odot \widetilde{G})(\theta) = \begin{cases} \dfrac{\widetilde{F}(\theta)\widetilde{G}(\theta)}{\text{hgt}(\widetilde{F} \cdot \widetilde{G})} & \text{if hgt}(\widetilde{F} \cdot \widetilde{G}) > 0 \\ 0 & \text{otherwise.} \end{cases} \quad (1)$$

where $\text{hgt}(\widetilde{F} \cdot \widetilde{G}) = \sup_{\theta \in \Theta} \widetilde{F}(\theta)\widetilde{G}(\theta)$ is the height of the product intersection of $\widetilde{F}$ and $\widetilde{G}$. This operation is associative; as shown in [6], it is the only normalized intersection operator having this property. The normalized product intersection can be extended to RFSs as follows. Let $(\Omega_i, \Sigma_i, P_i, \Theta, \Sigma_\Theta, \widetilde{X}_i)$, $i = 1, 2$, be two RFSs representing independent pieces of evidence. Their product intersection is defined as the RFS $(\Omega_1 \times \Omega_2, \Sigma_1 \otimes \Sigma_2, \widetilde{P}_{12}, \Theta, \Sigma_\Theta, \widetilde{X}_1 \oplus \widetilde{X}_2)$, where $\widetilde{X}_1 \oplus \widetilde{X}_2$ is the mapping from $\Omega_1 \times \Omega_2$ to $[0,1]^\Theta$ defined as $(\widetilde{X}_1 \oplus \widetilde{X}_2)(\omega_1, \omega_2) = \widetilde{X}_1(\omega_1) \odot \widetilde{X}_1(\omega_2)$, $\Sigma_1 \otimes \Sigma_2$ is the tensor product of $\Sigma_1$ and $\Sigma_2$, and $\widetilde{P}_{12}$ is the probability measure on $(\Omega_1 \times \Omega_2, \Sigma_1 \otimes \Sigma_2)$ obtained by conditioning the product measure $P_1 \times P_2$ by the fuzzy set of consistent pairs $(\omega_1, \omega_2)$, defined as $\widetilde{F}(\omega_1, \omega_2) = \text{hgt}\left(\widetilde{X}_1(\omega_1) \cdot \widetilde{X}_2(\omega_2)\right)$, i.e.,

$$\forall A \in \Sigma_1 \otimes \Sigma_2, \widetilde{P}_{12}(A) = \frac{\int_{\Omega_1} \int_{\Omega_2} A(\omega_1, \omega_2) \widetilde{F}(\omega_1, \omega_2) dP_2(\omega_2) dP_1(\omega_1)}{\int_{\Omega_1} \int_{\Omega_2} \widetilde{F}(\omega_1, \omega_2) dP_2(\omega_2) dP_1(\omega_1)}, \quad (2)$$

where $A(\cdot, \cdot)$ denotes the indicator function of $A$. The *degree of conflict* between the two pieces of evidence is defined as one minus the denominator in the right-hand side of (2). The product intersection of RFSs is commutative and associative. It extends both Dempster's rule for combining random sets, and the normalized product intersection (1) for combining possibility distributions.

### 2.2 Gaussian Random Fuzzy Numbers

*Gaussian Fuzzy Numbers* (GFNs) play the same role in quantitative possibility theory as Gaussian random variables (GRVs) in probability theory. They are defined as fuzzy subsets of $\mathbb{R}$ with membership function $x \mapsto \exp\left(-\frac{h}{2}(x-m)^2\right)$, where $m \in \mathbb{R}$ is the *mode* and $h \in [0, +\infty]$ is the *precision*. A GFN with mode $m$ and precision $h$ will be denoted by $\text{GFN}(m, h)$. GFNs are easily combined by the normalized product-intersection operator, as the following property holds: $\text{GFN}(m_1, h_1) \odot \text{GFN}(m_2, h_2) = \text{GFN}(m_{12}, h_1 + h_2)$, with $m_{12} = (h_1 m_1 + h_2 m_2)/(h_1 + h_2)$.

Let us now consider a GRV $M : \Omega \to \mathbb{R}$ with mean $\mu$ and variance $\sigma^2$. The mapping $\widetilde{X} : \Omega \to [0,1]^\mathbb{R}$ such that $\widetilde{X}(\omega) = \text{GFN}(M(\omega), h)$ defines a random fuzzy set called a *Gaussian random fuzzy number* (GRFN) with mean $\mu$, variance $\sigma^2$ and precision $h$. A GRFN can, thus, be seen as a GFN whose mode is uncertain and described by a Gaussian probability distribution. It is defined by a location parameter $\mu$, and two parameters $h$ and $\sigma^2$ corresponding, respectively, to possibilistic and probabilistic uncertainty. A GRV or a GFN is recovered when, respectively, $h = +\infty$ or $\sigma^2 = 0$. Formulas for the contour function $pl_{\tilde{X}} = Pl_{\tilde{X}}(\{x\})$ as well as for the lower and upper cumulative distribution functions (cdfs) $Bel_{\tilde{X}}((-\infty, x])$ and $Pl_{\tilde{X}}((-\infty, x])$ are given in [5].

As shown in [5], the family of GRFNs is closed under the product-intersection combination operation $\oplus$. Let $M_1 \sim N(\mu_1, \sigma_1^2)$ and $M_2 \sim N(\mu_2, \sigma_2^2)$ be two independent GRVs, and let $\tilde{X}_1 = \text{GFN}(M_1, h_1)$ and $\tilde{X}_2 = \text{GFN}(M_2, h_2)$ be corresponding GRFNs. To combine $\tilde{X}_1$ and $\tilde{X}_2$ by the product-intersection rule, we proceed as follows [5]:

1. We condition the joint probability distribution of $(M_1, M_2)$ by the fuzzy subset $\tilde{F}$ of $\mathbb{R}$ defined by $\tilde{F}(m_1, m_2) = \text{hgt}\left(\text{GFN}(m_1, h_1) \cdot \text{GFN}(m_2, h_2)\right) = \exp(-0.5\overline{h}(m_1 - m_2)^2)$, where $\overline{h} = h_1 h_2/(h_1 + h_2)$. This conditional distribution is normal with mean $\tilde{\boldsymbol{\mu}}$ and covariance matrix $\tilde{\boldsymbol{\Sigma}}$, whose expressions are given in [5].
2. The combined random fuzzy set $\tilde{X} = \tilde{X}_1 \oplus \tilde{X}_2$ is $\text{GFN}(M_c, h_1 + h_2)$, where $M_c \sim N(\mu_c, \sigma_c^2)$ with $\mu_c = \boldsymbol{h}^{*T} \tilde{\boldsymbol{\mu}}$ and $\sigma_c^2 = \boldsymbol{h}^{*T} \tilde{\boldsymbol{\Sigma}} \boldsymbol{h}^*$, where

$$\boldsymbol{h}^* = (h_1, h_2)^T/(h_1 + h_2)$$

is the vector of normalized precisions.

## 3 Extension to $n$ Dependent GRFNs

The formulas for the product intersection of two GRFNs mentioned in Sect. 2.2 were established under the assumption that the underlying GRVs are independent. In this section, we generalize these formulas to the combination of $n$ dependent GRFNs $\text{GFN}(M_i, h_i)$, $i = 1, \ldots, n$, where $(M_1, \ldots, M_n)$ has a multidimensional normal distribution with an arbitrary covariance matrix. After preliminaries exposed in Sect. 3.1, we prove our main result in Sect. 3.2.

### 3.1 Preliminaries

Let us first recall that a Gaussian fuzzy vector (GFV) with mode $\boldsymbol{m} \in \mathbb{R}^n$ and symmetric, positive semidefinite (PSD) precision matrix $\boldsymbol{H} \in \mathbb{R}^{n \times n}$ as a fuzzy subset of $\mathbb{R}^n$ with membership function

$$\boldsymbol{x} \mapsto \exp(-0.5(\boldsymbol{x} - \boldsymbol{m})^T \boldsymbol{H}(\boldsymbol{x} - \boldsymbol{m})).$$

It is denoted as $\text{GFV}(\boldsymbol{m}, \boldsymbol{H})$. The results derived in Sect. 3.2 are based on the following propositions, which are direct consequences of results about the product of univariate and multivariate normal densities proved in [1]. Propositions 1 and 2 generalize[1], respectively, Propositions 3 and 11 in [5].

---
[1] The rigorous proof of Proposition 2 cannot be given here for lack of space; it will be given in an extended version of this paper.

**Proposition 1.** Let $GFN(m_i, h_i)$, $i = 1, \ldots, n$, be $n$ GFNs.

1. The height of their product intersection is

$$\widetilde{F}(m_1, \ldots, m_n) = \exp\left[-\frac{1}{2}\left(\sum_{i=1}^{n} h_i m_i^2 - \frac{(\sum_{i=1}^{n} h_i m_i)^2}{\sum_{i=1}^{n} h_i}\right)\right]. \quad (3)$$

2. Their normalized product intersection is a GFN with precision $h = \sum_{i=1}^{n} h_i$ and mode $m = (1/h) \sum_{i=1}^{n} h_i m_i$.

**Proposition 2.** Let $GFV(\boldsymbol{m}_1, \boldsymbol{H}_1)$ and $GFV(\boldsymbol{m}_2, \boldsymbol{H}_2)$ be two GRVs. Assuming that $\boldsymbol{H}_1$ is positive definite and $\boldsymbol{H}_2$ is PSD,

1. The height of their product intersection is

$$\exp\left(-\frac{1}{2}(\boldsymbol{m}_1 - \boldsymbol{m}_2)^T \boldsymbol{H}_2 [\boldsymbol{I} + \boldsymbol{H}_1^{-1}\boldsymbol{H}_2]^{-1}(\boldsymbol{m}_1 - \boldsymbol{m}_2)\right), \quad (4)$$

where $\boldsymbol{I}$ is the $n \times n$ identity matrix;
2. Their normalized product intersection is a GFV with precision matrix $\boldsymbol{H} = \boldsymbol{H}_1 + \boldsymbol{H}_2$ and mode $\boldsymbol{m} = \boldsymbol{H}^{-1}(\boldsymbol{H}_1 \boldsymbol{m}_1 + \boldsymbol{H}_2 \boldsymbol{m}_2)$.

### 3.2 Main Result

Let us consider $n$ GRFNs $GFN(M_i, h_i)$, $i = 1, \ldots, n$, where $\boldsymbol{M} = (M_1, \ldots, M_n)$ has a multivariate normal distribution with mean $\boldsymbol{\mu}$ and covariance $\boldsymbol{\Sigma}$. We first assume this distribution to be non-degenerate, but this assumption will be relaxed later. To combine these $n$ GRFNs by the product intersection rule, we first need to condition the distribution of $\boldsymbol{M}$ by the fuzzy subset $\widetilde{F}$ of consistent tuples $m_1, \ldots, m_n$ given by (3). We first remark that (3) can be written in vector form as $\widetilde{F}(\boldsymbol{m}) = \exp\left[-\frac{1}{2}\boldsymbol{m}^T \boldsymbol{A} \boldsymbol{m}\right]$, where $\boldsymbol{m} = (m_1, \ldots, m_n)^T$ and $\boldsymbol{A}$ is the symmetric and PSD matrix

$$\boldsymbol{A} = \mathrm{diag}(\boldsymbol{h}) - \frac{\boldsymbol{h}\boldsymbol{h}^T}{\boldsymbol{1}^T \boldsymbol{h}},$$

where $\boldsymbol{h} = (h_1, \ldots, h_n)^T$ and $\boldsymbol{1} = (1, \ldots, 1)^T$. The conditional density of $\boldsymbol{M}$ given fuzzy event $\widetilde{F}$ is

$$f(\boldsymbol{m}|\widetilde{F}) = \frac{f(\boldsymbol{m})\widetilde{F}(\boldsymbol{m})}{\int f(\boldsymbol{m})\widetilde{F}(\boldsymbol{m}) d\boldsymbol{m}}. \quad (5)$$

From Proposition 2, $f(\boldsymbol{m}|\widetilde{F}) \propto \exp\left(-\frac{1}{2}(\boldsymbol{m}-\widetilde{\boldsymbol{\mu}})^T \widetilde{\boldsymbol{\Sigma}}^{-1}(\boldsymbol{m}-\widetilde{\boldsymbol{\mu}})\right)$ with

$$\widetilde{\boldsymbol{\Sigma}}^{-1} = \boldsymbol{\Sigma}^{-1} + \boldsymbol{A} \quad \text{and} \quad \widetilde{\boldsymbol{\mu}} = \widetilde{\boldsymbol{\Sigma}}(\boldsymbol{\Sigma}^{-1}\boldsymbol{\mu} + \boldsymbol{A}\boldsymbol{0}) = \widetilde{\boldsymbol{\Sigma}}\boldsymbol{\Sigma}^{-1}\boldsymbol{\mu}.$$

Writing $\widetilde{\boldsymbol{\Sigma}}^{-1} = \boldsymbol{\Sigma}^{-1}(\boldsymbol{I} + \boldsymbol{\Sigma}\boldsymbol{A})$, we get

$$\widetilde{\boldsymbol{\Sigma}} = (\boldsymbol{I} + \boldsymbol{\Sigma}\boldsymbol{A})^{-1}\boldsymbol{\Sigma} \quad \text{and} \quad \widetilde{\boldsymbol{\mu}} = (\boldsymbol{I} + \boldsymbol{\Sigma}\boldsymbol{A})^{-1}\boldsymbol{\mu}. \tag{6}$$

*Remark 1.* We have established (6) under the assumption that the distribution of $\boldsymbol{M}$ has a density (i.e., $\boldsymbol{\Sigma}$ is nonsingular). However, it can be shown that $\boldsymbol{I}+\boldsymbol{\Sigma}\boldsymbol{A}$ is nonsingular. As, when singular, $\boldsymbol{\Sigma}$ is the limit of a sequence of nonsingular covariance matrices, (6) remains true by continuity, even when the distribution of $\boldsymbol{M}$ is degenerate.

*Combination.* From Proposition 1, the product intersection of the $n$ GRFNs GFN$(M_i, h_i)$, $i=1,\ldots,n$ GRFN is the GRFN GFN $(M_c, \sum_{i=1}^n h_i)$ with

$$M_c = \frac{\sum_{i=1}^n h_i M_i}{\sum_{i=1}^n h_i} = \boldsymbol{h}^{*T}\boldsymbol{M} \sim N(\mu_c, \sigma_c^2), \tag{7}$$

where $\boldsymbol{h}^* = (h_1,\ldots,h_n)/\sum_{i=1}^n h_i$, $\mu_c = \boldsymbol{h}^{*T}\widetilde{\boldsymbol{\mu}}$ and $\sigma_c^2 = \boldsymbol{h}^{*T}\widetilde{\boldsymbol{\Sigma}}\boldsymbol{h}^*$.

*Degree of Conflict.* The degree of conflict between the $n$ GRFNs is defined as one minus the denominator on the right-hand side of (5). From (4) in Proposition 2, and noticing that $|\widetilde{\boldsymbol{\Sigma}}|/|\boldsymbol{\Sigma}| = |\widetilde{\boldsymbol{\Sigma}}\boldsymbol{\Sigma}^{-1}| = |\boldsymbol{I}+\boldsymbol{\Sigma}\boldsymbol{A}|^{-1}$, we get

$$\kappa = 1 - |\boldsymbol{I}+\boldsymbol{\Sigma}\boldsymbol{A}|^{-1/2}\exp\left(-\frac{1}{2}\boldsymbol{\mu}^T\boldsymbol{A}[\boldsymbol{I}+\boldsymbol{\Sigma}\boldsymbol{A}]^{-1}\boldsymbol{\mu}\right). \tag{8}$$

*Remark 2.* The equations given in [5] for the product intersection of two independent GRFNs and their degree of conflict can be recovered, respectively, from (7) and (8) when $n=2$ and $\boldsymbol{\Sigma}$ is diagonal.

## 4 Numerical Example

Let us consider two GRFNs $\widetilde{X}_1 = $ GFN$(M_1, h_1)$ and $\widetilde{X}_2 = $ GFN$(M_2, h_2)$ with $M_1 \sim N(1,4)$, $M_2 \sim N(3,1)$, $h_1 = 2$, and $h_2 = 1$. Their contour functions as well as their lower and upper cdfs are shown in Fig. 1a. Figures 2a and 2b show ellipses with 95% coverage probability for the unconditional distribution of random vector $(M_1, M_2)$ and its conditional distribution given $\widetilde{F}$ for, respectively, $\rho = 0.9$ and $\rho = -0.9$. The combined GRFNS for three values of the correlation coefficient $\rho \in \{-1, 0, 1\}$ are shown in Fig. 1b. It is clear that the assumed correlation coefficient strongly influences the result of the combination. Figures 3a and 3b show, respectively, the mean and standard deviation of the combined GRFN as functions of $\rho$. The standard deviation appears to be particularly sensitive to the value of $\rho$.

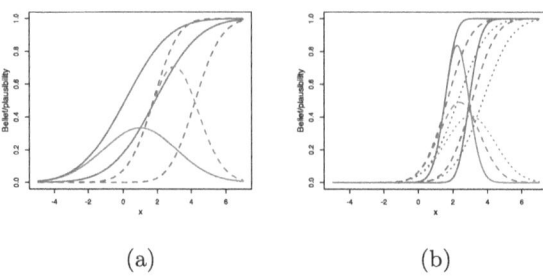

**Fig. 1.** Two GRFNs (a) and their combination (b) assuming $\rho = -1$ (solid lines), $\rho = 0$ (broken lines) and $\rho = 1$ (dotted lines). Each GRFN is represented by its contour function (red curve) and by its lower and upper cdfs (blue curves). (Color figure online)

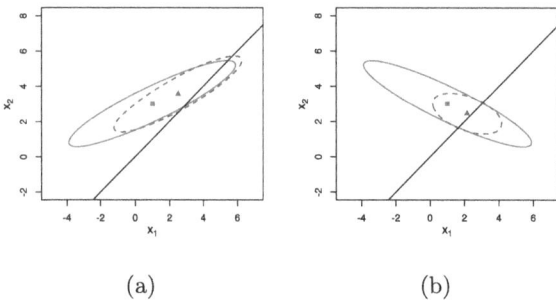

**Fig. 2.** 95% coverage probability ellipses for the unconditional distribution of random vector $(M_1, M_2)$ (solid red curve) and its conditional distribution given $\widetilde{F}$ (dashed blue curve), for $\rho = 0.9$ (a) and $\rho = -0.9$ (b). (Color figure online)

Figure 3c shows the degree of conflict as a function of $\rho$. It reaches a minimum value of 0.629 for $\rho = 0.625$. In real applications where the sources cannot be assumed to be independent but the precise value of the correlation is unknown, the value corresponding to the minimum conflict between the sources could be chosen. Figures 4a and 4b show, respectively, the minimum degree of conflict as a function of the distance $\Delta = |\mu_1 - \mu_2|$ between means and the corresponding correlation coefficient $\widehat{\rho}$ with, as before, $\sigma_1 = 2$, $\sigma_2 = 1$, $h_1 = 2$, and $h_2 = 1$. Figure 4a also displays the conflict for the product-intersection rule ($\rho = 0$) and the complete positive dependence rule ($\rho = 1$). Interestingly, we have $\widehat{\rho} = 1$ when the conflict is low and $\widehat{\rho} = -1$ when the conflict is high, while $\widehat{\rho}$ takes values between $-1$ and $1$ for intermediate distances.

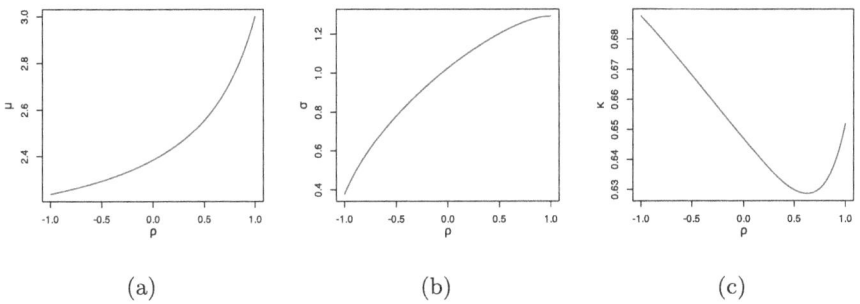

**Fig. 3.** Mean (a), standard deviation (b) and degree of conflict (c) as functions of $\rho$ for the combination of the two GRFNs shown in Fig. 1a.

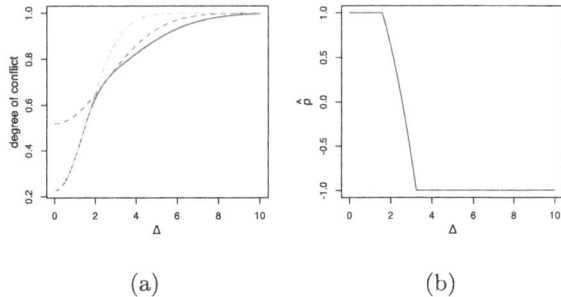

**Fig. 4.** (a): (a): Minimum degree of conflict (solid blue curve), degree of conflict for $\rho = 0$ (dashed red curve) and $\rho = 1$ (green dashed-dotted curve) vs. $\Delta = |\mu_1 - \mu_2|$. (b): Minimum-conflict correlation coefficient $\widehat{\rho}$ as a function of $\Delta$.(Color figure online)

## 5 Conclusions

We provided formulas for the combination of $n \geq 2$ GRFNs with an arbitrary correlation matrix. The results can be directly applied to the combination of transformed GRFNs and mixtures of GRFNs, as defined in [7]. In practice, the correlation matrix will rarely be known and it will need to be estimated. In the case where $n$ sources provide GRFNs and no ground truth is available, the correlation coefficients minimizing the conflict can be determined. This approach was exemplified in Sect. 4 in the case of two sources; it can be easily extended to $n$ sources using a suitable representation of the correlation matrix. In the supervised case such as, e.g., a regression task where $n$ ENNreg models [4] provide predictive GRFNs and ground truth is available, a loss function such as the generalized negative log-likelihood introduced in [4] can be minimized. Another problem, not addressed in this paper, is the consideration of the reliability or "relevance" of information sources in the combination. These and other research directions will be explored in future work.

# References

1. Bromiley, P.A.: Products and convolutions of Gaussian probability density functions. Technical Report 2003-003, TINA (2014). https://citeseerx.ist.psu.edu/document?repid=rep1&type=pdf&doi=ebf0adc76e49a9619c042792b26e0c599e1aef5c
2. Couso, I., Sánchez, L.: Upper and lower probabilities induced by a fuzzy random variable. Fuzzy Sets Syst. **165**(1), 1–23 (2011)
3. Denœux, T.: Belief functions induced by random fuzzy sets: a general framework for representing uncertain and fuzzy evidence. Fuzzy Sets Syst. **424**, 63–91 (2021)
4. Denœux, T.: Quantifying prediction uncertainty in regression using random fuzzy sets: the ENNreg model. IEEE Trans. Fuzzy Syst. **31**, 3690–3699 (2023)
5. Denœux, T.: Reasoning with fuzzy and uncertain evidence using epistemic random fuzzy sets: general framework and practical models. Fuzzy Sets Syst. **453**, 1–36 (2023)
6. Denœux, T., Kreinovich, V.: Algebraic product is the only "and-like"-operation for which normalized intersection is associative: a proof. In: Fifth International Conference on Artificial Intelligence and Computational Intelligence (AICI 2024), Hanoi, Vietnam, January 2024. https://hal.science/hal-04436177
7. Denœux, T.: Parametric families of continuous belief functions based on generalized Gaussian random fuzzy numbers. Fuzzy Sets Syst. **471**, 108679 (2023)
8. Nguyen, H.T.: On random sets and belief functions. J. Math. Anal. Appl. **65**, 531–542 (1978)
9. Shafer, G.: A Mathematical Theory of Evidence. Princeton University Press, Princeton, N.J. (1976)
10. Zadeh, L.A.: Fuzzy sets as a basis for a theory of possibility. Fuzzy Sets Syst. **1**, 3–28 (1978)
11. Zadeh, L.A.: Fuzzy sets and information granularity. In: Gupta, M.M., Ragade, R.K., Yager, R.R. (eds.) Advances in Fuzzy Sets Theory and Applications, pp. 3–18. North-Holland, Amsterdam (1979)

# A 3-Valued Logical Foundation for Evidential Reasoning

Chunlai Zhou

School of Information, Renmin University of China, Beijing, China
czhou@ruc.edu.cn

**Abstract.** The famous Cox's Theorem provides a two-valued logical justification of probability theory, which was crowned with *the Logic of Science* by Edwin Jaynes. In this paper, we similarly study a three-valued logical foundation for belief functions in this paper by focusing on the connection of Dempster's rule of combination and common knowledge. Dempster's rule allows one to combine belief functions from different sources and common knowledge is a mutual knowledge shared by an interacting group of agents. First, by interpreting belief functions as probabilities on knowledge, we prove that the combination of belief functions according to Dempster's rule is probabilities of common knowledge. Next, we explore a three-valued logical foundation of evidential reasoning. Belief (mass) functions are derived from any probability function over the set of all three-valued truth assignments. Moreover, we show that, in the Carnapian universe consisting of all atoms where each propositional letter or its negation appears only once, each three-valued assignment actually induces an $S5$ knowledge operator. Common knowledge is interpreted naturally as the combination of consistent three-valued assignments. Through these derived knowledge operators, we establish a natural connection between Dempster's rule of combination and common knowledge by reasoning in the three-valued logic.

**Keywords:** Evidential Reasoning · Belief Functions · three-valued logic

## 1 Introduction

Dealing with uncertainty is a fundamental issue for Artificial Intelligence [13]. Numerous approaches have been proposed, including evidential reasoning (also called Dempster-Shafer theory of belief functions or evidence theory). Ever since the pioneering works by Dempster [7] and Shafer [20], belief functions have become a standard tool in Artificial Intelligence for knowledge representation. On the other hand, reasoning about knowledge has become an active topic for investigation for researchers in such diverse fields as philosophy [15], economics [2], and artificial intelligence [12]. The classical model for knowledge is the so-called Kripke models or possible-world semantics. An agent is then said to *know* a fact $\phi$ if $\phi$ is true in all the worlds that he thinks possible. The modal logic system $S5$ is usually used as the logic for knowledge and is characterized by the class of Kripke models with accessibility relations being equivalence relations. When reasoning about knowledge of a group, it becomes useful to reason not just about an individual's state of knowledge, but also the knowledge of the group. For example, we might want to make statements such as "everyone in Group $G$ knows $\phi$". It turns out to be useful to be able to make even more complicated statements such as "everyone in $G$ knows that everyone in $G$ knows $\phi$", and "$\phi$ is common knowledge among the agents in $G$", where *common knowledge* is, informally, the infinite conjunction of the statements "everyone knows, everyone knows that everyone knows, everyone knows that everyone knows that everyone knows, $\cdots$". The semantics for common knowledge is simple and it can be characterized by the *coarsest refinement* of the equivalence relations for all individual agents. When common knowledge is expressed in terms of knowledge of individual agents, the expression is similar to the well-known formula of Dempster's rule of combination. This analogy has led to our first contribution on the connection between these two. We show that, if belief functions are represented as probabilities over the epistemic universe which consists of the possible worlds and knowledge of agents, then the combination of belief functions according to Dempster's rule is simply probabilities over common knowledge. Our second contribution is to explore the three-valued logical foundation of evidential reasoning by replacing the modal logic $S5$ with Kleene's three-valued logic for reasoning about knowledge. DS theory is founded on appending a third category "don't know" to the familiar dichotomy "it's true" or "it's false" and hence can be reasoned in three-valued logic. More precisely, a DS model provides three non-negative probabilities $(p, q, r)$ with $p + q + r = 1$ to the three categories of the modal triad "known to be true", "known to be false", and "don't know" associated with each assertion specified in the model [8]. With the deductive system for three-valued logic, we provide a *new* semantics for belief functions in terms of probabilities on three-valued valuations. Moreover, in the Carnapian space of two-valued (or classical) valuations, three-valued valuations provide a natural semantics for knowledge. In this semantics, common knowledge is reduced to the combination of consistent three-valued valuations and the Dempsterian combination is just probabilities on common knowledge.

The rest of the paper is organized as follows. In Sect. 2, we provide the basic background about the Dempster-Shafer theory. In Sect. 3, we present the modal logic system $S5$ for knowledge and reveal the link between Dempster's rule of combination and common knowledge. We elaborate on a three-valued logical foundation for evidential reasoning in Sect. 4. In Sect. 5 we discuss related work and conclude by pointing to future research. For a detailed introduction to belief function, one may refer to the classic [20].

## 2 Dempster's Rule and Common Knowledge

Narrowly construed, modal logic studies reasoning that involves the use of the expressions "necessarily" and "possibly". The most familiar logics in the modal family are constructed from a weak logic called **K** (after Saul Kripke) [4]. The set $\Phi_\Box$ of formulas of modal logics is constructed according to the syntax:

$\phi = p \mid \neg\phi \mid \phi \wedge \phi \mid \phi \vee \phi \mid \Box\phi$. Let $\Phi$ denote the set of formulas in propositional logic or those without the modality $\Box$. Let $\mathcal{P}$ denote the set of propositional letters. From now on, we assume that $\mathcal{P}$ is finite. The basic normal modal logic **K** results from adding the following to the principles of propositional logic: Necessitation Rule: If $A$ is a theorem of **K**, then so is $\Box A$; Distribution Axiom: $\Box(A \to B) \to (\Box A \to \Box B)$. The theoremhood of **K** is denoted by $\vdash_\mathbf{K}$. **K** is usually characterized by its Kripke semantics. Formula $\phi$ is *valid* (denoted as $\models \phi$) if it is satisfied at every possible-world of all Kripke models. The modal logic **K** is sound and complete with respect to the class of all Kripke models in the sense that $\vdash_\mathbf{K} \phi$ iff $\models \phi$ for all formulas $\phi \in \Phi_\Box$. Kripke semantics also provide a classical model for knowledge. In order to connect Kripke structures for knowledge to belief functions (usually on finite frames of discernment), without further notice, we assume in this paper that Kripke structures for knowledge are finite. Here we will use the modal logic system $S5$ to characterize knowledge of different agents. The system $S5$ is the above basic system **K** plus the following additional axioms: $\mathbf{T}: \Box\phi \to \phi$; $\mathbf{4}: \Box\phi \to \Box\Box\phi$; $\mathbf{B}: \phi \to \Box\neg\Box\neg\phi$. The logic $S5$ is characterized by the class of Kripke models with accessibility relations being equivalence relations. In other words, $\phi$ is theorem of $S5$ if and only if it is satisfied at every possible-world of all Kripke models with equivalence relations [12]. For readability, we use $K$ (sometimes with subscripts for different agents) as the modality of $S5$ instead of the usual $\Box$ and the set $\Phi_\Box$ is denoted as $\Phi_K$ accordingly. $K_i\phi$ is read "the agent $i$ knows $\phi$". In order to express different knowledge of a group of agents $\{1, 2, \cdots, n\}$, we augment the language $\Phi_K$ with the modal operators $E_G$ ("everyone on Group $G$ knows) and $C_G$ ("it is common knowledge among agents in $G$") for any subset $G$ of $\{1, 2, \cdots, n\}$. We can also give a Kripke semantics for common knowledge. For simplicity, we assume that there are only two agents 1 and 2. $E$ and $C$ are short for $E_{\{1,2\}}$ and $C_{\{1,2\}}$, respectively. Let $W$ be a set of states or possible worlds. $\mathcal{K}_1$ and $\mathcal{K}_2$ are two equivalence relations on $W$. $\mathcal{K}_i(i \in \{1,2\})$ is accessibility relation for agent $i$. Intuitively, $(s,t) \in \mathcal{K}_i$ if agent $i$ can't distinguish state $s$ from state $t$. Let $\pi_1$ and $\pi_2$ denote the partitions associated with the equivalence relations $\mathcal{K}_1$ and $\mathcal{K}_2$, and $\pi_C$ the *coarsest refinement* of both $\pi_1$ and $\pi_2$. In other words, $\pi_C$ is a *refinement* of both $\pi_1$ and $\pi_2$, i.e., for any two states $w_1$ and $w_2$, if $\mathcal{K}_C w_1 w_2$, then $\mathcal{K}_1 w_1 w_2$ and $\mathcal{K}_2 w_1 w_2$; If $\pi'$ is also a refinement of both $\pi_1$ and $\pi_2$, then $\pi'$ is a refinement of $\pi_C$. $\mathcal{K}_C$ denotes the corresponding equivalence relation of $\pi_C$. It models common knowledge $C$ in the sense that, for any formula $\phi$, $M, w \models C\phi$ if, for any state $w'$ such that $\mathcal{K}_C ww'$, $M, w' \models \phi$. Let $\mathcal{B}_1, \mathcal{B}_2$ and $\mathcal{B}_C$ be the Boolean algebras generated by the equivalence classes of $\pi_1, \pi_2$ and $\pi_C$, respectively. Let $M = \langle W, \mathcal{K}_1, \mathcal{K}_2, \mathcal{K}_C, \nu \rangle$ be a Kripke structure about knowledge of agents 1 and 2 where $\mathcal{K}_C$ is the equivalence relation for the common knowledge operator $C$. Then, for any formula $\phi \in \Phi$ and any state $w \in W$, $M, w \models C\phi$ iff there exist formulas $\phi_1$ and $\phi_2$ such that $M, w \models K_1\phi_1, M, w \models K_2\phi_2$ and $[[\phi_1]] \cap [[\phi_2]] \subseteq [[\phi]]$. For any $A \subseteq W$, define $e_i(A) := K_i(A) \setminus \bigcup_{A' \subsetneq A} K_i(A')$ for $i \in \{1,2\}$ where $K_i(A) = \{w : \pi_i(w) \subseteq A\}$ and $K_i(A') = \{w : \pi_i(w) \subseteq A'\}$. It is easy to see that $K_i(A) = \bigcup_{A' \subseteq A} e_i(A')$; If $A \neq A'$, then $e_i(A) \cap e_i(A') = \emptyset$; $W = \bigcup_{A \subseteq W} e_i(A)$. The set-functions $e_i$ partition $W$ into different components $e_i(A)(A \subseteq W)$. For any $A_1, A_2 \subseteq W$, $e_1(A_1) \cap e_2(A_2) \in \mathcal{B}_C$. Similarly, we can define $e_C(A) = K_C(A) \setminus \bigcup_{A' \subsetneq A} K_C(A')$ for any $A \subseteq W$ where $K_C(A) = \{w : \pi_C(w) \subseteq A\}$ and $K_C(A') = \{w : \pi_C(w) \subseteq A'\}$. Note that $e_C$ partitions $W$ into different components $e_C(A)(A \subseteq W)$ and $K_C(A) = \bigcup_{A' \subseteq A} e_C(A')$. This implies $[[C\phi]] = \bigcup_{[[\phi_1]] \cap [[\phi_2]] \subseteq [[\phi]]} [[K_1\phi_1]] \cap [[K_2\phi_2]]$. So $e_C([[\phi]]) = \bigcup_{[[\phi_1]] \cap [[\phi_2]] = [[\phi]]} [e_1([[\phi_1]]) \cap e_2([[\phi_2]])]$, and for any $A \subseteq [[\phi]]$, $e_C(A) = \bigcup_{A_1 \cap A_2 = A} [e_1(A_1) \cap e_2(A_2)]$. The structure of this formula looks like that of Dempster's rule of combination. Let $pr_i$'s be probability distributions over $\mathcal{B}_i$. These distributions together with the partitions $\pi_i$ determine set functions $m_i$: $m_i(A) := pr_i(e_i(A))$. It is easy to check that such defined $m_i$ are indeed mass functions and $bel_i(A) = pr_i(K_i(A))(= \sum_{A' \subseteq A} m_i(A'))$ are belief functions. For the above probability distributions $pr_i$'s on $\mathcal{B}_i$'s, their *convolution* $pr_1 \oplus pr_2$ is a probability function over the Boolean algebra $\mathcal{B}_C$ (associated with common knowledge) defined as: for any equivalence class $A$ of $\pi_C$ (or an atom of $\mathcal{B}_C$), $(pr_1 \oplus pr_2)(A) = \frac{(pr_1 \times pr_2)(\{e_1(A_1) \times e_2(A_2) : e_1(A_1) \cap e_2(A_2) = A\})}{K}$. where $K = (pr_1 \times pr_2)(\{e_1(A_1) \times e_2(A_2) : e_1(A_1) \cap e_2(A_2) \neq \emptyset\})$ is a normalization factor. By additivity, we can compute probability measure for any $A \in \mathcal{B}_C$. Note that, for any equivalence class $B$ of $\pi_C$, there is a set $B' \subseteq W$ such that $e_C(B') = B$. We can simply choose $B' = B \cup \{w\}$ for any $w \notin B$. The convolution $pr_1 \oplus pr_2$ together with the set function $e_C$ determines a mass function $m_C(A) := (pr_1 \oplus pr_2)(e_C(A))$ and a belief function $bel_C(A) := (pr_1 \oplus pr_2)(K_C(A))$.

**Theorem 1.** *For the above defined belief functions $bel_1, bel_2$ and $bel_C$, $bel_C = bel_1 \oplus bel_2$, i.e., $bel_C$ is the combination of $bel_1$ and $bel_2$ according to Dempster's rule.*

This theorem establishes a natural connection between Dempster's rule of combination and common knowledge. If we interpret belief functions as probabilities of knowledge, then the combined belief function according to Dempster's rule is nothing but the probability of the common knowledge.

## 3  Three-Valued Logical Foundation

Three-valued logics have been used for different purposes, depending on the meaning of the third truth-value. Among them, Kleene logic is typically assumed to deal with incomplete knowledge, with the third truth-value interpreted as *unknown*. In order to make the semantics of the three-valued logic transparent, we choose to interpret the three values $\{\mathbf{t}, \mathbf{f}, \mathbf{u}\}$ in terms of the power set of $\{0, 1\}$ as follows: $\mathbf{u} = \emptyset, \mathbf{t} = \{1\}, \mathbf{f} = \{0\}$. The meanings of the three values are self-evident: $\mathbf{t}$ represents "known to be true", $\mathbf{f}$ denotes "known to be false" and $\mathbf{u}$ designates "unknown". Their truth ordering is as follows: $\mathbf{f} \leq \mathbf{u} \leq \mathbf{t}$. From this ordering, we can define the lattice operations $\wedge$ and $\vee$:

| $\wedge$ | f | u | t |   | $\vee$ | f | u | t |   | $\neg$ |   |
|---|---|---|---|---|---|---|---|---|---|---|---|
| f | f | f | f |   | f | f | u | t |   | f | t |
| u | f | u | u |   | u | u | u | t |   | u | u |
| t | f | u | t |   | t | t | t | t |   | t | f |

Let **3** denote the lattice $\langle \{\mathbf{f}, \mathbf{t}, \mathbf{u}\}, \wedge, \vee \rangle$ and **2** the sublattice $\langle \{\mathbf{f}, \mathbf{t}\}, \wedge, \vee \rangle$. The meaning of these three "epistemic truth values" highly differs from the meaning of standard Boolean truth values since they are not intrinsic to propositions but are intended to reflect what an agent may have been informed about (regarding these propositions). Thus, interpreting a proposition $\phi$ as 0 (resp., 1) does not mean that $\phi$ is false (resp., true) but that the agent under consideration has some reasons to consider that $\phi$ is false (resp., true) or is told that $\phi$ is "false" (resp. "true"). The agent may have no reasons to consider $\phi$ as true and no reasons to consider it as false; in this situation, $\phi$ is given the epistemic truth value $\mathbf{u}$ (or $\emptyset$), which reflects a situation of ignorance. The agent may have some reasons to consider that $\phi$ is false and other reasons to consider that $\phi$ is true, and the epistemic truth value may be represented by $\{0, 1\}$ which reflects a situation of inconsistency. However, in this paper we will not consider this epistemic truth value. The epistemic nature in this interpretations agrees well with Shafer's emphasis of the epistemic nature of the set of possibilities on the frame of discernment in his theory (Page 36 in [20]) but under incomplete information. Recall that the language $\Phi$ includes the three connectives $\wedge$, $\vee$ and $\neg$ and a formula in $\Phi$ is formed by the following syntax: $\phi := p \mid \neg \phi \mid \phi_1 \wedge \phi_2 \mid \phi_1 \vee \phi_2$ where $p$ is a propositional letter. A *valuation* $v$ into the lattice **3** is a function from the set $\mathcal{P}$ of propositional letters into **3**. It is also called a *Kleene valuation* or an *assignment*. Let $\mathcal{L}$ denote the set of literals, i.e., $\mathcal{L} = \mathcal{P} \cup \{\neg p : p \in \mathcal{P}\}$. It is easy to see that $v$ can be extended to the set of formulas naturally as follows: $v(\neg \phi) = \neg v(\phi)$; $v(\phi \wedge \psi) = v(\phi) \wedge v(\psi)$; $v(\phi \vee \psi) = v(\phi) \vee v(\psi)$. Following Belnap [3], we simply say "$\phi$ is at least true" if $1 \in v(\phi)$; "$\phi$ is at least false" if $0 \in v(\phi)$. It follows immediately that $1 \in v(\phi)$ iff $0 \in v(\neg \phi)$, $0 \in v(\phi)$ iff $1 \in v(\neg \phi)$; $1 \in v(\phi \vee \psi)$ iff $1 \in v(\phi)$ or $1 \in v(\psi)$, $0 \in v(\phi \vee \psi)$ iff $0 \in v(\phi)$ and $0 \in v(\psi)$; $1 \in v(\phi \wedge \psi)$ iff $1 \in v(\phi)$ and $1 \in v(\psi)$, $0 \in v(\phi \wedge \psi)$ iff $0 \in v(\phi)$ or $0 \in v(\psi)$. A *Belnap structure* is a tuple $S = \langle S, \leq, v \rangle$ where $\langle S, \leq \rangle$ is a poset; $v$ is a valuation into **3** on the set of propositional letters satisfying the following *persistency* condition: for any $s_1, s_2 \in S$ and propositional letter $p$, if $s_1 \leq s_2$, then $1 \in v(s_1)(p)$ implies $1 \in v(s_2)(p)$ and $0 \in v(s_1)(p)$ implies $0 \in v(s_2)(p)$. It is a *Boolean structure* if $\leq$ is the identity relation and $v$ is a valuation into **2**. Belnap structures are usually defined for four-valued logics. But, since we express **3** in terms of four-valued logic, we also choose the semantics of four-valued logic here for uniformity. Two *support relations* between states and formulas are defined according to the above Kleene valuation. Note that, for any $s \in S$ and any formula $\phi$, $1 \in v(s)(\phi)$ iff $S, s \models_\mathbf{t} \phi$ and $0 \in v(s)(\phi)$ iff $S, s \models_\mathbf{f} \phi$. A formula $\psi$ is a **t**-*consequence* of a formula $\phi$ with respect to the class $\mathcal{B}$ of Belnap structures (denoted $\phi \models_\mathbf{t} \psi$) if, for any Belnap structure $S = \langle S, \leq, v \rangle$ and any $s \in S$, $S, s \models_\mathbf{t} \phi$ implies $S, s \models_\mathbf{t} \psi$ and is called **t**-*equivalence* of $\psi$ if further $\psi$ is a **t**-consequence of $\phi$. Recall that a Kleene valuation is a function $\nu : \mathcal{P} \rightarrow \mathbf{3}$ where $\mathcal{P}$ is the set of propositional letters. We consider a characterization of Kleene valuations in terms of positive and negative propositions as represented by the sets of "epistemically true" propositional letters and negated propositional letters respectively. More formally, a Kleene valuation $\nu$ can be characterized by an *orthopair* $(P_\nu, N_\nu) \in 2^\mathcal{P} \times 2^\mathcal{P}$ where $P_\nu = \{p \in \mathcal{P} : \nu(p) = \mathbf{t}\}$ and $N_\nu = \{p \in \mathcal{P} : \nu(p) = \mathbf{f}\}$ (or $\{p \in \mathcal{P} : \nu(\neg p) = \mathbf{t}\}$). Note that from the above definition, it holds immediately that $P_\nu \cap N_\nu = \emptyset$. Given an orthopair $(P, N) \in 2^\mathcal{P} \times 2^\mathcal{P}$, we denote the associated Kleene valuation by $\nu_{(P,N)}$. Two valuations $\nu_1$ and $\nu_2$ are *consistent* if $P_1 \cap N_2 = P_2 \cap N_1 = \emptyset$ where $(P_1, N_1)$ and $(P_2, N_2)$ are the orthopairs of $\nu_1$ and $\nu_2$ respectively. The underlying intuition behind the definition is that two valuations are consistent provided that, if a proposition letter is epistemically true according to one valuation it is not epistemically false according to the other, and vice versa. A consequence of this definition is that if a proposition is classified as being unknown in a given valuation then that proposition can't be the source of conflict with any other valuation. We can order valuations according to their relative semantic precision which is determined by the extent to which they tend to categorize formulas as being unknown. Let $\nu_1$ and $\nu_2$ be two valuations and $(P_1, N_1)$ and $(P_2, N_2)$ be their orthopairs respectively. We define the semantic precision as $\nu_1 \preceq \nu_2$ if $P_1 \subseteq P_2$ and $N_1 \subseteq N_2$. Given two *consistent* valuations $\nu_1$ and $\nu_2$ with associated orthopairs $(P_1, N_1)$ and $(P_2, N_2)$, the combination of $\nu_1$ and $\nu_2$ is the valuation $\nu_1 \oplus \nu_2 := \nu_{(P_1 \cup P_2, N_1 \cup N_2)}$. In other words, the orthopair of the combined valuation $\nu_1 \oplus \nu_2$ is $(P_1 \cup P_2, N_1 \cup N_2)$. In the case that $\nu_1$ and $\nu_2$ are inconsistent, $\oplus$ is undefined. Note that $\nu_1 \oplus \nu_2$ is well-defined. Since $\nu_1$ and $\nu_2$ are three-valued valuations, $P_1 \cap N_1 = P_2 \cap N_2 = \emptyset$. Moreover, their consistency implies that $P_1 \cap N_2 = N_1 \cap P_2 = \emptyset$. It follows hat $(P_1 \cup P_2) \cap (N_1 \cup N_2) = (P_1 \cap N_1) \cup (P_1 \cap N_2) \cup (P_2 \cap N_1) \cup (P_2 \cap N_2) = \emptyset$. One can show that $\nu_1 \oplus \nu_2$ is the least upper bound of $\nu_1$ and $\nu_2$ with respect to the precision ordering $\preceq$ in the sense of the following proposition. For consistent valuations $\nu_1$ and $\nu_2$, $\nu_1 \preceq \nu_1 \oplus \nu_2, \nu_2 \preceq \nu_1 \oplus \nu_2$; If $\nu_1 \preceq \nu_3$ and $\nu_2 \preceq \nu_3$, then $\nu_1 \oplus \nu_2 \preceq \nu_3$. From now on, we mainly focus on the *canonical Belnap structure* $\langle S_\mathbf{3}, \leq_\mathbf{3}, \nu_3 \rangle$ where $S_\mathbf{3} = \{\nu : \nu \text{ is a Kleene valuation on } \mathcal{P}\}$; For any two $\nu_1, \nu_2 \in S_\mathbf{3}, \nu_1 \leq_\mathbf{3} \nu_2$ if $\nu_1 \preceq \nu_2$; for any propositional letter $p$, $\nu_3(\nu)(p) = \mathbf{t}$ if $p \in P_\nu$, $\mathbf{f}$ if $p \in N_\nu$, $\mathbf{u}$ otherwise. One can show that $S_\mathbf{3}, \nu \models \phi$ iff $\bigwedge_{p \in P_\nu} \wedge \bigwedge_{q \in N_\nu} \neg q \vdash \phi$. For any formula $\phi \in \Phi$, $[[\phi]]_\mathbf{t} := \{\nu \in S_\mathbf{3} : S_\mathbf{3}, \nu \models_\mathbf{t} \phi\}$, $[[\phi]]_\mathbf{f} = \{\nu \in S_\mathbf{3} : S_\mathbf{3}, \nu \models_\mathbf{f} \phi\}$ and $[[\phi]]_\mathbf{u} = \{\nu : \nu_3(\nu)(\phi) = \mathbf{u}\}$. Since we assume that $\Phi$ is finite, $S_\mathbf{3}$ is also finite. For any formula $\phi$, we identify $\phi$ as the set of all formulas that is $\mathbf{t}$-equivalent to $\phi$, i.e., $\{\psi : \psi \dashv\vdash \phi\}$. Let $\bigwedge [[\phi]]_\mathbf{t}$ denote the formula $\bigwedge_{\nu \in [[\phi]]_\mathbf{t}} (\bigwedge \{p : p \in P_\nu\} \wedge \bigwedge \{\neg p : p \in N_\nu\})$. It is easy to see that $\phi \dashv\vdash \bigwedge [[\phi]]_\mathbf{t}$. So the set of non-$\mathbf{t}$-equivalent formulas is finite and is a Boolean algebra with the associated lattice operators. Let $\nu_1$ and $\nu_2$ be two consistent Kleene valuations. For any formula $\phi \in \Phi$, $S_\mathbf{3}, \nu_1 \oplus \nu_2 \models_\mathbf{t} \phi$ iff $\exists \phi_1, \phi_2 \in \Phi$ such that $S_\mathbf{3}, \nu_1 \models_\mathbf{t} \phi_1$, $S_\mathbf{3}, \nu_2 \models_\mathbf{t} \phi_2$ and $\phi_1 \wedge \phi_2 \vdash \phi$. We view uncertainty as being epistemic in nature, resulting from a lack of knowledge concerning either the state of the world to which propositions refer, or the underlying definitions of concepts used in propositions. In the following, we assume that this uncertainty is quantified by a probability measure $pr$ on $S_\mathbf{3}$ the set of Kleene valuations.

**Definition 1.** Let $pr$ be a probability measure on $S_\mathbf{3}$ so that $pr(\nu)$ is the agent's subjective belief that $\nu$ is the true Kleene valuation for $\mathcal{P}$. For any formula $\phi \in \Phi$, we define

- $bel(\phi) := pr([[\phi]]_\mathbf{t}) = pr\{\nu \in S_\mathbf{3} : S_\mathbf{3}, \nu \models_\mathbf{t} \phi\}$;
- $pl(\phi) := 1 - pr\{\nu \in S_\mathbf{3} : S_\mathbf{3}, \nu \models_\mathbf{t} \neg\phi\}$.

It is easy to see that $pl(\phi) = 1 - bel(\neg\phi) = pr([[\phi]]_\mathbf{t}) + pr([[\phi]]_\mathbf{u}) = bel(\phi) + pr(\{\nu : \nu_\mathbf{3}(\nu)(\phi) = \mathbf{u}\})$. In other words, the degree of plausibility in $\phi$ is the belief degree in $\phi$ plus the degree of the unknown cases.

**Lemma 1.** *Such defined bel and pl are indeed belief and plausibility functions over $S_3$, respectively.*

Carnap's logical approach to probability starts with the construction of possible worlds that encompasses all valid states of a systems of interest. First all propositions of relevance to the system $p_1, \cdots, p_n$ are considered. All possible conjunctions of the type $p_1 \wedge \neg p_2 \wedge \cdots \wedge p_n$ are considered. A propositional letter or its negation is called a *literal* and a conjunction of literals where every proposition letter in $\mathcal{P}$ appears once and only once is called an *atom*. There is a one-to-one correspondence between the set $\mathcal{A}$ of atoms and the set $\mathcal{VL}_2$ of all two-valued truth assignments over $\mathcal{P}$. For an atom $\alpha$, $P_\alpha$ and $N_\alpha$ denote the sets of propositional letters in $\mathcal{P}$ that occurs in $\alpha$ positively and negatively, respectively. Both $P_\alpha$ and $N_\alpha$ determines a valuation $\nu_\alpha$ in the sense that, if $p \in P_\alpha$, then $\nu_\alpha(p) = 1$ and, if $p \in N_\alpha$, then $\nu_\alpha(p) = 0$. Conversely, a two-valued truth assignment $\nu$ defines an atom $\alpha_\nu = \bigwedge_{\nu(p)=1} p \wedge \bigwedge_{\nu(q)=0} \neg q$. The Carnapian universe $S_\mathbf{2}$ is a collection of all possible states of the system and consists of all possible worlds identified as all atoms over $\mathcal{P}$.

For each agent $i(\in \{1,2\})$, his knowledge about the Carnapian universe $S_\mathbf{2}$ can be represented by a subset $\mathcal{P}_i$ of $\mathcal{P}$ which consists of all propositional letters that he knows for sure. This implies that the values of other propositional letters, i.e., those not in $\mathcal{P}_i$, are *unknown*. Such a $\mathcal{P}_i$ induces an equivalence relation $\mathcal{K}_i$ on $S_\mathbf{2}$: for any two atoms $\alpha$ and $\alpha' \in S_\mathbf{2}$,

$$\alpha \mathcal{K}_i \alpha' \text{ if } P_\alpha \cap \mathcal{P}_i = P_{\alpha'} \cap \mathcal{P}_i \text{ and } N_\alpha \cap \mathcal{P}_i = N_{\alpha'} \cap \mathcal{P}_i$$

Intuitively, $\alpha \mathcal{K}_i \alpha'$ means that, if we restrict the knowledge to those in $\mathcal{P}_i$, we are unable to distinguish $\alpha$ from $\alpha'$. Each equivalence class $[\alpha]_i = \{\alpha' : \alpha \mathcal{K}_i \alpha'\}$ can be identified with a three-valued valuation $\nu_\alpha^i$:

$$\nu_\alpha^i(p) = \begin{cases} \mathbf{t} \text{ if } p \in P_\alpha \cap \mathcal{P}_i, \\ \mathbf{f} \text{ if } p \in N_\alpha \cap \mathcal{P}_i, \\ \mathbf{u} \text{ otherwise }. \end{cases}$$

Now we combine the knowledge of agents 1 and 2. Let $\mathcal{P}_{12}$ denote the union of $\mathcal{P}_1$ and $\mathcal{P}_2$. Such a $\mathcal{P}_{12}$ induces an equivalence relation $\mathcal{K}_C$ on $S_\mathbf{2}$: for any two atoms $\alpha$ and $\alpha' \in S_\mathbf{2}$,

$$\alpha \mathcal{K}_C \alpha' \text{ if } P_\alpha \cap \mathcal{P}_{12} = P_{\alpha'} \cap \mathcal{P}_{12} \text{ and } N_\alpha \cap \mathcal{P}_{12} = N_{\alpha'} \cap \mathcal{P}_{12}$$

It is easy to check that $\mathcal{K}_C$ is the coarsest refinement of both $\mathcal{K}_1$ and $\mathcal{K}_2$ and hence it models the modal operator $K_C$ of common knowledge of agents 1 and

2. Similarly, each equivalence class $[\alpha]_{12} = \{\alpha' : \alpha \mathcal{K}_C \alpha'\}$ can be identified with a three-valued valuation $\nu_\alpha^{12}$:

$$\nu_\alpha^{12}(p) = \begin{cases} \mathbf{t} \text{ if } p \in P_\alpha \cap \mathcal{P}_{12}, \\ \mathbf{f} \text{ if } p \in N_\alpha \cap \mathcal{P}_{12}, \\ \mathbf{u} \text{ otherwise}. \end{cases}$$

Actually the coarsest refinement for $\mathcal{K}_C$ is equivalent to the combination of three-valued valuations restricted to $\mathcal{P}_1$ and $\mathcal{P}_2$. Assume that $[\alpha]_{12}$ is an equivalence class of $\mathcal{K}_C$ and there are two equivalence classes $[\alpha_i]_i$'s of $\mathcal{K}_i (i \in \{1,2\})$ such that $[\alpha]_{12} = [\alpha_1]_1 \cap [\alpha_2]_2$. This equality is equivalent to the combination formula $\nu_{\alpha_1}^1 \oplus \nu_{\alpha_2}^2 = \nu_\alpha^{12}$.

Each formula $\phi$ in $\Phi$ can be identified as the set of all atoms $\alpha$ in $S_2$ such that $\alpha \to \phi$ is a tautology (in classical propositional logic) and $K_i \phi$ is identified as the set of all three-valued valuations $\nu$ such that $\bigwedge_{p \in P_\nu} p \wedge \bigwedge_{q \in N_\nu} \neg q \to \phi$ is a tautology. Formally define $[\phi] = \{\alpha \in \mathcal{A} : \alpha \to \phi \text{ is a tautology }\}$. It follows that $K_i([\phi]) = \{\alpha \in \mathcal{A} : \bigwedge_{p \in P_\alpha \cap \mathcal{P}_i} p \wedge \bigwedge_{q \in N_\alpha \cap \mathcal{P}_i} \neg q \to \alpha \text{ is a tautology }\}$.

Let $\mathcal{B}_1$ and $\mathcal{B}_2$ be the Boolean algebras generated by the equivalence classes associated with the equivalence relations $\mathcal{K}_1$ and $\mathcal{K}_2$, respectively. Let $pr_1$ and $pr_2$ be probability distributions on $\mathcal{B}_1$ and $\mathcal{B}_2$. With these probability distributions, each $\mathcal{K}_i$ determines a mass function and a belief function. Define

- $bel_i([\phi]) = pr_i(K_i[\phi])$ and
- $m_i([\phi]) = pr_i(e_i([\phi]))$ where $e_i([\phi]) := K_i[\phi] \setminus \bigcup_{A \subsetneq [\phi]} K_i(A))$.

It is easy to check that $bel_i$ is a belief function and $m_i$ is a mass function. The proof is similar to that of Lemma 1. Similarly we can define $K_C([\phi]), e_C([\phi]), bel_C([\phi])(= (pr_1 \oplus pr_2)(K_C([\phi])))$ and $m_C([\phi])(= (pr_1 \oplus pr_2)(e_C([\phi])))$ and show that $bel_C$ is a belief function and $m_C$ is a mass function on $\mathcal{B}_C$, the Boolean algebra generated by the equivalence classes of $\mathcal{K}_C$.

**Corollary 1.** *For any formula $\phi \in \Phi$ and $w \in S_2$, $S_2, w \models C\phi$ iff there are formulas $\phi_1, \phi_2 \in \Phi$ such that $S_2, w \models K_1 \phi_1$, $S_2, w \models K_2 \phi_2$ and $\phi_1 \wedge \phi_2 \to \phi$ is a tautology.*

**Theorem 2.** *On the Carnapian universe, $bel_C = bel_1 \oplus bel_2$ in thse sense that, for any formula $\phi$, $bel_C(\phi) = (bel_1 \oplus bel_2)(\phi)$*

*Proof.* It follows directly from the above corollary. The proof is similar to that of Theorem 1. □

## 4 Related Works and Conclusions

Ruspini [19] employed epistemic logic S5, which is developed to deal with problems of representation and manipulation of the states of knowledge of rational agents, to generalize Carnap's space of possible worlds, or universe. Our approach here is based on Ruspini's work and further studies the connection of Dempster's rule and common knowledge.

Our work on three-valued logic is mainly adapted from [5,17]. They introduced a bipolar framework for combining agents' beliefs so as to enable them to reach a common shared position or viewpoint. They adopted a bipolar truth model for propositional logic characterized by lower and upper valuations on the sentences of the languages. As they pointed out, their lower and upper valuations are equivalent to Kleene's three-valued valuations. Our valuation combining operator $\oplus$ is actually the optimistic combination operator in [17]. Also they introduced lower and upper measures quantifying epistemic uncertainty. In [10,18], they discussed the relationship between non-classical logics and epistemic logic $S5$. Dempster [8] proposed a three-valued semantics for the Dempster-Shafer theory of probabilistic reasoning. In this semantics, every meaningful formal assertion is associated with a triple $(p,q,r)$ where $p$ is the probability "for" the assertion, $q$ is the probability "against" the assertion, and $r$ is the probability "don't know". He also sketched a calculus for this semantics. But none of them discuss the connection of Dempster's rule with common knowledge. Smets [21] offered a Cox-style justification of the normative representation of quantified beliefs with belief functions but the background logic is classical propositional logic. In addition to the above mentioned works, some others also try to provide a logical semantics for evidential reasoning [9,11,14,22–24]. We need to understand more about formulas in three-valued logic which are consistent with evidential reasoning.

**Acknowledgments.** We want to thank three reviewers for critical and helpful comments.

# References

1. Anderson, B., Belnap, N.: Entailment: the Logic of Relevance and Necessity. Princeton University Press, New Jersey (1975)
2. Aumann, R.: Agreeing to disagree. Ann. Statist **4**, 1236–1239 (1976)
3. Belnap, N.: A useful four-valued logic. In: Dunn, J.M., Epstein, G. (eds.) Modern uses of multi-valued logic, pp. 8–37. Reidel, Dordrecht (1977)
4. Blackburn, P., de Rijke, M., Venema, Y.: Modal Logic. Cambridge Tracts in Theoretical Computer Science, 53. Oxford University Press (2000)
5. Ciucci, D., Dubois, D., Lawry, J.: Borderline vs. unknown: comparing three-valued representations of imperfect information. Int. J. Approx. Reasoning **55**(9), 1866–1889 (2014)
6. Cox, R.: Probability, frequency and reasonable expectation. Am. J. Phys. **14**(1), 1–13 (1946)
7. Dempster, A.P.: Upper and lower probabilities induced by a multivalued mapping. Annals of Math. Stat. **38**, 325–339 (1967)
8. Dempster, A.P.: The dempster-shafer calculus for statisticians. Int. J. Approx. Reasoning **48**(2), 365–377 (2008)
9. Dubois, D., Godo, L., Prade, H.: An elementary belief function logic. J. Appl. Non-Class. Logics **33**(3-4), 365–377 (2023)
10. Fagin, R., Halpern, J., Vardi, M.: A nonstandard approach to the logical omniscience problem. Artif. Intell. **79**(2), 203–240 (1995)

11. Frittella, S., Majer, O., Nazari, S.: Toward Updating Belief Functions over Belnap-Dunn Logic. BELIEF **261–272**, 2022 (2022)
12. Halpern, J., Fagin, R., Moses, Y., Vardi, V.: Reasonign about Knowledge. MIT Press, MA, USA (1995)
13. Halpern, J.: Reasoning about Uncertainty. MIT Press, Cambridge (2005)
14. Harmanec, D., Klir, G.J., Resconi, G.: On modal logic interpretation of dempster-shafer theory of evidence. Int. J. Intell. Syst. **9**(10), 941-951 (1994)
15. Hintikka, J.: Knowledge and Belief. Cornell University Press, Ithaca, NY (1962)
16. Kleene, S.: Introduction to Metamathematics. D. van Nostrand Company, New York (1952)
17. Lawry, J., Dubois, D.: A bipolar framework for combining beliefs about vague propositions. In: Brewka, G., Eiter, T., McIlraith, S.A. (eds.), KR 2012, AAAI Press (2012)
18. Levesque, H.: A logic of implicit and explicit belief. In: AAAI, pp. 198–202 (1984)
19. Ruspini, E.H.: Epistemic logics, probability, and the calculus of evidence. In: McDermott, J.P. (ed.), Proceedings of the 10th International Joint Conference on Artificial Intelligence. Milan, Italy, 23–28 August pp. 924–931. Morgan Kaufmann (1987)
20. Shafer, G.: A Mathematical Theory of Evidence. Princeton University Press, Princeton, N.J. (1976)
21. Smets, P.: The normative representation of quantified beliefs by belief functions. Artif. Intell. **92**(1–2), 229–242 (1997)
22. Zhou, C.: Logical Foundations of Evidential Reasoning with Contradictory Information. In: Bimbó, K. (ed.) J. Michael Dunn on Information Based Logics. Outstanding Contributions to Logic, vol. 8, pp. 213–246. Springer, Cham (2016). https://doi.org/10.1007/978-3-319-29300-4_12
23. Zhou, C.: Belief functions on distributive lattices. Artif. Intell. **201**, 1–31 (2014)
24. Zhou, C., Qin, B., Du, X.: Plato's cave in the dempster-shafer land: the link between Pignistic and plausibility transformations. In: IJCAI-17, pp. 4676–4682 (2017)
25. Zhou, C., Qin, B., Li, D., Xu, Z., Du, X.: Basic utility theory for belief functions. In: ECAI 2020, pp. 2648–2655 (2020)

# Accelerated Dempster Shafer Using Tensor Train Representation

Duc P. Truong[(✉)], Erik Skau, Cassandra L. Armstrong, and Kari Sentz

Los Alamos National Laboratory, Santa Fe, USA
dptruong@lanl.gov

**Abstract.** We propose a tensor train based data structure to accelerate the calculation of Dempster-Shafer operations such as belief and Dempster's rule of combination. This approach relies on the fact that the matrix representation of these operators possess rank-1 tensor network decompositions, allowing for far more efficient calculations in tensor train format. Numerical experiments demonstrate the superior performance of the proposed method in computing Dempster-Shafer quantities.

**Keywords:** Dempster-Shafer · Tensor Decomposition · Tensor Train

## 1 Introduction

Dempster-Shafer Theory (DST) is a method of imprecise probabilities that allows for the assignment of "probabilities" to sets or intervals [4,8]. This is a generalization of probability that represents uncertainty with lower and upper probability bounds (*belief* and *plausibility*, respectively) [9,31]. The difference between these upper and lower bounds, called *imprecision*, offers a second order characterization of uncertainty when there is incomplete or conflicting information [37]. DST has proven to be highly useful in practical applications when there is significant uncertainty that cannot be represented with traditional probabilities on singletons [6,29,38], and more recently has been breaking ground in artificial intelligence and machine learning [1,10,12–14,20].

One of the critical challenges practitioners face in applying DST lies in the exponential computational complexity resulting from the requirement for computation over the power set. There is significant precedent of approximate solutions for DST. As discussed in [28], there are four main approaches to approximation: simplification approaches where sets are simplified with a probability-based, consonant-based or logic-based approximation [5,16,22,36]; sampling-based techniques (Monte Carlo, Latin hypercube) techniques where random sampling methods are used to estimate belief and plausibility [19,24,25,39,40]; hierarchical approaches [11,17,30]; and heuristic approaches [18,34].

A new approach to accelerating the computation of DST is possible by representing Dempster Shafer structures with multidimensional arrays called *tensors*.

This enables the simplification of basic DST operations with a simple nonrecursive form of tensor decomposition called *tensor train (TT) decomposition*. [27] The novelty of TT decomposition is that a multidimensional tensor can be approximated by the product of a series of low-dimensional tensors with prescribed error. This has been shown to be both efficient and stable with many and various computational applications, such as: computing high-dimensional integrals [3]; solving high dimensional PDEs [2,7,23,35]; approximating density estimation [26]; or accelerating calculations in probabilistic models [15].

In this work, we propose a TT-based approach to compute DST objects more efficiently. This approach transforms the matrix representation of DST operators [33] into low-rank tensor-network structures. These representations is exact, and requires much less storage than the matrix representations. Utilizing this tensor-network structure also enables to approximate BPAs in TT-format, and to further speed up the calculation of DST objects. Moreover, the accuracy of performing DST calculation in TT format is well controlled by the BPA TT approximation error. Due to the space limitation, these error relationship is now shown here in this paper. Numerical experiments suggest that, if BPAs can be highly compressed in the TT format, then performing calculation for DST objects in TT format will be significantly more efficient, with well controlled accuracy. We denote vectors by lowercase bold letters, e.g., $\mathbf{v}$, matrices by bold capital letters, e.g., $\mathbf{A}$, and tensors (of order three and higher) by script letter, e.g., $\mathcal{X}$. The outline of the paper is as follows. In Sect. 2, an overview of Dempster-Shafer theory is given. Next, we describe the formulation of DST operations as low rank tensor networks in Sect. 3. We then formulate the DST calculation in which BPAs are approximately given in TT formats in Sect. 4. In Sect. 5, numerical experiments for belief and DS rules of combination shows the advantages of the computations in TT format.

## 2 Dempster-Shafer Theory Basics

Here we provide a brief overview of DST with highlights of some basic DST operations [8,9,31] and their extension into the Transferable Belief Model (TBM) [32]. The operations we describe here are *belief* and *Dempster's rule of combination* with their connection to *implicability* and the *conjunctive join*, respectively.

**Definition 1.** *The* Frame Of Discernment *(FOD) is the collection of singletons, or mutually exclusive elements, denoted by $\Omega$. This is also called the sample space or the universal set in probability.*

**Definition 2.** *An* element *is a member of the powerset of the frame of discernment. If $|\Omega| = N$, then there are $2^N$ elements and are referred to with $x \in 2^\Omega$.*

**Definition 3.** *A* Basic Probability Assignment *(BPA) is a function $m : 2^\Omega \to \mathbb{R}$ that satisfies $0 \leq m(x) \leq 1$ and $\sum_{x \in 2^\Omega} m(x) = 1$.*

In Dempster Shafer theory, BPAs are often defined with the additional constraint, $m(\emptyset) = 0$. The Transferable Belief Model (TBM) modifies DST [32] so mass can be assigned to the empty set to represent conflict, $m(\emptyset) \geq 0$. This work concentrates on the TBM generalization, but can easily be adapted to DST.

**Definition 4.** A *focal element* $A$ is an element where $m(A) > 0$.

We can think of the BPA as a probability-like mass that is assigned over the powerset. This allows for the assignment of a probability-like mass to sets with a cardinality greater than 1 in contrast to assigning masses to only singletons as in probability theory. There are many alternative functions that can encode interchangable information as a BPA. *Belief (bel)*, *implicability (b)*, and *commonality (q)* functions satisfy different constraints [31,33] and can be related to BPAs with the following equations:

$$\begin{aligned}
\text{bel}(x) &= \sum_{y \mid \emptyset \neq y \subseteq x} m(y) & m(x) &= \sum_{y \mid y \subseteq x} (-1)^{|x|-|y|} \text{bel}(x) \\
\text{bel}(\emptyset) &= 0 & m(\emptyset) &= 1 - \text{bel}(\Omega) \\
b(x) &= \sum_{y \mid y \subseteq x} m(y) & m(x) &= \sum_{y \mid y \subseteq x} (-1)^{|x|-|y|} b(x) \qquad (1) \\
q(x) &= \sum_{y \mid x \subseteq y} m(y) & m(x) &= \sum_{y \mid x \subseteq y} (-1)^{|y|-|x|} q(x) \ .
\end{aligned}$$

Belief (*bel*) of an element $x$ is the sum of all masses of nonempty subsets of $x$. Commonality of an element is the sum of all of the masses assigned to elements to which $x$ is a subset. These relations make it clear that belief represents a lower probability bound, and commonality represents all probability that could be given to any element of $x$ with equal likelihood [31].

Dempster's rule of combination [9] is the first rule of combination that combines independent BPAs.

**Definition 5.** *Dempster's rule of combination*, $\oplus^{DS}$, operates on two BPAs, $m_1$ and $m_2$, defined by

$$(m_1 \oplus^{DS} m_2)(x) = \frac{\sum_{y,z \mid y \cap z = x} m_1(y) m_2(z)}{1 - K} \text{ where } K = \sum_{y,z \mid y \cap z = \emptyset} m_1(y) m_2(z) ,$$

$$(m_1 \oplus^{DS} m_2)(\emptyset) = 0.$$
(2)

This is a consensus rule that has the effect of removing all conflict through the normalization factor $1 - K$. For a different approach to conflict, authors have developed alternative rules of combination, one of which is the conjunctive join [32]. The conjunctive join is a bilinear operator that combines two independent BPAs to generate a new BPA without the normalizing function.

**Definition 6.** The *conjunctive join*, $\oplus$, operates on two BPAs, $m_1$ and $m_2$, defined by

$$(m_1 \oplus m_2)(x) = \sum_{y,z \mid y \cap z = x} m_1(y) m_2(z) \ . \qquad (3)$$

Rather than performing this combination directly on BPAs, it is often computationally advantageous to compute an equivalent operation on commonality functions. On commonality functions this is done with element-wise multiplication, denoted with $\odot$, so $q(m_1 \oplus m_2) = q(m_1) \odot q(m_2)$.

## 3 Tensor Operators

While the measures such as BPA, implicability/belief, plausibility, and commonality are traditionally defined as a functional on subsets of the frame of discernment $m, b, p, q \in 2^\Omega \to \mathbb{R}$, they can equivalently be represented as vectors, $\mathbf{m}, \mathbf{b}, \mathbf{p}, \mathbf{q} \in \mathbb{R}^{2^N}$. Represented as vectors, the transformations between the different DST objects are represented as matrices computed from recursive Kronecker multiplications. This construction highlights that these operators act on singletons independently, and can motivate the use of the Möbius transform [21,33].

Here we investigate the advantages of representing DST objects as tensors instead of functions or vectors. In this setting, the operators transforming between DST objects are naturally represented as low rank tensor networks, which provide even more compression capability. A tensor $\bigotimes^N \mathbb{R}^2$ can encapsulate the same information as a BPA, implicability, orcommonality function with $N$ singletons by associating each dimension of the tensor with a different singleton. The natural mapping between an element $x$ and a tensor index given an ordering of the singletons, is tensorindex$(x) = x_0, x_1, \ldots, x_{N-1}$ where $x_l$ is the Boolean indicator if the $l$-th singleton is in the set $x$, or equivalently as the coefficients of the vector index binary expansion. We use $\mathbf{m}, \mathbf{b}, \mathbf{q} \in \bigotimes^N \mathbb{R}^2$ to represent tensor forms of BPA, implicability, and commonality respectively, where $\bigotimes$ is tensor product defined in the Appendix of [35]. With tensor representations, the transformations from BPAs to implicability and commonality are posed as multi-linear tensor functions with both tensor inputs and outputs.

**Definition 7.** The *implicability/commonality tensor* for $N$ singletons, $\mathcal{B}_N/\mathcal{Q}_N : \bigotimes^N \mathbb{R}^2 \to \bigotimes^N \mathbb{R}^2$ is defined by,

$$\mathcal{B}_N = \mathcal{B}_{N-1} \otimes \mathcal{B}_1, \quad \mathcal{B}_1 = \begin{bmatrix} 0 & 0 \\ 0 & 1 \end{bmatrix}, \quad \mathcal{Q}_N = \mathcal{Q}_{N-1} \otimes \mathcal{Q}_1, \quad \mathcal{Q}_1 = \begin{bmatrix} 1 & 1 \\ 0 & 1 \end{bmatrix}. \quad (4)$$

Immediately following Definition 7 the identities $\mathcal{B}_N = \bigotimes^N \mathcal{B}_1$, $\mathcal{Q}_N = \bigotimes^N \mathcal{Q}_1$ describes the operations of computing implicability and commonality from a BPA as a tensor decomposition. These decompositions highlight how each singleton is represented by an individual axis. The implicability can be computed from a BPA tensor $\mathbf{m}$ with three singletones, $N = 3$, by applying $\mathcal{B}_1$ along every axis independently (similar to commonality):

$$\begin{aligned} \mathbf{b}(A, B, C) &= \sum_{a,b,c} \mathcal{B}_3(a, A, b, B, c, C) \mathbf{m}(a, b, c) \\ &= \sum_{a,b,c} \mathcal{B}_1(a, A) \mathcal{B}_1(b, B) \mathcal{B}_1(c, C) \mathbf{m}(a, b, c) \end{aligned} \quad (5)$$

In Eq. (5), the summation over three singletons results in $2^3$ summands for each of the $2^3$ elements, generalizing to the complexity of $\mathcal{O}(2^{2N})$. The operator decomposition applied in the Equation (5) allows for the reordering of the summation which reduces the complexity to $\mathcal{O}(N2^N)$ in Equation (6):

$$\mathbf{b}(A,B,C) = \sum_a \mathcal{B}_1(a,A) \left( \sum_b \mathcal{B}_1(b,B) \left( \sum_c \mathcal{B}_1(c,C) \mathbf{m}(a,b,c) \right) \right) \quad (6)$$

Similarly, the conjunctive join tensor operator is a tensor defined to act on two tensor BPAs, $\mathcal{M}_1$ and $\mathcal{M}_2$, producing a tensor result:

**Definition 8.** The conjunctive join tensor operator for $N$ singletons, $\mathcal{C}_N : \bigotimes^N \mathbb{R}^2 \times \bigotimes^N \mathbb{R}^2 \to \bigotimes^N \mathbb{R}^2$ is defined by,

$$\mathcal{C}_N = \mathcal{C}_{N-1} \otimes \mathcal{C}_1 \text{ where } \mathcal{C}_1 = \begin{bmatrix} 1 & 1 & 0 & 0 \\ 1 & 0 & 0 & 1 \end{bmatrix}. \quad (7)$$

Immediately following the recursive definition, we have the decomposition of the conjunctive join tensor $\mathcal{C}_N = \bigotimes^N \mathcal{C}_1$.

$$(\mathbf{m}_1 \oplus \mathbf{m}_2)(\alpha, \beta, \gamma) = \sum_{aAbBcC} \mathcal{C}_3(a,A,\alpha,b,B,\beta,c,C,\gamma) \mathbf{m}_1(a,b,c) \mathbf{m}_2(A,B,C)$$

$$= \sum_{aA} \mathcal{C}_1(a,A,\alpha) \left( \sum_{bB} \mathcal{C}_1(b,B,\beta) \left( \sum_{cC} \mathcal{C}_1(c,C,\gamma) \mathbf{m}_1(a,b,c) \mathbf{m}_2(A,B,C) \right) \right) \quad (8)$$

This decomposition allows for more efficient conjunctive join computations, which can be further aided by commonality. Through the commonality, the conjunctive join can be computed with

$$(\mathbf{m}_1 \oplus \mathbf{m}_2)(A,B,C)$$
$$= \sum_a \mathcal{Q}_1^{-1}(a,A) \left( \sum_b \mathcal{Q}_1^{-1}(b,B) \left( \sum_c \mathcal{Q}_1^{-1}(c,C) \mathbf{q}_1(a,b,c) \mathbf{q}_2(a,b,c) \right) \right). \quad (9)$$

## 4 BPA Approximations Using Tensor Train

Here, we investigate the advantages of approximating a tensor BPA with tensor train decomposition [27]. TT decomposition can approximate high order tensors with a train of order three core tensors. A TT approximation of a BPA tensor $\mathbf{m} \in \bigotimes^N \mathbb{R}^2$ takes the form

$$\mathbf{m}(i_0, i_1, \ldots, i_{N-1}) \approx \sum_{\alpha_0, \alpha_1, \ldots, \alpha_N}^{r_0, r_1, \ldots, r_N} \prod_{l \in (0, \ldots, N-1)} \mathcal{G}_{l+1}(\alpha_l, i_l, \alpha_{l+1})$$

where $\alpha_l$ for $l = 0, \ldots, N$ is summed over the TT rank $r_l$ which specifies the dimensions of each core $\mathcal{G}_l \in \mathbb{R}^{r_{l-1} \times 2 \times r_l}$ with $r_0 = r_N = 1$. TT approximation can be computed with a sequence of truncated singular value decomposition (SVD). At each stage, a single rank and dimension are factored off of all the remaining dimensions by applying a SVD to reshape the remaining tensor [27]. A desired error can be prescribed beforehand which determines the truncation dimension of the singular value decomposition which become the TT ranks $r_l$. It is beneficial to use TT format when $r = \max\{r_0, r_1, \ldots r_N\}$ stays relatively small. With BPA in TT format, the implicability computations for three singletons (Equation (6)) can be further expressed in TT format as

$$\mathbf{b}(A,B,C) = \sum_a \mathcal{B}_1(a,A) \left( \sum_b \mathcal{B}_1(b,B) \left( \sum_c \mathcal{B}_1(c,C) \mathbf{m}(a,b,c) \right) \right)$$

$$\approx \sum_a \mathcal{B}_1(a,A) \left( \sum_b \mathcal{B}_1(b,B) \left( \sum_c \mathcal{B}_1(c,C) \sum_{\alpha_0,\ldots,\alpha_3} \prod_{l=1,2,3}^{s=a,b,c} \mathcal{G}_{l+1}(\alpha_l, s, \alpha_{l+1}) \right) \right)$$

$$\approx \sum_{\alpha_0,\ldots,\alpha_3} \left( \sum_a \mathcal{B}_1(a,A) \mathcal{G}_1(\alpha_0, a, \alpha_1) \right) \ldots$$

$$\ldots \left( \sum_b \mathcal{B}_1(b,B) \mathcal{G}_2(\alpha_1, b, \alpha_2) \right) \left( \sum_c \mathcal{B}_1(c,C) \mathcal{G}_3(\alpha_2, c, \alpha_3) \right).$$

$$\approx \sum_{\alpha_0,\ldots,\alpha_3} \mathcal{G}'_1(\alpha_0, A, \alpha_1) \mathcal{G}'_2(\alpha_1, B, \alpha_2) \mathcal{G}'_3(\alpha_2, C, \alpha_3).$$

(10)

The advantage of computing implicability in the TT-format is that the calculation is split into a series of products between $\mathcal{B}_1$ and the cores $\mathcal{G}_i$ of $\mathbf{m}$. This allows to compute the implicability of much larger BPAs. A similar analysis can be done for commonality, and plausibility can be computed indirectly from implicability.

A similar expansion can be done for the conjunctive join of two $2 \times 2 \times 2$ BPA tensors $\mathcal{X}$ and $\mathcal{Y}$. Let $\mathcal{X}$ and $\mathcal{Y}$ have TT decompositions $\mathcal{X} \approx \mathcal{G}^x_1 \mathcal{G}^x_2 \mathcal{G}^x_3$ and $\mathcal{Y} \approx \mathcal{G}^y_1 \mathcal{G}^y_2 \mathcal{G}^y_3$, then the conjunctive join of $\mathcal{X}^{TT}$ and $\mathcal{Y}^{TT}$ can be computed with

$$\mathcal{G}'_1 = \sum_{a^x, a^y} \mathcal{C}_1(a^x, a^y, A) \mathcal{G}^x_1(\alpha^x_0, a^x, \alpha^x_1) \mathcal{G}^y_1(\alpha^y_0, a^y, \alpha^y_1)$$

$$\mathcal{G}'_2 = \sum_{b^x, b^y} \mathcal{C}_1(b^x, b^y, B) \mathcal{G}^x_2(\alpha^x_1, b^x, \alpha^x_2) \mathcal{G}^y_1(\alpha^y_1, b^y, \alpha^y_2)$$

$$\mathcal{G}'_3 = \sum_{c^x, c^y} \mathcal{C}_1(c^x, c^y, C) \mathcal{G}^x_3(\alpha^x_2, c^x, \alpha^x_3) \mathcal{G}^y_3(\alpha^y_2, c^y, \alpha^y_3)$$

(11)

$$(\mathcal{X} \oplus \mathcal{Y})(A,B,C) \approx \sum_{\alpha_0,\ldots,\alpha_3} \mathcal{G}'_1(\alpha_0, A, \alpha_1) \mathcal{G}'_2(\alpha_1, B, \alpha_2) \mathcal{G}'_3(\alpha_2, C, \alpha_3).$$

## 5 Numerical Experiments

To experimentally demonstrate these computational benefits we compare the performance of different ways to compute belief and Dempster's rule of combination. The methods being compared are (1) Vector - using matrix-vector operations on the BPA vector; (2) FMT - using fast Mobius Transform, (3) Tensor - using the decomposition of the operators applied to BPA tensor, (4) TT - using decomposition of the operators applied to the tensor train format of BPA tensor. The synthetic BPAs in the tensor format are generated from the tensor train structure with various prescribed ranks. Given a BPA tensor, a TT representation with a higher rank will better approximate the BPA. Also, it is known that the performance of TT algorithms are strongly correlated with the TT-ranks. Consequently, we examine the performance at four values of TT-ranks, i.e. $r = 1, 50, 100, 150$, in these experiments.

Figure 1-*Left* shows the plot for the elapsed time of different methods in computing belief as the number of singletons varies. The vector data format is shown to take the largest amount of time with very steep scaling and could not perform the calculation with 16 singletons due to the large amount of memory being required. FMT and tensor approaches scale comparably across the number of singletons, with the tensor approach being a bit faster, but demonstrating the same complexity. The TT approach on the rank-1 TT BPA shows superior performance with a very flat scaling, and is about 1000 times faster then FMT and Tensors with 25 singletons. We also measured the elapsed time to compute the TT format of a BPA tensor; with the decomposition time decompt+TT, the TT approach is comparable with FMT and Tensor. This suggests the TT approach will be favorable when multiple DST operations are performed consecutively from a TT decomposition. Figure 1-*Middle* shows the comparison of Tensor and TT-approaches on four different TT-rank BPA approximations. It is immediately evident that the scaling of the computations of BPAs already in TT-format is vastly superior to computations starting from Tensor format. Even with the relatively high TT-ranks at 150, the TT-approach is shown to be much less expensive. Figure 1-*Right* shows the accuracy of different approaches in computing belief, using FMT as the reference. From the figure, we see the vector and Tensor agree to floating point precision. The TT analysis has been broken into two errors, first the TT approximation error incurred from approximating the BPA with a TT, the TT Belief error is the error in the resulting TT decomposition of belief. Note that the belief error is less than the approximation error.

Figure 2-*Left* shows the elapsed time of different methods for computing Dempster's rule of combination where the number of singletons varies. All methods computing Dempster's rule of combination use commonality except BPA-Vector which computes the rule of combination directly using Definition 5. The BPA-Vector approach is the most inefficient approach. Commonality-Vector demonstrates a better complexity than BPA-Vector, but not as good as FMT, Tensor, and decomp+TT which scale comparably. The TT approach on the rank-1 TT BPA is far superior, but does not include the time needed to compute

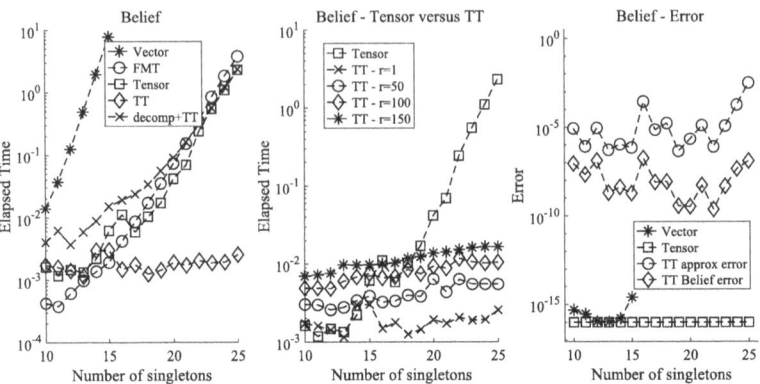

**Fig. 1.** Elapsed time and errors of different methods in computing belief versus the number of singletons.

the TT decompositions. Figure 2-*Middle* shows the comparison of Tensor and TT-approaches on different TT-Rank BPA approximations. The TT-approaches scale much better than the Tensor approach. Even with the relatively high TT-ranks of 150 the TT-approach is shown to be much less expensive. Figure 2-*Right* shows the accuracy of different approaches in computing Dempster's rule of combination using the FMT as reference. As expected, the error of BPA-vector, commonality-vector, and tensor are around machine precision. The TT approximation error, and the resulting TT Rule Of Combination (ROC) error are of a similar magnitude.

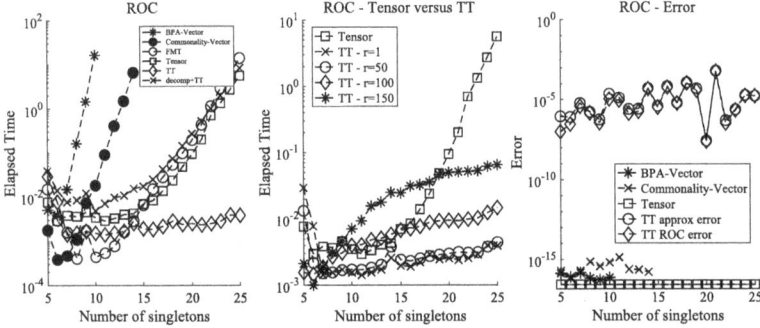

**Fig. 2.** Elapsed time and errors of different methods in computing Dempster's rule of combination versus the number of singletons.

**Acknowledgments.** The research presented in this article was supported by the Laboratory Directed Research and Development program of Los Alamos National Laboratory under project number 20220799DI.

# References

1. Abdelkhalek, R., Elouedi, Z.: Hybrid artificial immune recognition system with improved belief classification process. In: BELIEF 2022, pp. 307–316 (2022)
2. Adak, D., Truong, D.P., Manzini, G., Rasmussen, K.Ø., Alexandrov, B.S.: Tensor network space-time spectral collocation method for time dependent convection-diffusion-reaction equations. arXiv preprint arXiv:2402.18073 (2024)
3. Alexandrov, B., Manzini, G., Skau, E.W., Truong, P.M.D., Vuchov, R.G.: Challenging the curse of dimensionality in multidimensional numerical integration by using a low-rank tensor-train format. Mathematics **11**(3), 534 (2023)
4. Augustin, T., Coolen, F.P., De Cooman, G., Troffaes, M.C.: Introduction to Imprecise Probabilities, vol. 591. Wiley, New York (2014)
5. Barnett, J.: Computational methods for a mathematical theory of evidence. In: International Journal for Computational Methods, IJCM-81, Vancouver, CA, pp. 868–875 (1981)
6. Chehibi, M., Ferchichi, A., Riadh Farah, I.: An intelligent system for managing uncertain temporal flood events. In: BELIEF 2022, pp. 153–162 (2022)
7. Danis, M.E., Truong, D., Boureima, I., Korobkin, O., Rasmussen, K., Alexandrov, B.: Tensor-train WENO scheme for compressible flows. arXiv preprint arXiv:2405.12301 (2024)
8. Dempster, A.: New methods for reasoning towards posterior distributions based on sample data. Ann. Math. Stat. **37**(2), 355–374 (1966)
9. Dempster, A.: Upper and lower probabilities induced by a multi-valued mapping. Ann. Math. Stat. **38**(2), 325–339 (1966)
10. Denœux, T.: A neural network classifier based on dempster shafer theory. IEEE Trans. Syst. Man Cybern.-Part A **30**(2), 131–150 (2000)
11. Denœux, T.: Inner and outer approximation of belief structures using a hierarchical clustering approach. Internat. J. Uncertain. Fuzziness Knowl.-Based Syst. **9**(04), 437–460 (2001)
12. Denœux, T.: NN-EVCLUS: neural network-based evidential clustering. Inf. Sci. **572**, 297–330 (2021)
13. Denœux, T.: An evidential neural network model for regression based on random fuzzy numbers. In: BELIEF 2022, pp. 44–53. BELIEF (2022)
14. Denœux, T., Kanjanatarakul, O., Sriboonchitta, S.: EK-NNclus: a clustering procedure based on the evidential K-nearest neighbor rule. Knowl.-Based Syst. **88**, 57–69 (2015)
15. Dolgov, S., Anaya-Izquierdo, K., Fox, C., Scheichl, R.: Approximation and sampling of multivariate probability distributions in the tensor train decomposition. Stat. Comput. **30**, 603–625 (2020)
16. Dubois, D., Prade, H.: LP consonant approximations of belief functions. In: IEEE Trans. Fuzzy Systems, vol. 22, pp. 420–436 (2014)
17. Gordon, J., Shortliffe, E.: A method of managing evidential reasoning in a hierarchical hypothesis space. Artif. Intell. **26**, 323–357 (1985)
18. Harmanec, D.: Faithful approximation of belief functions. In: Uncertainty in Artificial Intelligence, 1999, Stockholm, Sweden, pp. 271–278 (1999)
19. Helton, J., Johnson, J., Oberkampf, W., Storlie, C.: A sampling-based computational strategy for the representation of epistemic uncertainty in model predictions with evidence theory. Technical Report, Sandia National Laboratories SAND2006-5557 (2006)

20. Hoarau, A., Martin, J.C.D., La Galle, Y.: Imperfect labels with belief functions for active learning. In: BELIEF 2022, pp. 44–53 (2022)
21. Kennes, R.: Computational aspects of the Mobius transformation of graphs. IEEE Trans. Syst. Man Cybern. **22**(2), 201–223 (1992)
22. Lowrence, J., Garvey, T., Strat, T.: A framework for evidential reasoning systems. In: Proceedings of the Fifth National Conference of the American Association for Artificial Intelligence, 1986, Hungary Budapest, pp. 896–903 (1986)
23. Manzini, G., Truong, P., Vuchkov, R., Alexandrov, B.: The tensor-train mimetic finite difference method for three-dimensional maxwell's wave propagation equations. Math. Comput. Simul. **210**, 615–639 (2023)
24. Moral, S., Wilson, N.: Fast markov-chain algorithms for the calculating Dempster-Shafer belief. In: Proceedings of the Twelveth European Conference on Artificial Intelligence, 1994, Hungary Budapest, pp. 672–676 (1994)
25. Moral, S., Wilson, N.: Markov-chain Monte-Carlo algorithms for the calculation of dempster-shafer belief. In: AAAI-94 Proceedings, pp. 269–274 (1994)
26. Novikov, G.S., Panov, M.E., Oseledets, I.V.: Tensor-train density estimation. In: Uncertainty in Artificial Intelligence, pp. 1321–1331. PMLR (2021)
27. Oseledets, I.V.: Tensor-train decomposition. SIAM J. Sci. Comput. **33**(5), 2295–2317 (2011)
28. Sarabis-Jamab, A., Araabi, B.N.: Information-based evaluation of approximation methods in Dempster-Shafer theory. Internat. J. Uncertain. Fuzziness Knowl.-Based Syst. **24**(4), 503–535 (2016)
29. Sentz, K., Scott, F.: Combination of evidence in Dempster Shafer theory. Technical Report, Sandia National Laboratories SAND2000-0835 (2002)
30. Shafer, G., Logan, R.: Implementing Dempster's rule for hierarchical evidence. Artif. Intell. **33**, 271–298 (1987)
31. Shafer, G.: A Mathematical Theory of Evidence. Princeton University Press, Princeton (1976). http://www.jstor.org/stable/j.ctv10vm1qb
32. Smets, P.: The combination of evidence in the transferable belief model. IEEE Trans. Pattern Anal. Mach. Intell. **12**(5), 447–458 (1990)
33. Smets, P.: The application of the matrix calculus to belief functions. Int. J. Approximate Reasoning **31**(1–2), 1–30 (2002)
34. Tessem, B.: Approximations for efficient computation in the theory of evidence. Artif. Intell. **61**, 315–329 (1993)
35. Truong, D.P., Ortega, M.I., Boureima, I., Manzini, G., Rasmussen, K.Ø., Alexandrov, B.S.: Tensor networks for solving the time-independent Boltzmann neutron transport equation. J. Comput. Phys. **507**, 112943 (2024)
36. Voorbraak, F.: A computationally efficient approximation of Dempster-Shafer theory. Int. J. Man Mach. Stud. **30**, 525–536 (1989)
37. Walley, P.: Statistical Reasoning with Imprecise Probabilities. Chapman and Hall (1991)
38. Wang, S., Liu, Z., Zhang, Z., Yang, L.: Heterogeneous image fusion for target recognition based on evidence reasoning. In: BELIEF 2022, pp. 153–162 (2022)
39. Wickramarathne, T.L., Premaratne, K., Murthi, M.N.: Monte-Carlo approximation for Dempster-Shafer belief theoretic algorithms. In: Proceedings of the International Conference on Information, Chicago, USA, pp. 1–11 (2011)
40. Wilson, N.: Algorithms for Dempster-Shafer theory. In: Handbook of Defensible Reasoning and Uncertainty Management, pp. 421–475 (2000)

# Author Index

**A**
Afifi, Sohaib 197
Armstrong, Cassandra L. 283

**B**
Bi, Shuhui 234
Bossé, Éloi 161

**C**
Cella, Leonardo 121, 189

**D**
Dang, Viet-Hung 98
de Campos, Cassio 31
Deng, Yong 161
Denneulin, Sébastien 3
Denœux, Thierry 40, 49, 264
Devanne, Maxime 87
Dezert, Jean 87
Dubois, Jean-Christophe 207, 244

**F**
Feng, Mengling 49, 78

**G**
Geletu, Mihreteab Negash 87
Giurgi, Dănuţ-Vasile 87
Gong, Chaoyu 13
Gross-Amblard, David 244
Guiziou, Loïc 3

**H**
Hoarau, Arthur 207
Huang, Ling 49, 78
Huang, Tao 58
Huynh, Van-Nam 98

**J**
Jing, Liping 58
Josso-Laurain, Thomas 87

**K**
Krak, Thomas 31

**L**
Lauffenburger, Jean-Philippe 87
Le Gall, Yolande 207, 244
Lefèvre, Éric 197
Lepskiy, Alexander 216
Liu, Chuanqi 22
Liu, Liping 255
Liu, Yang 131
Liu, Zechao 22
Liu, Zhekun 58
Liu, Zhunga 22, 68
Luo, Hao 161

**M**
Ma, Liyao 234
Martin, Ryan 111, 121, 140
Montalván Hernández, David Ricardo 31

**N**
Ning, Liangbo 22

**P**
Pichon, Frédéric 150, 197

**R**
Ramasso, Emmanuel 3
Ramel, Sébastien 150
Ruan, Yucheng 78

**S**
Sentz, Kari 283
Shenoy, Prakash P. 225
Skau, Erik 283
Su, Zhi-gang 13

**T**
Tanaydin, Tugba  180
Thibaud, Sébastien  3
Thierry, Constance  207, 244
Tian, Hongpeng  68
Tran, Anh-Tu  98
Truong, Duc P.  283
Turhan, Hasan Ihsan  180

**V**
Vu, Tuan-Anh  197

**W**
Wang, Rui  58
Wang, Sihan  13
Wang, Yiyang  234
Williams, Jonathan P.  111, 131

**X**
Xing, Yucheng  49
Xu, Qianyi  78

**Z**
Zhang, Haifei  171
Zhang, Zuowei  22, 68
Zhao, Xiong  234
Zhou, Chunlai  273
Zhou, Qianli  161
Zuo, Jingwei  68

**SPRINGER NATURE**

## GPSR Compliance

*The European Union's (EU) General Product Safety Regulation (GPSR) is a set of rules that requires consumer products to be safe and our obligations to ensure this.*

*If you have any concerns about our products, you can contact us on ProductSafety@springernature.com*

In case Publisher is established outside the EU, the EU authorized representative is:

Springer Nature Customer Service Center GmbH
Europaplatz 3
69115 Heidelberg, Germany

The manufacturer's authorised representative in the EU is Springer Nature Customer Service Centre GmbH, Europaplatz 3, 69115 Heidelberg, Germany. If you have any concerns regarding our products, please contact ProductSafety@springernature.com

Printed and bound by CPI Group (UK) Ltd, Croydon, CR0 4YY

26/03/2026

02078963-0005